Quantum Biosensing in Medical Diagnostics

Quantum Biosensing in Medical Diagnostics

Edited by J. G. Manjunatha

WILEY

This edition first published 2026

Registered Offices
John Wiley & Sons, Inc., 111 River Street, Hoboken, NJ 07030, USA
John Wiley & Sons Ltd, New Era House, 8 Oldlands Way, Bognor Regis, West Sussex, PO22 9NQ, UK

For details of our global editorial offices, customer services, and more information about Wiley products visit us at www.wiley.com.

The manufacturer's authorized representative according to the EU General Product Safety Regulation is Wiley-VCH GmbH, Boschstr. 12, 69469 Weinheim, Germany, e-mail: Product_Safety@wiley.com.

Wiley also publishes its books in a variety of electronic formats and by print-on-demand. Some content that appears in standard print versions of this book may not be available in other formats.

Library of Congress Cataloging-in-Publication Data Applied for

Hardback ISBN: 9781394338979

Cover Design: Wiley
Cover Image: © KanStockPng/stock.adobe.com

Set in 9.5/12.5pt STIXTwoText by Straive, Chennai, India

Contents

List of Contributors

Prashanth S. Adarakatti
Department of Chemistry, SVM Arts, Science and Commerce College
ILKAL – 587125 Affiliated to Bagalkot University
Jamakhandi, Karnataka
India

Vinayak Adimule
Department of Chemistry, Angadi Institute of Technology and Management (AITM)
Belagavi, Karnataka
India

Arasan Saroja Anakath
Department of Computer Science and Engineering, Saveetha School of Engineering
Saveetha Institute of Medical and Technical Sciences
Chennai, Tamil Nadu
India

and

Department of Analytical Chemistry
Faculty of Pharmacy
Ankara University
Ankara
Türkiye

Pandurangan Anandan
Department of Physics
Thiru Kolanjiappar Government Arts College
Virudhachalam, Tamil Nadu
India

Dhivya Antony
Department of Chemistry
St. Joseph's College of Engineering
Chennai, Tamil Nadu
India

Balamurugan Arumugam
Department of Chemistry
National Sun Yat-sen University
Taiwan
Republic of China

and

PG & Research Department of Chemistry
Thiagarajar College, Affiliated to Madurai Kamaraj University
Madurai, Tamil Nadu
India

Siddaramanna Ashoka
Department of Chemistry, School of Applied Sciences
REVA University
Bengaluru, Karnataka
India

Nazife Aslan
Chemistry Department, Polatlı Science and Arts Faculty
Ankara Hacı Bayram Veli University
Ankara
Turkey

Subrata Barick
Department of Medicinal Chemistry
National Institute of Pharmaceutical Education and Research Raebareli (NIPER-R)
Lucknow, Uttar Pradesh-226002
India

Praveen Barmavatu
Department of Mechanical Engineering
Faculty of Engineering
Universidad Tecnológica Metropolitana
Santiago
Chile

Sandeep Chandrashekharappa
Department of Medicinal Chemistry
National Institute of Pharmaceutical Education and Research Raebareli (NIPER-R)
Lucknow, Uttar Pradesh
India

Po-Ling Chang
Department of Chemistry
National Sun Yat-sen University
Taiwan
Republic of China

Tanmay Chatterjee
Department of Chemistry (Science and Humanities)
PES University
Bengaluru, Karnataka
India

Jayeeta Chattopadhyay
Amity Institute of Applied Sciences
Amity University Jharkhand
Ranchi, Jharkhand
India

Raril Chenthattil
Department of Chemistry
Amrita Vishwa Vidyapeetham
Kollam, Kerala
India

Fatih Ciftci
Department of Biomedical Engineering
Fatih Sultan Mehmet Vakıf University
Istanbul
Turkey

and

Biomedical Electronic Design Application and Research Center (BETAM)
Fatih Sultan Mehmet Vakıf University
Istanbul
Turkey

Sonu Gandhi
BRIC-National Institute of Animal Biotechnology (NIAB)
Hyderabad, Telangana
India

and

BRIC-Regional Centre for Biotechnology (RCB)
Faridabad, Haryana (NCR Delhi)
India

Savita Hanaji
Department of Engineering Science and Humanities
KLE Technological University, Dr. M. S. Sheshgiri Campus
Belagavi, Karnataka
India

Cigdem Kanbes-Dindar
Chemistry Department, Polatlı Science and Arts Faculty
Ankara Hacı Bayram Veli University
Ankara
Turkey

and

Department of Analytical Chemistry
Faculty of Pharmacy
Ankara University
Ankara
Turkey

Rangappa Keri
Centre for Nano and Material Sciences (CNMS), Jain (Deemed-to-be) University
Bangalore, Karnataka
India

Sathish Kumar Ponnaiah
Department of Chemical Engineering
Pohang University of Science and Technology (POSTECH)
Pohang 37673
Republic of Korea

Brikshadipa Mandal
Department of Medicinal Chemistry
National Institute of Pharmaceutical Education and Research Raebareli (NIPER-R)
Lucknow, Uttar Pradesh-226002
India

Nikiwe Mhlanga
Department of Science Technology and Innovation(DSTI)/Mintek Nanotechnology Innovation Centre (NIC), Advanced Materials Division, Mintek
Randburg
South Africa

Aswathy S. Murali
Department of Chemistry
Amrita Vishwa Vidyapeetham
Kollam, Kerala
India

Belakatte P. Nandeshwarappa
Department of PG Studies and Research in Chemistry
Shivagangothri, Davangere University
Davanagere, Karnataka
India

Santosh Nandi
Department of Mechanical Engineering
Faculty of Engineering
Universidad Tecnológica Metropolitana
Santiago
Chile

Nireekshana Nandigam
BRIC-National Institute of Animal Biotechnology (NIAB)
Hyderabad, Telangana
India

Shankramma S. Nesargi
Department of Chemistry
KLE Technological University, Dr. M. S. Sheshgiri Campus
Belagavi, Karnataka
India

Raghu Ningegowda
Department of Chemistry, Jyoti Nivas College Autonomous
Bangalore, Karnataka
India

Tara Sankar Pathak
Chemistry Department
Surendra Institute of Engineering and Management
Siliguri
West Bengal
India

Sayee Kannan Ramaraj
PG & Research Department of Chemistry
Thiagarajar College, Affiliated to Madurai
Kamaraj University
Madurai, Tamil Nadu
India

Gnanesh Rao
Department of Biochemistry
Bangalore University
Bangalore, Karnataka
India

Jyotirmayee Sahoo
BRIC-National Institute of Animal
Biotechnology (NIAB)
Hyderabad, Telangana
India

and

BRIC-Regional Centre for Biotechnology
(RCB)
Faridabad, Haryana (NCR Delhi)
India

Shridevi Salagare
Department of Chemistry and Research
Centre, NMKRV College for Women
Bengaluru, Karnataka
India

Begüm Sarac
Department of Biomedical Engineering
Fatih Sultan Mehmet Vakıf University
Istanbul
Turkey

Beena Saraswathyamma
Department of Chemistry, Amrita Vishwa
Vidyapeetham
Kollam, Kerala
India

Monima Sarma
Department of Chemistry (Science and
Humanities)
PES University
Bengaluru, Karnataka
India

Hülya Silah
Faculty of Science, Department of
Chemistry
Bilecik Seyh Edebali University
Bilecik
Türkiye

Sree Lekshmi
Department of Chemistry, Amrita Vishwa
Vidyapeetham
Kollam, Kerala
India

Anuja Sudarsanan
Department of Chemistry, Amrita Vishwa
Vidyapeetham
Kollam, Kerala
India

Priya Tiwari
Department of Medicinal Chemistry
National Institute of Pharmaceutical
Education and Research Raebareli
Lucknow, Uttar Pradesh
India

Bengi Uslu
Faculty of Pharmacy
Department of Analytical Chemistry
Ankara University
Ankara
Turkey

Muneshwar Harsha Vardhan
BRIC-National Institute of Animal Biotechnology (NIAB)
Hyderabad, Telangana
India

and

BRIC-Regional Centre for Biotechnology (RCB)
Faridabad, Haryana (NCR Delhi)
India

Leona R. Varghese
Department of Chemistry, Amrita Vishwa Vidyapeetham
Kollam, Kerala
India

Chinnapiyan Vedhi
Department of Chemistry, V.O. Chidambaram
Thoothukudi, Tamil Nadu
India

Seydanur Yücer
Department of Biomedical Engineering
Fatih Sultan Mehmet Vakıf University
Istanbul
Turkey

Notes on Contributors

Dr. Sonu Gandhi (Corresponding Author) is a scientist at the National Institute of Animal Biotechnology, Hyderabad, India, and her work is mainly focused on biosensor development and theranostic applications. Dr. Gandhi did her PhD at IMTECH, Chandigarh, and her postdoc at IFOM-IEO, Milan, Italy. She worked as visiting scientist at the University of Washington, Seattle, United States. Dr. Gandhi received the SMC Bronze Medal from the Society of Materials Chemistry; the Sri Om Prakash Sharma Award from IABS; the Distinguished Women Scientist Award from ISCB; the ICAR Panjabrao Deshmukh Award; the SERB-Women Excellence Award; the SERB-Early Career Research Award; the DBT-BioCARe Women Scientist Award; the Best Paper Award from CITAC-Portugal; the Tech Exhibition Award from IKP-BIRAC, DBT, and DST; and the Best Young Investigator Award from IIT-BHU. Recently, she has been elected as a Member of the National Academy of Sciences, India (NASI); Young Associate of the Indian Academy of Sciences (IASc), Bengaluru; Member of INYAS; Fellow of the Royal Society of Chemistry (RSC), Cambridge, United Kingdom; Women in STEM from Asia in the Australian Academy of Sciences (AAS); and member of the American Chemical Society (ACS), USA. Dr. Gandhi has published more than 100 publications in peer-reviewed journals, and her research has been highlighted in PIB, Govt. of India; Nature India News; IISc Bengaluru News; and several newspapers and media.

Jyotirmayee Sahoo is currently pursuing her doctoral research at the National Institute of Animal Biotechnology (NIAB), Hyderabad. Her research is centered on the development of sensor technologies for detecting microplastic degradation. Integrating a novel approach that facilitates the degradation of microplastics prior to detection, with an aim to enhance environmental monitoring and remediation strategies.

Muneshwar Harshavardhan is a Ph.D. student at the National Institute of Animal Biotechnology, Hyderabad. His research area focuses on tissue engineering, where he is developing scaffolds for real-time wound monitoring and accelerated wound healing.

Nireekshana Nandigam She is pursuing a master's in biotechnology at Vellore Institute of Technology, Vellore. Currently working as a dissertation trainee at the National Institute of Animal Biotechnology, where she works on the expression and purification of recombinant proteins.

Dr. Balamurugan Arumugam earned his Ph.D. (2022) in Chemistry from Madurai Kamaraj University, Tamil Nadu, India. Subsequently, he worked as a postdoctoral research fellow in the Laboratory of Bio-Analytical Chemistry, Department of Chemistry, National Sun Yat-sen University, Taiwan. His research interests include bioanalytical chemistry, biomedical technologies, preparation of functional nanomaterials, hybrid materials, and composites for electrochemical sensors, biosensors, photocatalysis, and catalytic applications.

Prof. Po-Ling Chang obtained his Ph.D. from National Taiwan University. Currently, he works as a professor and leads the Laboratory of Bio-Analytical Chemistry in the Department of Chemistry at the National Sun Yat-sen University, Taiwan. His research focuses on bioanalytical chemistry, biomedical technologies, biosensors, super-resolution microscopy, nanocapillary/capillary electrophoresis with laser-induced fluorescence, cell organelle bioimaging, DNA electro-osmotic flow mechanisms, etc.

Dr. Sathish Kumar Ponnaiah obtained his Ph.D. (2020) in Chemistry from Madurai Kamaraj University, Tamil Nadu, India. Now, he is a postdoctoral researcher in the Department of Chemical Engineering, Pohang University of Science and Technology, Republic of Korea. His research interest is the synthesis of functional nanomaterials for electrochemical sensors, biosensors, energy storage, photocatalysis, and water splitting applications.

Prof. Sayee Kannan Ramaraj obtained his Ph.D. degree from Madurai University, Madurai, Tamil Nadu, India. He is currently working as an associate professor and head in the Department of Chemistry, Thiagarajar College, Madurai, 625009, Tamil Nadu, India. His research focuses on the synthesis of nanomaterials for electrochemical sensors, biosensors, energy devices, photocatalysis, and water-splitting applications. Professor Sayee Kannan Ramaraj has published more than 163 research articles in international peer-reviewed journals.

Jayeeta Chattopadhyay is a professor and deputy director (R&D) at Amity University Jharkhand, Ranchi, Jharkhand, India. She completed her Ph.D. in New Energy Engineering from Seoul National University of Science and Technology, South Korea. She has published over 60 peer-reviewed papers, contributed book chapters, and holds two international patents. Her notable work includes the development of various hollow spherical nanostructures, which have been utilized as potential electrocatalysts in water electrolysis.

Tara Sankar Pathak is the former head of the Department of Basic Science and Humanities, Surendra Institute of Engineering and Management, Siliguri, West Bengal, India. He is a distinguished academician and researcher with over two decades of experience in teaching and scientific research. He holds an M.Sc. from the Indian Institute of Technology (IIT) Kharagpur and an M.Tech. from the National Institute of Technology (NIT) Jamshedpur. He earned his Ph.D. from the National Chemical Laboratory (CSIR-NCL), Pune, affiliated with Pune University.

Dr. Monima Sarma is a faculty member in the Department of Chemistry (Science and Humanities) at PES University, Bengaluru, India. Her research focuses on luminescent materials for organic electronics (OLEDs) and bioimaging studies, nanobiophotonics, device structure optimization, supramolecular chemistry, sensors, nonlinear optics, etc.

Dr. Tanmay Chatterjee is a faculty member in the Department of Chemistry (Science and Humanities) at PES University, Bengaluru, India. His research focuses on artificial intelligence and machine learning (AI&ML) for OLED material design and screening, predictive analysis of device performance metrics, device structure optimization, and manufacturing process control; thermally activated delayed fluorescence (TADF) materials as emitters and hosts in OLEDs; fluorescent probes for imaging studies; etc.

Dr. Dhivya Antony, assistant professor, Department of Chemistry, St. Joseph's College of Engineering, Chennai, 600119, Tamil Nadu, India, for writing the manuscript; **Dr. Arasan Saroja Anakath**, associate professor, Department of Computer Science and Engineering, Saveetha School of Engineering, Saveetha Institute of Medical and Technical Sciences, Chennai, Tamil Nadu, India, for participating in improving the drawing quality; **Dr. Pandurangan Anandan**, assistant professor, Department of Physics, Thiru Kolanjiappar Government Arts College, Virudhachalam, 606001, India, contributed to reviewing and editing; and **Dr. Chinnapiyan Vedhi**, associate professor, Department of Chemistry, V.O. Chidambaram, Thoothukudi, 628008, Tamil Nadu, India, contributed significantly to this manuscript in reviewing and editing.

Dr. Hülya Silah is currently an assistant professor at the Department of Chemistry, Faculty of Science, Bilecik Seyh Edebali University, Turkey. Her research focuses on analytical chemistry, electrochemical sensor development, and environmental applications such as water purification and adsorption. She has published peer-reviewed articles covering a wide range of topics, including nanomaterial-based sensors, drug and pesticide analysis, and the removal of pollutants from aqueous media.

Prof. Dr. Bengi Uslu is currently a full professor at the Department of Analytical Chemistry, Faculty of Pharmacy, Ankara University, Turkey. Her areas of interest are electrochemistry, biosensors, nanosensors, biotechnology, drug analysis, bioanalytical method development, redox mechanisms of drugs, and analytical method validation. She has authored or co-authored more than 200 peer-reviewed full papers and more than 180 oral or poster presentations at scientific conferences. She has more than 60 book chapters, 6000 citations of research papers, and an h-index of 42.

Seydanur Yücer is an undergraduate education in biomedical engineering at Fatih Sultan Mehmet Vakıf University. She is working as a scholarship holder on the project "Franz Diffusion Mathematical Modelling of Transdermal Bioactive Biomaterials" conducted by Dr. Ciftci. The young researcher is involved in many studies under the supervision of Dr. Ciftci.

Begüm Sarac is an undergraduate education in biomedical engineering at Fatih Sultan Mehmet Vakıf University. She is working as a scholarship holder on the project "Franz

Diffusion Mathematical Modelling of Transdermal Bioactive Biomaterials" conducted by Dr. Ciftci. The young researcher is involved in many studies under the supervision of Dr. Ciftci.

Fatih Ciftci is the head of biomedical engineering at Fatih Sultan Mehmet Vakıf University. He is also an editorial board member of Nature Scientific Reports. He is the founder of BioriginAI Research Group and the head of biomedical electronic design application and research center.

Brikshadipa Mandal is a postgraduate student pursuing her M.S. (Pharm) in Medicinal Chemistry at the National Institute of Pharmaceutical Education and Research, Raebareli (NIPER-R). Her research work mainly focuses on the synthesis of small heterocyclic molecules for therapeutic purposes.

Subrata Barick is a Ph.D. candidate in the Department of Medicinal Chemistry at the National Institute of Pharmaceutical Education and Research, Raebareli (NIPER-R). He earned his M.Pharm in pharmaceutical chemistry from Central University of Punjab. His research area focuses on the design and synthesis of biologically active compounds.

Santosh Nandi is a postdoctoral research fellow in the Department of Mechanical Engineering, Faculty of Engineering, Universidad Tecnológica Metropolitana, Chile. His research focuses on the thermal behavior of nanocomposites, nanocomposites for supercapacitor applications, advancements in electrical and dielectric properties, and liquid crystals for optical applications.

Dr. Shankramma S. Nesargi is an experienced academician with 17 years of teaching experience and one year of industrial experience. She specializes in advanced functional nanomaterials for energy, environmental, and electrochemical applications. Her research focuses on the synthesis and characterization of metal oxide nanomaterials and nanocomposites for energy storage, photocatalysis, and sensing applications. She has expertise in techniques such as SEM, XRD, UV–Visible spectroscopy, FTIR, and electrochemical analysis (CV, EIS, GCD). She is actively involved in mentoring students and contributing to applied materials research.

Vinayak Adimule is a world's top 2% listed scientist and professor at Angadi Institute of Technology and Management, Belagavi, India. His research interests include advancement in nanocomposites for supercapacitors, electrical and optical properties, liquid crystals, covalent organic frameworks, and polymer nanocomposites for dielectric and electro-optical applications.

S. S. Kerur is a professor at KLE Technological University, Belagavi, India. Her research focuses on the synthesis of nanocomposites, fundamental characterizations such as UV–visible spectroscopy and FTIR, and their applications in photocatalysis.

Savita Hanaji is a professor at KLE Technological University, Belagavi, India. Her work is centered on the development and advancement of wavelet-based numerical techniques, including Legendre wavelet methods, Jacobi wavelet integration, and Gegenbauer wavelet approaches.

Rangappa Keri is a world's top 2% listed scientist and Jain (Deemed-to-be University), Bangalore, India. He authored around 150 research articles. His work focuses on the development of organic molecules for anticancer, antibacterial, neuroprotective compounds, acetylcholinesterase (AChE) inhibitors, nanocomposites for supercapacitors, sensors, liquid crystal nanocomposites, etc.

Praveen Barmavatu is an assistant professor at the Department of Mechanical Engineering, Faculty of Engineering, Universidad Tecnologica Metropolitana, Chile, USA. His work focuses on heat exchangers and PCM on the thermal performance and energy conversion and storage applications.

Dr. Shridevi Salagare was born in Raibag, Karnataka, India, on July 31, 1988. Received her M.Sc. in physical chemistry from Gulbarga University, Gulbarga, India. She has completed B.Ed. from Rani Channamma University, Belagavi. She obtained her Ph.D. in the PG Department of Chemistry and Research Centre, NMKRV College for Women, Jayanagar, affiliated with Bangalore University. Her area of research work includes synthesis of nanomaterials, MOFs, nanocomposites, electrochemistry of chemically modified electrodes, and quantification of nitrite and its application toward real environmental samples.

Dr. Siddaramanna Ashoka graduated with M.Sc. and Ph.D. degrees in chemistry from Bangalore University. He was a postdoctoral fellow at various institutes in India and abroad: the Warsaw University of Technology, Warsaw, Poland; the University of Quebec, Canada; and the Indian Institute of Science, India. He was awarded the Indo-Quebec postdoctoral international merit scholarship. He has published 100 research articles in reputed international journals. He has one national patent on fluoride and arsenic removal from drinking water. He has an h-index of 31 with 3537 citations. He has written five book chapters published in Elsevier and RSC publications. Four of his research articles have been selected as among the top 25 hottest articles among Elsevier publications. His research areas of interest include synthesis and applications of nanostructured materials for energy storage and conversion, wastewater treatment, and electrocatalysis. Currently, he is working as an assistant professor at REVA University, Bengaluru, India.

Dr. Prashanth S. A., born on September 30, 1989 in Hubli, Karnataka, is a dedicated chemist and academician currently serving as an assistant professor of chemistry at SVM Arts, Science, and Commerce College, Ilkal, affiliated with Bagalkot University, Jamkhandi. He earned his B.Sc. and M.Sc. degrees from Karnatak University, Dharwad, and completed his Ph.D. in chemistry from Bangalore University in 2017. He pursued postdoctoral research at the Indian Institute of Science (IISc), Bengaluru. He has published over 55 peer-reviewed research papers and contributed more than 30 book chapters with publishers like the Royal Society of Chemistry (RSC) and Elsevier. He has worked as an academic visitor at JNCASR in the group of Bharat Ratna Prof. C. N. R. Rao and as a visiting fellow at Manchester Metropolitan University, UK. A reviewer for several international journals, his research interests include electrochemical sensors, electroanalytical chemistry, additive manufacturing, and materials chemistry for energy storage.

Dr. Nikiwe Mhlanga is a principal research scientist at Mintek in the Advanced Materials Division, which is home to the DSTI/Mintek Nanotechnology Innovation Centre. She holds a Ph.D. degree in chemistry from the University of Johannesburg. Dr. Mhlanga's current research interest focuses on surface-enhanced Raman spectroscopy (SERS) probes for the detection and quantification of infectious diseases such as malaria. She excels in both theoretical (simulation) and experimental understanding of various materials, particularly at the nanoscale.

Dr. Çiğdem Kanbeş-Dindar is a lecturer at the Department of Chemistry, Faculty of Science and Letters, Polatlı Campus, Ankara Hacı Bayram Veli University. She is also working as a postdoctoral researcher at the Department of Analytical Chemistry, Faculty of Pharmacy, Ankara University, where she contributes to various national and international research projects. In addition, she is the cofounder of CHIDO Technology, a company established to develop innovations based on cold plasma technologies. Her research interests focus on cold plasma-based nanomaterials, electrochemical and optical sensors for applications in agriculture, healthcare, and defense. She has served as a principal investigator and researcher in numerous funded projects and has published several peer-reviewed scientific papers.

Prof. Dr. Nazife Aslan is the Head of the Department of Chemistry at the Faculty of Science and Letters, Ankara Hacı Bayram Veli University, Polatlı Campus. Her research focuses on biosensors, surface modification techniques, and nuclear analytical methods. As a senior academic, she has played a pivotal role in shaping the department's research strategy and academic programs. Prof. Aslan has led and collaborated on various multidisciplinary projects and has contributed extensively to the scientific literature in her field.

Gnanesh Rao has earned his Master of Science (M.Sc.) in biochemistry from Bangalore University, and B.Sc. (Biochemistry, Botany, Microbiology) from Government Science College, Bangalore. He has 6 years of experience in the field of research and development in the domain of nanotechnology, synthetic biology, cell biology, molecular biology, and virology. He worked as a project assistant in the Institute for Stem Cell Science and Regenerative Medicine (InStem), Bengaluru; the Institute for Biotechnology and Applied Bioinformatics (IBAB), Bengaluru; and a research associate at private CROs. He works in the cross-sectional areas of research and regulatory in the fields of therapeutics, biomedical devices, nutraceuticals, alternative protein, and cosmetics. He has authored and coauthored several research articles and book chapters with recognized publishers.

Priya Tiwari is a Ph.D. candidate in the Department of Medicinal Chemistry at the National Institute of Pharmaceutical Education and Research, Raebareli (NIPER-R). She earned her Master of Pharmacy (M. Pharm) in Pharmaceutical Chemistry from Guru Ghasidas Vishwavidyalaya (a Central University), Bilaspur, and B. Pharm from Uttar Pradesh University, Medical Sciences, Saifai, Etawah. Her research centers on the design, synthesis, and biological evaluation of novel bioactive compounds, with a particular focus on antitubercular, anticancer, and larvicidal drug discovery.

She has developed expertise in computer-aided drug design (CADD), synthetic organic chemistry, structure–activity relationship (SAR) studies, and biochemical screening methodologies. She has authored and co-authored several high-impact research articles and book chapters with internationally recognized publishers, including Springer Nature and the Royal Society of Chemistry (RSC). She is also a recipient of a research grant from AMRIT and holds two granted patents (one Indian and one South African), highlighting the novelty and translational potential of her work. Through her ongoing doctoral research and prior academic endeavors; she has consistently demonstrated a strong commitment to translational medicinal chemistry aimed at addressing critical unmet therapeutic needs.

Dr. Raghu Ningegowda is an Assistant Professor in the Department of Chemistry at the Jyoti Nivas College Autonomous, Bangalore, INDIA, 560095. He earned his Ph.D. in Chemistry from Mysore University, Mysore (2016). Dr. Raghu Ningegowda brings over 21 years of extensive experience in both industrial and academic research, complemented by more than seven years of teaching excellence. His scientific expertise spans a wide range of therapeutic areas, with a strong emphasis on rational drug design, heterocyclic chemistry, and medicinal chemistry-driven lead optimization. He has consistently demonstrated innovation and leadership in advancing drug discovery research. His work integrates fundamental chemistry with translational applications, contributing to the development of novel bioactive molecules with therapeutic potential. He has published over 22 peer-reviewed research articles in reputed international journals and is the inventor/co-inventor of more than five national and international patents, underscoring his strong innovation-driven research. In addition to his research accomplishments, he is actively engaged in mentoring young researchers, completing two minor research projects, and one major research project is ongoing. His professional vision emphasizes the integration of organic chemistry, green catalyst, and pharmaceutical sciences to accelerate the discovery of next-generation therapeutic agents.

Belakatte P. Nandeshwarappa is a professor in the Department of Studies in Chemistry at Davangere University, where he also serves as coordinator of the Department of Criminology and Forensic Science. He obtained his M.Sc. (2001) and Ph.D. in Organic Chemistry (2007) from Kuvempu University, followed by postdoctoral research at the Institute of Fine Chemicals, East China University of Science and Technology, Shanghai. With nearly 21 years of research and 18 years of postgraduate teaching experience, he has also worked in the pharmaceutical industry as a research associate at Apotex Pharma Chem. India Pvt. Ltd., Bangalore. His research focuses on synthetic organic and medicinal chemistry, with notable contributions to heterocyclic compounds.

A prolific scholar, he has authored 12 books, edited 8 volumes, contributed 30 book chapters, and published 50 research papers in reputed journals. He has guided 3 doctoral students and 65 postgraduate students and served as examiner and adjudicator for many universities. His academic service includes roles as BOE Chairman, BOS/BOE member, and organizer of seminars, conferences, and training programs. Recognized with multiple honors and awards, he has actively introduced innovative teaching methodologies and departmental programs, reflecting his commitment to advancing research, academic excellence, and professional development in the chemical sciences.

Dr. Sandeep Chandrashekharappa is an assistant professor in the Department of Medicinal Chemistry at the National Institute of Pharmaceutical Education and Research, Raebareli (NIPER-R). He earned his Ph.D. in Chemistry from Kuvempu University, Shimoga (2014), and subsequently pursued his postdoctoral research at the Institute for Stem Cell Science and Regenerative Medicine (InStem), Bengaluru.

He brings over 15 years of extensive experience in both industrial and academic research, complemented by more than 5 years of teaching excellence. His scientific expertise spans a wide range of therapeutic areas, with a strong emphasis on rational drug design, heterocyclic chemistry, and medicinal chemistry-driven lead optimization. A recipient of the prestigious **Gandhian Young Technological Innovation (GYTI) Young Scientist Award**, Dr. Chandrashekharappa has consistently demonstrated innovation and leadership in advancing drug discovery research. His work integrates fundamental chemistry with translational applications, contributing significantly to the development of novel bioactive molecules with therapeutic potential.

He has published over **105** peer-reviewed research articles in reputed international journals and is the inventor/co-inventor of more than **49** national and international patents, underscoring his strong innovation-driven research. In addition to his research accomplishments, he is actively engaged in mentoring young researchers, which is proven by his recent **Gyanodaya Award** for his excellence and fostering collaborations across interdisciplinary domains, as well as getting **six grants** for his research work, among them one completed and five ongoing. His professional vision emphasizes the integration of organic chemistry, chemical biology, and pharmaceutical sciences to accelerate the discovery of next-generation therapeutic agents.

Raril Chenthattili is a dynamic postdoctoral fellow in the Department of Chemistry at Amrita Vishwa Vidyapeetham, Amritapuri campus. He earned his Ph.D. in chemistry from Mangalore University. His work centers on the innovative design and development of electrochemical sensors for detecting biologically active molecules, alongside pioneering advancements in energy storage devices. With over 50 peer-reviewed publications to his name, he is making significant strides in the sensor science community and continues to contribute impactful research at the intersection of chemistry, health, and sustainable energy.

Sree Lekshmi is a Ph.D. candidate in the Department of Chemistry at Amrita Vishwa Vidyapeetham, Amritapuri campus. Her research focuses on the fabrication of advanced electrochemical sensors using functionalized metal sulfides, with applications in the detection of biologically and environmentally relevant molecules.

Aswathy S. Murali is a Ph.D. graduate from Amrita Vishwa Vidyapeetham, Amritapuri campus. Her research focuses on the design and development of functionalized materials for sensing applications across diverse domains, including environmental, biological, and industrial fields. Her work reflects a strong commitment to innovation in material science and sensor technology.

Leona R. Varghese is a Ph.D. candidate in the Department of Chemistry at Amrita Vishwa Vidyapeetham, Amritapuri campus. She holds an M.Sc. in analytical chemistry from Mahatma Gandhi University. Her research focuses on the design and development of nanostructured materials for electrochemical biosensors, along with the exploration of advanced electrode materials for high-performance supercapacitors.

Anuja Sudarsanan is a Ph.D. candidate in the Department of Chemistry at Amrita Vishwa Vidyapeetham, Amritapuri campus. She holds an M.Sc. in chemistry from the University of Kerala. Her research focuses on the development of electrocatalysts for water splitting, with an emphasis on the kinetics and mechanisms of the hydrogen evolution reaction (HER) in both acidic and alkaline media. She is also actively involved in exploring advanced materials for energy storage applications, aiming to contribute to sustainable energy technologies.

Dr. Beena Saraswathyamma is an associate professor in the Department of Chemistry at Amrita Vishwa Vidyapeetham, Amritapuri campus. She earned her Ph.D. in 2011 from Cochin University of Science and Technology (CUSAT), Kerala, with a specialization in electrochemical sensors for metal ion detection. As part of the DST Indo-Polish collaborative project, she also gained valuable international research experience at the Polish Academy of Sciences, Poland. Her current research interests include the synthesis of metal-organic frameworks (MOFs), transition metal dichalcogenides (TMDs), 2D materials, and nanocomposites, with applications in electrochemical sensors, biosensors, and energy storage devices.

Preface

Quantum science is rapidly transforming the landscape of modern technology, and its influence is now extending into the field of medical diagnostics through the emerging domain of quantum biosensing. The convergence of quantum mechanics and biosensing technologies offers an extraordinary opportunity to improve the sensitivity, specificity, and speed of disease detection. By leveraging phenomena such as quantum entanglement, tunneling, and coherence, quantum biosensors are enabling breakthroughs in the identification of biomolecular interactions, the detection of trace-level biomarkers, and the monitoring of dynamic biological processes in real time.

This book, *Quantum Biosensing in Medical Diagnostics*, provides a timely and comprehensive overview of the principles, materials, and applications of quantum biosensors in the context of human health. It brings together contributions from leading researchers and practitioners across disciplines, including physics, chemistry, engineering, and biomedical sciences. Together, these chapters illuminate the fundamental mechanisms underlying quantum sensing techniques, such as nitrogen-vacancy centers in diamonds, quantum dots, superconducting circuits, and photonic systems, while highlighting their integration with biological systems for diagnostic purposes. As we move toward an era of precision medicine, the demand for rapid, noninvasive, and highly accurate diagnostic tools is greater than ever. Quantum biosensing holds the potential to detect diseases such as cancer, neurodegenerative disorders, and infectious diseases at their earliest stages—well before conventional methods can. Beyond early detection, quantum-enabled platforms offer promising routes for real-time patient monitoring, therapeutic guidance, and point-of-care testing. The preparation of this book was motivated by a growing need to consolidate the latest advancements and foster interdisciplinary dialogue among quantum scientists, biosensor developers, and clinical researchers. It is our hope that this volume will serve not only as a reference for established experts but also as a catalyst for students and early-career scientists entering this exciting field. By expanding the boundaries of what is technically possible, quantum biosensing stands to make a profound impact on the future of healthcare and disease management.

We invite the readers to explore the evolving landscape of quantum biosensors and to envision the transformative role these technologies will play in the next generation of medical diagnostics.

February 2nd, 2026
Madikeri

By Editor
Dr. J. G. Manjunatha

1

Introduction to Quantum Biosensing: A New Frontier in Diagnostics

Jyotirmayee Sahoo[1,2], *Muneshwar Harsha Vardhan*[1,2], *Nireekshana Nandigam*[1], and *Sonu Gandhi*[1,2]

[1] *BRIC-National Institute of Animal Biotechnology (NIAB), Hyderabad, Telangana, India*
[2] *BRIC-Regional Centre for Biotechnology (RCB), Faridabad, Haryana (NCR Delhi), India*

1.1 Introduction

Biosensing technology has undergone significant evolution since its outset in the mid-twentieth century, driven by advancements in biochemistry, engineering, and nanotechnology, leading to transformative impacts across various domains, including clinical treatment, pharmacy, healthcare sectors, environmental monitoring, and food safety [1]. The earliest biosensor, innovated by Clark and Lyons in 1962 for glucose detection, employed enzyme-based recognition systems to selectively identify target analytes, integrating biological recognition elements with electrochemical transducers [2, 3] to yield measurable signals as shown in Figure 1.1 [5]. This groundbreaking discovery set the stage for the advancement of complex biochemical sensing systems, paving the way for the evolution of modern biosensing technologies. Advancing further, second-generation biosensors incorporated mediated interactions, improving sensitivity and specificity by employing redox mediators to enhance electron transfer at electrode surfaces [6]. Furthermore, third-generation biosensors have greatly benefited from advancements in nanotechnology, where nanoscale materials and advanced fabrication methods have led to the creation of highly sensitive biosensors [7]. These innovations enable the detection of single molecules, expanding their use in clinical diagnostics and personalized medicine. Various sensing technologies, including optical, electrochemical, and piezoelectric detection, have expanded biosensing capabilities. Each method provides distinct advantages in sensitivity, portability, and speed of analysis, making biosensors more versatile and efficient. The following timeline map illustrates the evolution of biosensors and shows the key advancements from their inception to modern innovations, as shown in Figure 1.1.

Moving forward, in the early twenty-first century (around 2010–2011), a new sensing approach emerged—quantum biosensing [8]. This pioneering advancement in diagnostic technology uses the principles of quantum mechanics (QMs). This plays a vital role in

Quantum Biosensing in Medical Diagnostics, First Edition. Edited by J. G. Manjunatha.

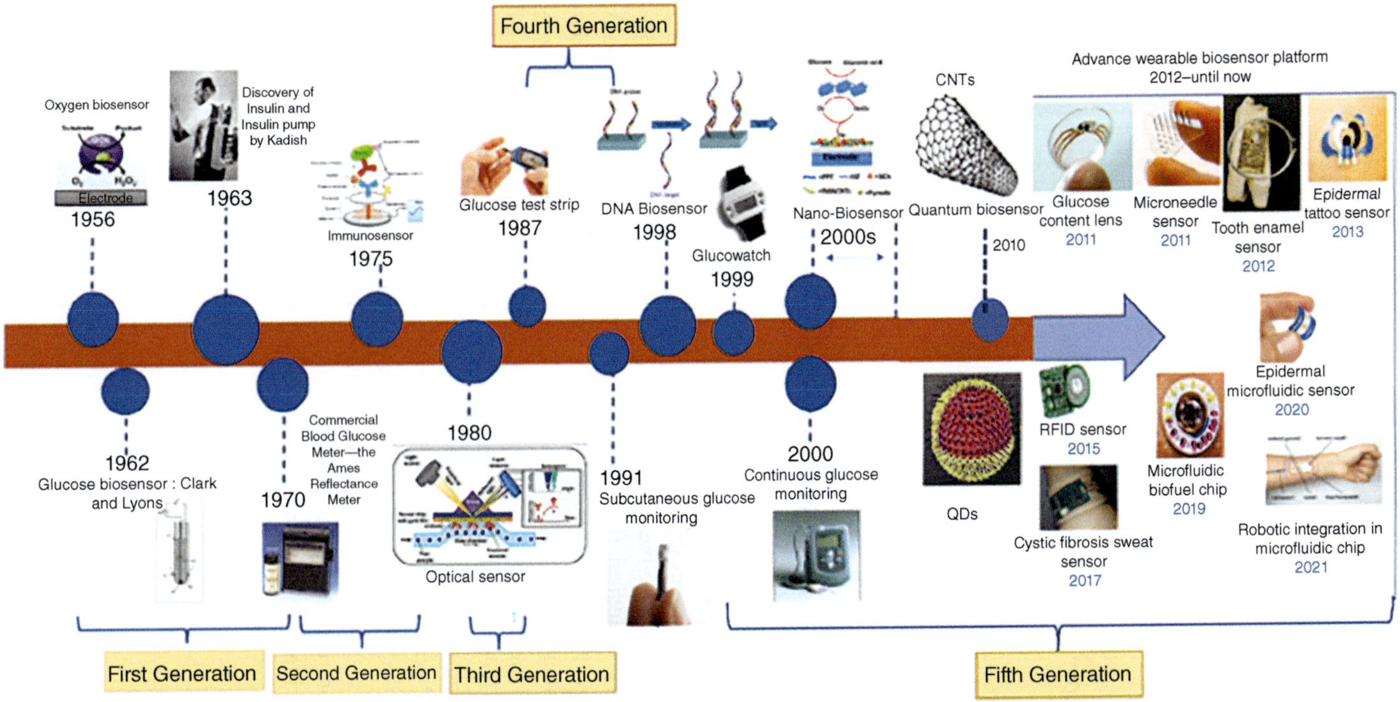

Figure 1.1 Evolution of biosensor technology, progressing from first-generation glucose biosensors to fifth-generation advanced wearable and microfluidic-integrated biosensors. Malik et al. [4] / with permission of Elsevier / CC BY 4.0.

Figure 1.2 The role of quantum biosensing across various fields of biology and its contribution to the advancement of biosensor technology. [10] / with permission of Twishmay.

biological systems, influencing phenomena like superposition, entanglement, coherence, and tunneling in biomolecular interactions and energy transfer. Furthermore, its applications span quantum photonics, information science, and complexity in biology, driving innovations in biosensing, neurobiology, and enzymatic regulation as detailed in Figure 1.2. By integrating QMs with biomedicine, quantum biosensors (QBs) enable the precise detection and characterization of subtle biological changes that conventional sensors often fail to identify, thereby transforming the landscape of biosensing and diagnostics. Furthermore, QBs exhibit a significant reduction in background noise compared to classical sensors, enhancing their ability to differentiate biophysical signals from background interference with exceptional sensitivity and specificity. Numerous QBs have been developed for diagnosing various diseases, yet their incorporation with next-generation sensing technologies may represent a groundbreaking advancement, offering exciting possibilities for future applications.

QB has come up as a transformative means in medical diagnostics, taking advantage of the principles of QMs to achieve unequaled precision and sensitivity in detecting biological analytes such as DNA, RNA proteins, and small biomolecules. These biosensors show quantum confinement effects, size-based optical properties, and exceptional electronic conductivity, making them highly responsive to micro-environmental changes induced by biomolecular interactions [11–13]. Upon interaction with target biomolecules, quantum materials exhibit measurable alterations in surface plasmon resonance (SPR), fluorescence intensity, electrical conductivity, etc., facilitating the transduction of biological signals into quantifiable outputs. Based on the transduction mechanism, various detection techniques are employed to interpret and generate the response, which is further explained in the subsequent section.

QBs serve as highly sensitive and precise tools with applications extending beyond medicine to agriculture and environmental monitoring. For instance, Chen et al. [14] explored the surface modifications and optical attributes of quantum dots (QDs) for detecting pollutants in aquaculture, focusing on their distinctive optical properties, including narrow emission bandwidths, broad excitation spectra, adjustable emission wavelengths, and resistance to photobleaching. Similarly, Safarpour et al. [15] developed a nanobiosensor for detecting plant pathogenic fungi by biofunctionalizing cadmium telluride QDs (CdTe-QDs) with a specific antibody against *Protostrongylus betae*, attaching the Glutathione-S-transferase (GST) antibody to thioglycolic acid-modified QDs. The formation of donor-acceptor complexes using rhodamine dye enabled the immunosensor to achieve 100% sensitivity and specificity with excellent stability and reproducibility. QDs have also proven to be an exceptional material for constructing biosensors with remarkable specificity and sensitivity for antibiotic detection. Additionally, their widespread integration into the detection of veterinary drug residues and pesticides underscores their significant impact on nanoscience-driven analytical advancements. The transition of quantum sensors from laboratory research to real-world applications is evident in the emergence of innovative commercial ventures in this domain. QBs, operating at atomic-scale dimensions with exceptional coherence properties, provide unprecedented spatial sensitivity and accuracy. Although assessing their full potential remains challenging, these quantum technologies hold immense promise for advancing biological research and diagnostics.

This introductory chapter provides a foundational understanding of biosensors by explaining their fundamental principles and working mechanisms. It then explores the concepts of QMs and the technological components integral to QB. This chapter also provides an overview of the diverse applications of QMs in medical diagnostics, showing their potential to revolutionize disease detection and monitoring with unparalleled precision. However, the scope of QMs extends far beyond medical diagnosis, playing a crucial role in various scientific and industrial fields, including environmental monitoring, food safety, and advanced material science. The subsequent chapters of this book will delve deeper into these applications in detail and their real-world implications.

1.2 Foundations of Quantum Mechanics in Biosensing

1.2.1 Basics of Biosensing Technologies

Biosensing technology combines biological recognition elements with physicochemical transducers to detect and quantify target analytes with high specificity and sensitivity.

A biosensor consists of three key components: a bioreceptor, a transducer, and a signal processor [16–18]. The bioreceptor, such as an aptamer, nucleic acid, antibody, or enzyme, selectively interacts with the target molecule, triggering a measurable biochemical or chemical event. The transducer then converts this event into a quantifiable signal, which may be optical, electrochemical [19], piezoelectric, or thermal. Finally, the signal processor amplifies and interprets the signal, providing a readable output corresponding to the analyte concentration. Moreover, biosensors employ various detection mechanisms tailored for specific applications, leading to different types of biosensors (Figure 1.3). Electrochemical biosensors detect biochemical reactions by measuring changes in current, voltage, or impedance at an electrode surface [20–22]. Optical biosensors utilize fluorescence, absorbance, or SPR to enable highly sensitive, real-time biomolecular detection without requiring labels or chemical modifications [23]. Piezoelectric biosensors, based on quartz crystal microbalance (QCM) technology, detect mass or mechanical changes upon analyte binding, providing high sensitivity [24, 25].

Recent advancements in nanotechnology have significantly enhanced the performance of biosensors. The incorporation of nanomaterials such as graphene, gold nanoparticles (AuNPs), and QDs has improved sensitivity, miniaturization, and real-time detection capabilities. For example, graphene-based biosensors exhibit excellent electrical conductivity and high surface area, which makes them ideal for electrochemical sensing applications [23]. Similarly, QDs, due to their size-dependent photoluminescence, enable multiplexed detection of multiple analytes in a single assay. The following sections provide an in-depth exploration of the fundamental principles of QMs that govern the optical and electronic properties of QDs, elucidating their role in biosensing applications.

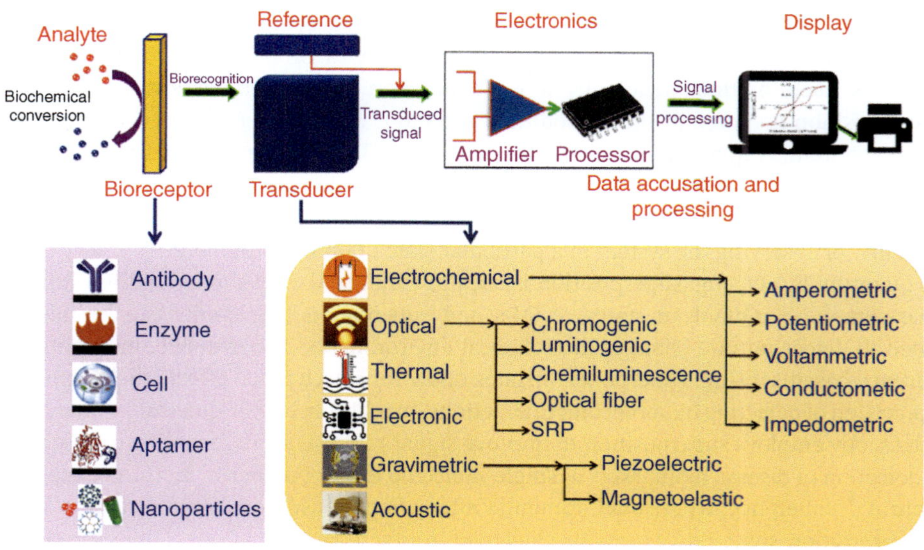

Figure 1.3 Overview of biosensor principle, where analyte undergoes biochemical recognition via bioreceptor, generating a signal transduced into electrical, optical, thermal, electronic, gravimetric, or acoustic responses. Mokhtari et al. [9]/ with permission of Elsevier / CC BY 4.0.

1.2.2 Principles of Quantum Mechanics

QMs is a fundamental theory in physics that governs the behavior of matter and energy at atomic and subatomic scales, where particles exhibit both particle-like and wave-like properties depending on how they are observed [26]. The foundation of QMs is built upon several key principles, including wave-particle duality, superposition, entanglement, and quantum tunneling. Wave-particle duality states that elementary particles, such as photons and electrons, can behave as discrete particles and continuous waves. Superposition is another fundamental concept, which asserts that a quantum system, until measured, can simultaneously exist in multiple states. Quantum entanglement is a phenomenon where two or more particles become correlated such that the state of one instantly influences the other, regardless of the distance between them [27]. Each of these principles plays an essential role in modern quantum technologies, including quantum biosensing, quantum computing, and cryptography. Quantum biosensing refers to the application of these quantum phenomena in biological detection and diagnostic systems [28]. By leveraging quantum properties, quantum biosensing enables highly sensitive, selective, and rapid detection of biomolecules, pathogens, and disease markers at ultralow concentrations. The integration of QDs, spin-based sensors, and quantum interference techniques has revolutionized biosensor technology, pushing the boundaries of detection limits and analytical performance [29]. In the following sections, the fundamental principles of QMs relevant to biosensor development will be explored in detail, highlighting their impact on emerging diagnostic technologies and next-generation biosensing platforms.

1.2.2.1 Quantum Superpositions

Quantum superposition, a basic principle of QMs, states that a quantum system, such as an electron, photon, or atom, can exist simultaneously in a combination of multiple states [30]. Unlike classical systems, which exist in well-defined states at any given time, quantum superposition allows particles to occupy a probabilistic distribution of states, described mathematically by a wave function. The mathematical foundation of superposition is rooted in the Schrodinger equation, which governs the evolution of a quantum system. A quantum state (Ψ) can be expressed as a linear combination of basis states: $|\psi\rangle = c1|\psi 1\rangle + c2|\psi 2\rangle + \cdots + cn|\psi n\rangle$, where c1, c2, …, c_n are probability amplitudes associated with each possible state [31]. The square of these amplitudes represents the probability of observing the system in a particular state upon measurement.

In quantum biosensing, superposition is used to enhance the sensitivity and resolution of measurements. Quantum states of QDs and superconducting qubits can be manipulated to detect minute changes in biological environments. For instance, in quantum coherence-based sensing, superposition states allow for interference effects that improve the precision of detecting biomolecular interactions. In addition to this, quantum-enhanced spectroscopy employs superposition to improve signal-to-noise ratios, enabling ultrasensitive detection of disease biomarkers at single-molecule levels. Furthermore, the concept of quantum decoherence, where environmental interactions cause a quantum system to lose its superposition state, poses a major challenge in practical applications. Understanding and controlling decoherence is essential for maintaining the integrity of quantum superposition in biosensing applications, ensuring accurate and reliable detection of biological targets.

1.2.2.2 Entanglement

The phenomenon of quantum entanglement takes place when two or more particles become intrinsically correlated, such that the state of one particle instantaneously influences the state of the other, regardless of the physical distance between them [32]. This unique property, which defines classical intuitions, has profound implications in biosensing and medical imaging, particularly enhancing the sensitivity and accuracy of techniques such as positron emission tomography (PET) [33]. Traditional PET imaging relies on detecting gamma-ray photons emitted during positron-electron annihilation but suffers from resolution limitations and background noise. By utilizing entangled photon pairs, PET imaging can achieve superior spatial and temporal resolution, reducing noise and significantly improving signal fidelity. This advancement enables the detection of early-stage disease biomarkers, which is crucial for conditions like cancer [34, 35], neurodegenerative disorders [36], and metabolic syndromes. Beyond imaging, quantum entanglement is also being explored in biosensors for real-time molecular detection. Quantum-enhanced fluorescence imaging, entanglement-assisted spectroscopy, and spin-based magnetic field sensing have demonstrated exceptional sensitivity in tracking biomolecular interactions at the single-molecule level. However, practical implementation faces challenges such as quantum decoherence in biological environments, which can rapidly degrade entangled states. Currently, research is focused on developing quantum error correction techniques, noise-resistant entangled states, and cryogenic quantum sensing platforms to overcome these limitations.

1.2.2.3 Quantum Tunneling

Another pivotal principle of QMs is quantum tunneling, which allows particles to pass through energy barriers [37]. This phenomenon arises due to the wave-like nature of quantum particles, described by the Schrodinger equation, which allows for nonzero probability amplitudes extending beyond an energy barrier. Mathematically, the probability of tunneling is governed by the transmission coefficient in the time-independent Schrodinger equation, which is exponentially dependent on the barrier width and height [27]. Electron tunneling sensors work by measuring minute variations in electrical conductance when a target biomolecule interacts with the sensor surface. These sensors utilize nanoscale junctions, such as metal-insulator-metal (MIM) and single-molecule transistor structures, where electron flow across an insulating barrier is modulated by molecular interactions [38]. Changes in tunneling current can indicate the presence of specific biomolecules, allowing for highly sensitive detection. Furthermore, quantum tunneling-based biosensors can detect viral RNA or antigenic proteins at femtomolar concentrations, significantly surpassing the sensitivity of conventional polymerase chain reaction (PCR) or enzyme-linked immunosorbent assay (ELISA) techniques. In addition to this, in label-free biosensing, tunneling-based field-effect transistors (FETs) and molecular break-junction sensors allow real-time monitoring of biomolecular interactions, eliminating the need for fluorescent or enzymatic labeling, which can introduce artifacts and lower detection fidelity [39].

1.2.3 Technological Components of Quantum Biosensing

QBs use the principles of QMs to achieve precise and highly sensitive detection of biomolecules. Their functionality is based on the ability to transduce biomolecular interactions into quantifiable signals, including variations in electrical conductivity, shifts in SPR, and fluorescence modulation. Such signal transduction mechanisms significantly enhance sensitivity and specificity, making QBs highly advantageous for medical diagnostics and early disease detection. A diverse range of quantum-engineered platforms contributes to the development of these biosensors, including QDs, nitrogen-vacancy (NV) centers in diamonds, quantum photonics, and light-matter interactions [28]. These technologies play a crucial role in advancing biosensing applications by enabling fast detection, multiplexed analysis, and single-molecule resolution. The subsequent sections will explore these key technological components in detail, elucidating their underlying mechanisms, practical applications, advantages, and the challenges associated with their implementation in real-world biosensing systems.

1.2.3.1 Quantum Dots

QDs are semiconductor nanocrystals that exhibit unique optical and electronic properties [40] due to their nanoscale dimensions, ranging from 2 to 10 nm. These properties arise from the quantum confinement effect, where the electronic characteristics of the material are influenced by its size and shape, allowing for diverse photoluminescence and high stability against photobleaching. The applications of QDs extend to bioimaging and medicine, where they serve as fluorescent biolabels, sensors, and drug delivery agents; quantum information, facilitated by their bright, narrowband emission and potential use in lasing and heterostructures; communications, where QDs contribute to advanced optical and electronic signaling; machine vision, aiding in enhanced imaging and recognition technologies; augmented reality, enabling improved interactive displays; cameras and displays, where their size-dependent absorption and emission properties enhance imaging quality; illumination, improving light sources due to their efficient luminescence; and energy harvesting, where they contribute to photovoltaic cells and solar concentrators. These wide-ranging applications stem from the unique quantum confinement effects and tunable electronic properties of QDs, as shown in Figure 1.4. It possesses a high surface area-to-volume ratio, making it suitable for bioconjugation with various biomolecules, including antibodies, nucleic acids, and peptides, thereby enhancing sensitivity and specificity in biosensing applications [42].

Although a universally accepted classification framework for QDs has yet to be established, the scientific community has recently categorized them into two broad classes: semiconductor quantum dots (SQDs) and carbon-based nanomaterials [43, 44], especially carbon quantum dots (CQDs). SQDs are primarily composed of metallic and other inorganic elements, exhibiting size-dependent electronic and optical properties due to quantum confinement effects. These materials, often synthesized from compounds such as cadmium selenide (CdSe), indium phosphide (InP), and lead sulfide (PbS) [45], demonstrate high quantum yield and tunable emission spectra, making them highly effective in optoelectronic and biosensing applications. In contrast, CQDs encompass a diverse range of carbon-based nanomaterials, including graphene quantum dots (GQDs) [46], carbon nanodots (CNDs) [47], and polymer dots (PDs). The following flowchart outlines the classification of QDs. Apart from their semiconductor counterparts, CQDs offer inherent biocompatibility, reduced cytotoxicity, and strong photostability, making them particularly suitable for biomedical applications like drug delivery, bioimaging, and fluorescence-based

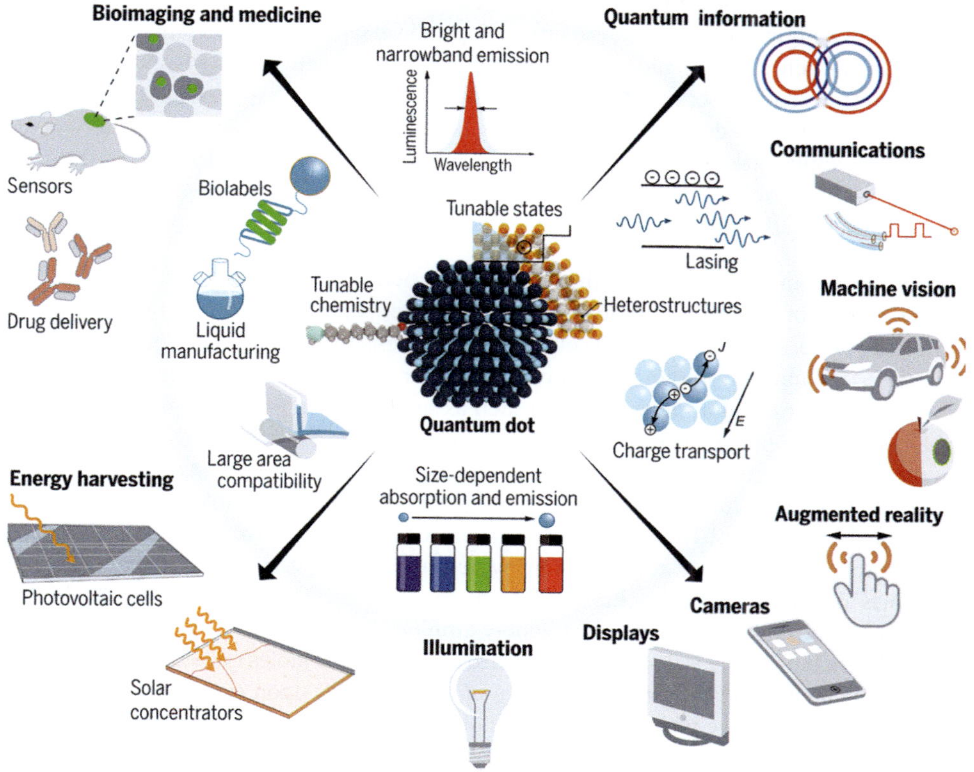

Figure 1.4 Diverse applications of semiconductor quantum dots, which extend up to bioimaging and medicine (fluorescent biolabels, sensors, drug delivery), quantum information (narrowband emission, lasing, heterostructures), biosensing (optical and electronic signaling), machine vision (enhanced imaging, recognition), augmented reality (interactive displays), cameras and displays, and energy harvesting (photovoltaics, solar concentrators). García de Arquer et al. [41] / American Association for the Advancement of Science / CC BY 4.0.

biosensing [48]. The distinct physicochemical properties of both categories have paved the way for their integration into advanced quantum biosensing technologies, enabling highly sensitive and selective detection of biomolecules at the molecular level.

1.2.3.2 Nitrogen-Vacancy (NV) Centers in Diamonds

The nitrogen-vacancy (NV), an intriguing defect in diamond crystal structures, has gained notable attention in recent years due to its exceptional properties and potential applications in quantum technologies, particularly in biosensor development [49, 50]. An NV center is formed when a nitrogen atom replaces a carbon atom in the diamond lattice and an adjacent carbon atom is missing, leading to a vacancy [51]. This defect gives the diamond unique quantum properties, such as the ability to exhibit optically detected electron spin resonance at room temperature [52]. NV centers utilize superposition in electron spin states to detect minute variations in magnetic fields, enabling highly precise mapping of biomolecular interactions and intracellular dynamics [77]. The highly sensitive nature of the electron spin in NV centers allows these diamonds to function as effective quantum bits (qubits) for quantum information processing and versatile nanoscale sensors. Furthermore, the NV centers in diamonds possess remarkable magnetic sensitivity, which stems from their spin states that can be manipulated and read optically [53]. These characteristics enable NV centers to detect minute changes in their environment, making them ideal for various applications in biosensing technologies. The emitted fluorescence from the NV center when excited by light serves as a powerful tool for quantitative measurement, further enriching their utility in biological sensing applications.

1.2.3.3 Quantum Photonics and Light-Matter Interaction

The principles of quantum photonics and light–matter interactions play a pivotal role in advancing biosensor technologies by enabling highly sensitive and precise detection mechanisms [54]. Quantum photonics involves the generation, control, and manipulation of photons exhibiting quantum phenomena such as superposition and entanglement, allowing for enhanced optical sensing and quantum-level interrogation of biological systems [55]. On the other hand, light-matter interaction encompasses the fundamental processes that occur when photons interact with matter, ranging from atoms and molecules to complex biomolecular structures. These interactions give rise to various optical phenomena, including absorption, scattering, and emission, which are governed by quantum mechanical principles dictating energy exchange and electronic state transitions within the interacting material [56]. One of the most widely employed techniques showing light-matter interaction in biosensing is fluorescence spectroscopy, which enables molecular-level investigations by measuring the emission spectra of biomolecules upon excitation by incident photons. This technique provides critical insights into cellular dynamics, protein interactions, and nucleic acid hybridization, making it vitally important in biomedical research and diagnostics. Another easy-advanced biosensing approach, SPR sensing, exploits the resonant oscillations of conduction electrons at the interface between a metal and a dielectric material. The resonance condition is highly sensitive to refractive index changes caused by the binding of biomolecules, making SPR a powerful tool for real-time, label-free detection of biomolecular interactions. This technology has been extensively utilized in diverse biomedical applications, including blood glucose monitoring, early-stage cancer biomarker detection, and pathogen identification. As quantum photonics and light-matter interactions continue to be refined, their integration into biosensors is expected to further revolutionize diagnostic platforms, enabling ultra-sensitive, high-resolution, and multiplexed detection systems. Future advancements may include the incorporation of entangled photon pairs for enhanced signal-to-noise ratio, quantum-enhanced Raman spectroscopy for single-molecule analysis, and nanoplasmonic structures for adjustable biosensing platforms, thereby crossing the boundaries of precision in real-time disease diagnostics.

1.3 Applications of Quantum Biosensing in Diagnostics

The efficacy of disease management is based on early and precise diagnosis, which necessitates a comprehensive understanding of its pathophysiology and progression. At the onset of the twenty-first century, the emergence of quantum biosensing has revolutionized medical diagnostics, enabling ultrasensitive, high-specificity, and real-time detection of biomarkers at the molecular level. By integrating quantum-enhanced sensing technologies with traditional diagnostic methodologies, researchers have unlocked unparalleled capabilities in oncology, neurology, cardiovascular disease diagnostics, infectious disease surveillance, and pharmaceutical research. These advancements leverage quantum phenomena such as quantum confinement, entanglement, and superposition, offering miniaturized, non-invasive, and highly efficient biosensing platforms. The following sections delve into the multifaceted applications of QBs, examining their technological foundations, diagnostic advantages, and future potential in reshaping the landscape of precision medicine and biomedical research.

1.3.1 Fluorescent Probes in Medical Diagnostics

One of the most significant contributions of quantum biosensing in diagnostics is the use of fluorescent quantum probes, which surpass conventional fluorescent dyes and organic fluorophores in terms of brightness, photostability, and tunability. A fluorescent probe fundamentally consists of three key components: a recognition unit that selectively binds to the target analyte, a fluorophore responsible for light emission, and a connector or linker that facilitates interaction between the recognition unit and the fluorophore, as illustrated in Figure 1.5a [57]. A diverse array of fluorescent probes has been engineered for the diagnosis of complex neurological disorders, including Alzheimer's disease (AD), stroke, depression, Wilson's disease, epilepsy, and Parkinson's disease (PD).

For instance, in 2022, Kim and colleagues designed Probe 5, an NIR fluorescent probe specifically tailored for detecting peroxynitrite ($ONOO^-$) [60], a reactive nitrogen species implicated in oxidative stress and neurodegeneration. Probe 5 is made up of diamino-substituted rhodol dye as the fluorophore, while the 4-aminophenol moiety serves as the selective $ONOO^-$ recognition unit. This molecular design enabled *in situ* imaging of $ONOO^-$ in AD mouse models, allowing Kim et al. to establish that $ONOO^-$ functions as a potential biomarker for AD, as shown in Figure 1.5b. The ability to visualize and quantify such reactive species in living biological systems underscores the significance of fluorescent quantum probes in advancing biomedical imaging, early disease diagnostics, and precision medicine applications.

1.3.2 Applications of Quantum Biosensing in Oncology

Cancer remains one of the leading causes of mortality worldwide, with an alarming 19.3 million new cases and nearly 10 million cancer-related deaths reported in the twenty-first century. A major contributing factor to high cancer mortality is late-stage diagnosis, which significantly limits therapeutic efficacy and survival rates. Consequently, early and accurate detection of cancer-specific biomarkers is crucial for improving patient prognosis. In this context, QBs have emerged as highly sensitive diagnostic tools, offering exceptional precision in detecting circulating tumor cells (CTCs), extracellular vesicles, tumor-specific genetic mutations, and oncogenic biomarkers [61]. Recent advancements in nanomaterial-based QB have demonstrated the potential of GQDs in electrochemical biosensing for early cancer detection. Notably, Ganganboina et al. (2019) leveraged

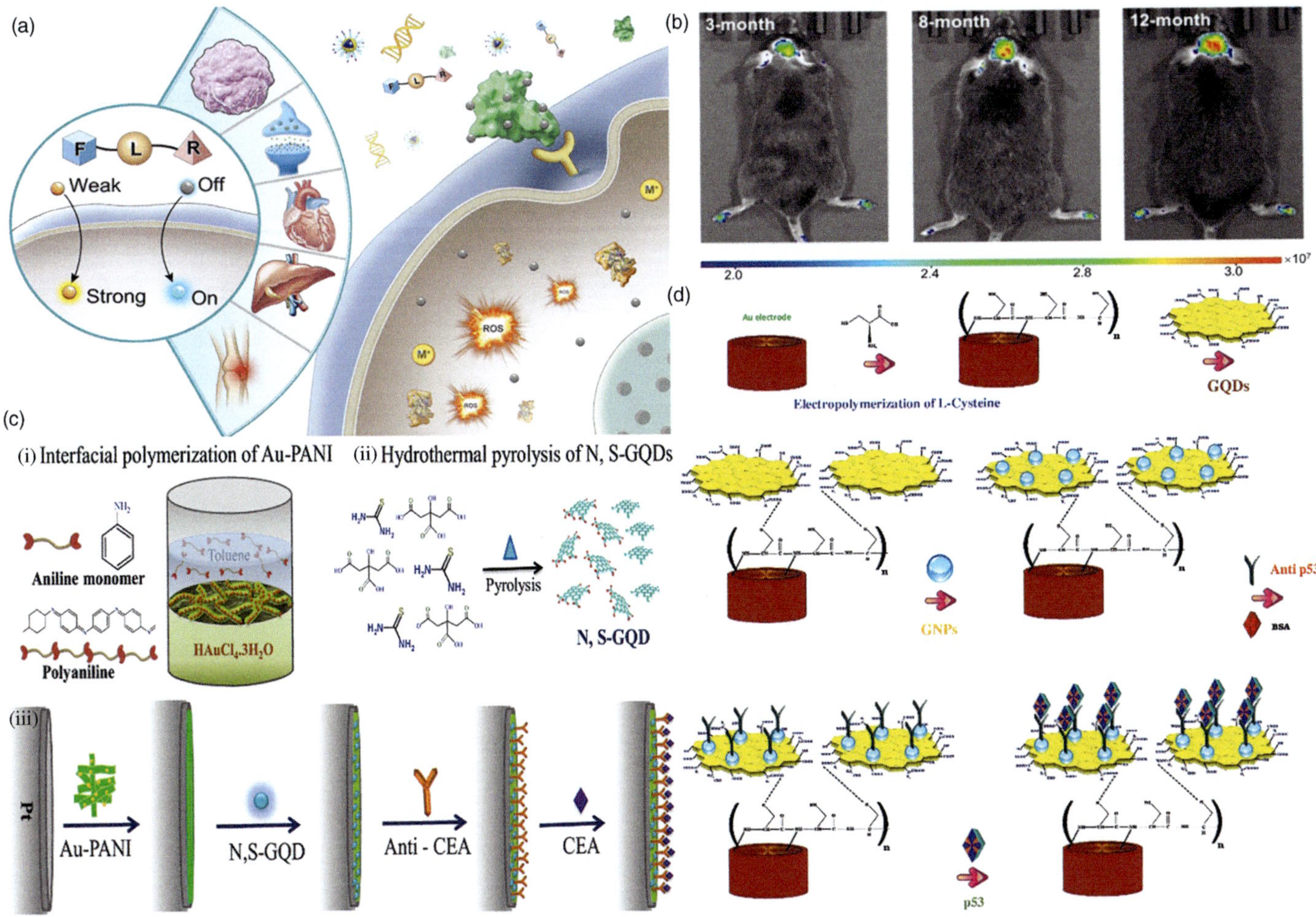

Figure 1.5 (a) Overview of fluorescent probes for medical diagnostics. (b) *in vivo* fluorescence imaging of AD mice at 3, 8, and 12 months via tail injection of Probe 5, indicating age-related elevation of $ONOO^-$ in the brain. Wang et al. [57] / with permission of American Chemical Society / CC BY 4.0. (c) (i) Synthesis of Au-PANI through polymerization, (ii) hydrothermal pyrolysis of N, S-GQD, and (iii) development of a label-free CEA electrochemical impedimetric immunosensor. Ganganboina and Doong [58] / Springer Nature / CC BY 4.0. (d) Label-free electrochemical immunosensor for detecting the p53, incorporating ternary signal amplification through P-Cys, GQDs, and GNPs to enhance sensitivity and performance. Hasanzadeh et al. [59] / with permission of Elsevier.

the inherent synergy between GQDs and electrochemical biosensors to develop an ultrasensitive detection platform. They employed N, S-doped GQDs@Au-polyaniline (N, S-GQDs@Au-PANI) nanowires as depicted in Figure 1.5c to detect carcinoembryonic antigen (CEA), a well-established biomarker, achieving an exceptionally low limit of detection (LoD) of 0.01 ng/mL. This approach underscores the potential of quantum-enhanced biosensing platforms in real-time cancer biomarker detection [58]. Similarly, Hasanzadeh et al. developed a P-Cys/GQDs/GNPs nanocomposite-based biosensor for the electrochemical detection of suppressor protein p53, as shown in Figure 1.5d, a key regulator of genomic stability. The integration of QDs and AuNPs enhanced the electrochemical response, enabling ultra-sensitive p53 quantification [59].

1.3.3 Quantum Biosensors for Cardiovascular Disease Detection

Cardiovascular diseases (CVDs) constitute a broad spectrum of disorders affecting the heart and vascular system, including coronary artery disease, heart failure, stroke, and hypertension. A primary underlying cause of these conditions is atherosclerosis, characterized by the progressive accumulation of lipid plaques within arterial walls, leading to vascular occlusion, impaired blood circulation, and an elevated risk of myocardial infarction and stroke [62]. Because of the asymptomatic nature of early-stage CVDs, their timely diagnosis is crucial for effective intervention and disease management. QBs have revolutionized cardiovascular diagnostics by enabling the ultrasensitive detection of key cardiac biomarkers, such as troponins, C-reactive protein (CRP), and procalcitonin (PCT), at exceptionally low concentrations. Among their most significant applications is the detection of cardiac troponin I (cTnI), a highly specific biomarker for myocardial infarction. To achieve this, photoelectrochemical biosensors incorporating CdSe/CdS/ZnS core-shell QDs in conjunction with TiO_2 heterostructures have been synthesized (Figure 1.6a), demonstrating enhanced photocurrent responsiveness, a broad dynamic detection range (10 pg/mL to 0.2 ng/mL), and remarkable chemical stability. Furthermore, QD-based immunochromatographic strips have been designed for the rapid and precise detection of CRP and PCT, both of which serve as critical indicators of myocardial inflammation and cardiovascular disease risk. These label-free, high-sensitivity diagnostic platforms not only facilitate early disease detection but also offer a non-invasive, portable, and cost-effective alternative to conventional biochemical assays [66]. As advancements in quantum biosensing technology continue to evolve, its integration into clinical cardiology holds immense potential for transforming point-of-care diagnostics, real-time monitoring, and personalized treatment strategies in CVD management.

1.3.4 Quantum Biosensing in Infectious Disease Diagnostics

Infectious diseases driven by highly adaptive pathogens account for nearly 15% of global mortality [67]. The transmission routes could be direct contact, contaminated substrates, vector mediation, and aerosolized dispersal. The increasing threat of antimicrobial resistance necessitates the development of advanced biosensing technologies that are crucial for early diagnosis, outbreak control, and therapeutics. Quantum-enabled biosensors can

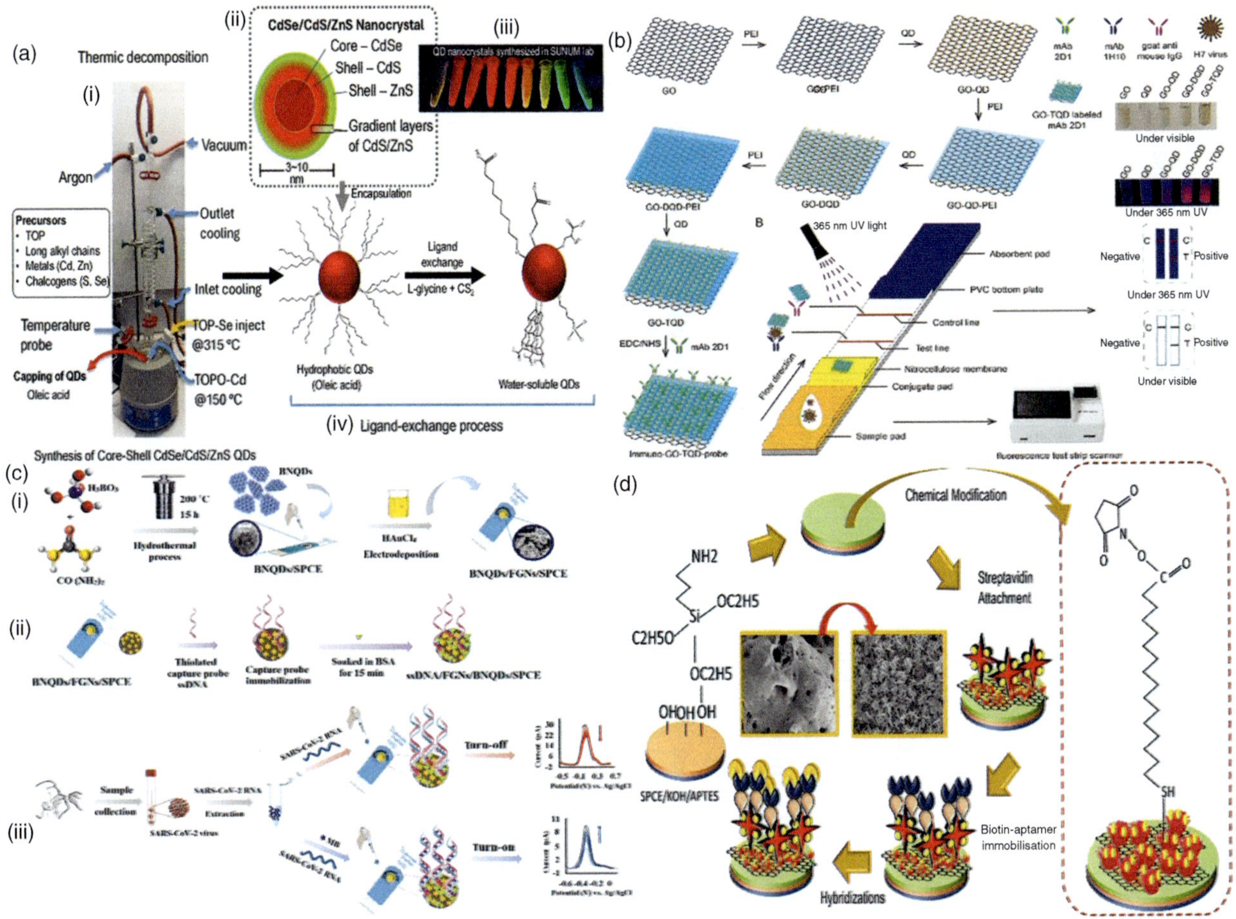

Figure 1.6 (a) Synthesis of core-shell CdSe/CdS/ZnS quantum dots (QDs) for cardiac troponin detection of cardiovascular disease monitoring. Memon et al. [62] / with permission of Elsevier. (b) GO-TQD nanoprobe and GO-TQD LFA were designed for the direct detection of H_7AIVs, enabling sensitive and specific analysis of avian influenza virus. Fu et al. [63] / with permission of Elsevier / CC BY 4.0. (c) Electrochemical biosensor was developed for the sensitive and specific detection of SARS-CoV-2. Hatamluyi et al. [64] / with permission of Elsevier. (d) An LSG/MoS_2/N-GQDs/L-Ag NPs hybrid was synthesized on a modified surface to facilitate aptamer–troponin I interaction, enhancing detection sensitivity and specificity. Vasudevan et al. [65] / with permission of Elsevier.

detect viral RNA [68–70], bacterial antigens, and inflammatory cytokines with ultrafast response times and minimal sample volumes. These technologies are being applied to pandemic preparedness, point-of-care testing, and antimicrobial resistance monitoring, ensuring early and accurate detection of infectious agents. Notable breakthroughs include graphene oxide multilayer quantum dot-based immunochromatographic strips (Figure 1.6b), which detect the H7N9 avian influenza virus with an exceptionally low detection limit of 0.016 ng/mL for hemagglutinin protein [63]. Similarly, a boron nitride quantum dot-based biosensor, integrated with gold nanostructures, has shown 100% specificity and sensitivity for SARS-CoV-2 detection, eliminating the need for nucleic acid amplification and significantly outperforming conventional RT-PCR methods [64]. Additionally, fluorescent biosensors leveraging Förster resonance energy transfer (FRET) between QDs and two-dimensional metal-organic frameworks, as shown in Figure 1.6c, have been successfully employed to detect the *Mycobacterium tuberculosis* IS6110 gene fragment, achieving a detection limit as low as 35 pM [71].

1.3.5 Nanoparticles in Quantum Biosensing

Nanoparticles are revolutionizing quantum biosensing by enabling highly sensitive and precise biomolecular detection. The integration of nanoparticles in quantum biosensing has further enhanced detection efficiency, with AuNPs, GQDs, and plasmonic nanostructures enabling highly sensitive, multiplexed, and miniaturized biosensors. Materials like QDs and AuNPs stand out due to their unique optical and electronic properties, such as fluorescence and SPR, which significantly enhance signal transduction in biosensors. These nanoparticles amplify sensor signals, allowing for real-time, high-resolution monitoring in medical diagnostics and environmental analysis, paving the way for more accurate and efficient detection methods [18]. In a recent study, Vasudevan et al. developed a biosensor using a quadruplet hybrid of 3D laser-scribed graphene (LSG), molybdenum disulfide (MoS_2), nitrogen-doped graphene quantum dots (N-GQDs), and lignin-based silver nanoparticles (L-AgNPs) to detect troponin I, as depicted in Figure 1.6d, a critical biomarker for myocardial injury. The integration of L-AgNPs improved biomarker recognition and binding stability, offering a cost-effective and environmentally friendly alternative to traditional gold nanoparticles [65]. Similarly, Gharehaghaji et al. designed a biosensor incorporating self-assembled Prussian blue (PB) nanoparticles on MXene quantum dots, supported by AuNPs, for the detection of ovarian cancer antigen (CA-125). The AuNPs in this system enhanced conductivity and facilitated the immobilization of CA-125 antibodies, ensuring highly specific and efficient biomarker capture. These cutting-edge nanoparticle-based biosensors demonstrate remarkable sensitivity, stability, and real-time detection capabilities, making them promising tools for early disease diagnosis and clinical applications [72].

1.3.6 Quantum Optical Sensors for Biomedical Applications

Finally, quantum optical sensors are also gaining prominence in biomedical applications, utilizing quantum-enhanced light-matter interactions to achieve superior detection limits in biospectroscopy, biophotonics, and precision diagnostics. Quantum optical sensors are transforming biomedical applications by offering high precision and real-time detection of biological molecules. By utilizing quantum principles like fluorescence enhancement and SPR [73, 74], these sensors significantly improve sensitivity in disease diagnostics. With advancements in microfluidics and artificial intelligence (AI) integration, they are proving to be valuable tools for personalized medicine, continuous health monitoring, and early disease detection [23]. Zhang et al. developed an optical fiber biosensor to detect acetylcholine, a crucial neurotransmitter, using pH-sensitive nitrogen-doped carbon QDs (N-CQDs) [75]. The sensor works by measuring changes in fluorescence—when acetylcholine is broken down by acetylcholinesterase, it causes a drop in pH, which in turn reduces the fluorescence intensity (Figure 1.7c). This simple yet highly sensitive approach makes it a valuable tool for monitoring neurotransmitter activity [78]. Similarly, Loyez et al. designed an optical fiber-based SPR (OF-SPR) sensor for detecting HER2, a key biomarker for breast cancer (Figure 1.7a,b). Their sensor uses a sandwich assay with aptamers and antibodies, which significantly boosts the signal, making detection much

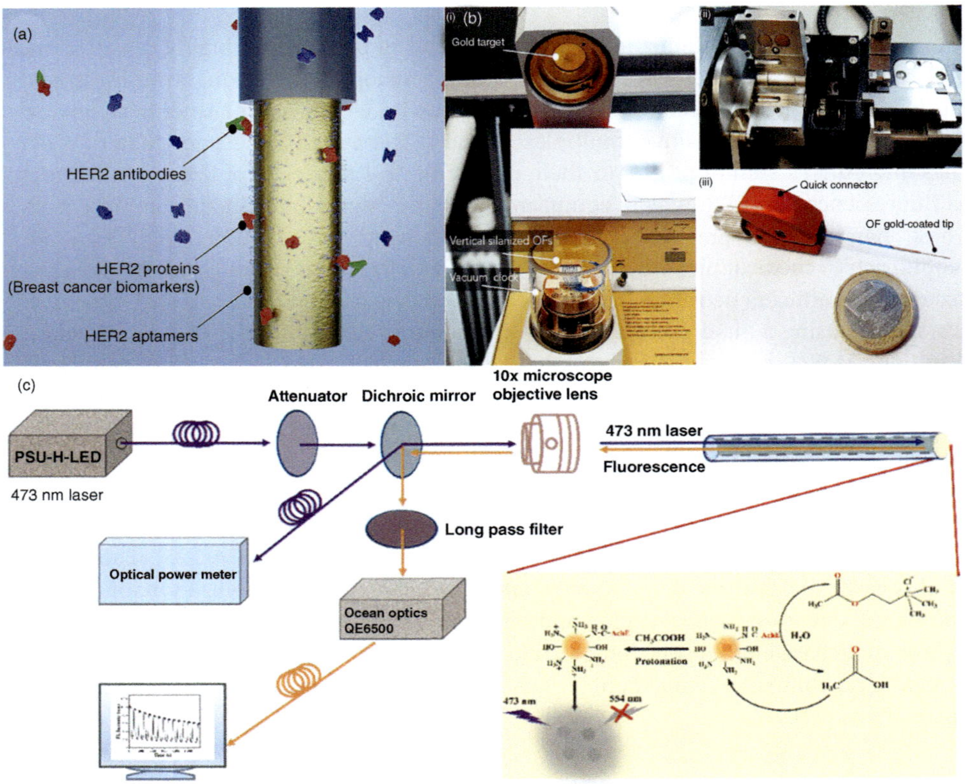

Figure 1.7 (a,b) Detection of the breast cancer biomarker HER2 using a sandwich-type surface plasmon resonance optical biosensor. Loyez et al. [76] / with permission of Elsevier. (c) Optical fiber biosensor for acetylcholine detection based on pH-sensitive fluorescent carbon quantum dots. Zhang et al. [78] / with permission of Elsevier.

more precise. By carefully optimizing the thickness of the gold coating on the optical fiber, they achieved ultra-sensitive detection, making this technology a promising candidate for early cancer diagnosis and improved patient care [76].

Furthermore, various types of materials have been utilized to enhance detection range and sensitivity, as summarized in Table 1.1. Beyond biomedical applications, QB is emerging as a powerful tool in environmental monitoring and public health surveillance. Quantum sensors can detect airborne pathogens, waterborne contaminants, and toxic heavy metals at extremely low concentrations, playing a crucial role in epidemiological studies and pollution monitoring. The integration of nanoparticles in QB has further enhanced detection efficiency, with gold nanoparticles, GQDs, and plasmonic nanostructures enabling highly sensitive, multiplexed, and miniaturized biosensors. However, the scope of QMs in diagnostics extends far beyond these applications. The subsequent

Table 1.1 Various types of quantum materials used in medical diagnosis.

S. No.	Material	Analyte	LOD	References
1	NV Centers in Diamonds	mi-RNA	2.6 aM	[78]
2	Quantum Dots	Serum albumin	0.2129 μM	[79]
3	Carbon Nanotubes	Paraoxon-ethyl	10 nM	[80]
4	Quantum Photonics	Glucose	1 mM	[81]
5	Single Electron Transistors	Dibutyl phthalate	0.08 fM	[82]
6	Copper Nanoclusters	Kanamycin	1.07 nM	[83]
7	Graphene Quantum Dots	Peroxide	20 nM	[84]
8	Silicon Quantum Dots	Virus	0.08 nM	[85]
9	Carbon Quantum Dots	Pyocyanin	37 nM	[86]
10	Gold Nanoclusters	micro-RNA 21	0.7 pM	[87]
11	Platinum Nanoclusters	Glutathione	0.37 μM	[88]
12	Copper Indium Sulfide Quantum Dots	Cholesterol	0.222 mM	[89]
13	Zinc Sulfide Quantum Dots	Bilirubin	1.8 μM	[90]
14	Cadmium Quantum Dots	DNA	1 fM	[91]
15	Silicon Nanowires	Prostate-specific antigen	0.01 pM	[92]
16	Gold Polyaniline Nanowires	Carcinoembryonic antigen	0.01 nM	[58]
17	Zinc Oxide Nanowires	Glycosylated hemoglobin A1c	61.8 nM	[93]
18	Boron Nitride Quantum Dots	Cardiac troponin	0.0005 nM	[94]
19	Molybdenum Disulfide Quantum Dots	Prostate-specific antigen	0.01 pM	[95]
20	Quantum Raman Spectroscopy	Folic acid	9 nM	[96]

chapters will provide an in-depth exploration of its broader potential and emerging innovations in the field.

1.4 Role of Artificial Intelligence in Quantum Biosensing

Recently, AI, 3D bioprinting [97], and the Internet of Things (IoT) have emerged as cutting-edge technologies that have been integrated to enhance conventional methodologies. These advancements have significantly transformed modern healthcare, facilitating the implementation of digital prescriptions, point-of-care diagnostics, and personalized drug monitoring systems, thereby improving efficiency, accessibility, and patient outcomes [98]. This convergence fosters a decentralized healthcare model, which is connected to IoT and has reduced the burden on medical professionals [99]. AI plays a pivotal role in precision medicine, offering breakthroughs in disease prevention, early diagnosis, and targeted treatment [28]. The success of AI-driven healthcare depends on its equitable, robust, and ethically responsible deployment, particularly in domains such as AI-assisted diagnostics, treatment optimization, wearable and implantable biosensors, clinical decision support, and bioinformatics [100]. Future IoT-based medical systems are expected to incorporate Green IoT, emphasizing energy efficiency, waste reduction, and minimal environmental impact. The emergence of the Internet of Medical Things (IoMTs) further revolutionizes healthcare by integrating software-driven, network-connected biomedical devices for real-time health monitoring [101]. Electrochemical biosensors, with their rapid analysis times, cost-effectiveness, and adaptability, are poised to play a crucial role in precision medicine and improving patient outcomes.

The rapid expansion of biomedical data demands high-performance computing tools such as cloud-based computing, GPUs, and TPUs to efficiently process complex QB datasets [102]. High-performance computing (HPC) clusters and distributed computing enable real-time data interpretation, feature extraction, and pattern recognition, accelerating diagnostics. GPU acceleration enhances DL models for image analysis and signal processing, while TPUs optimize large-scale AI-driven medical applications with improved energy efficiency [103]. machine learning (ML) techniques, including regression analysis and classification, play a crucial role in biosensor data analysis, enabling nanobiosensors to extract meaningful insights through advanced pattern recognition. ML reduces background noise, false positives, and cross-sensitivity, improving biosensor accuracy and detection limits [104]. Additionally, ML-driven advancements in fluorescent nanodiamonds (FNDs) for intracellular magnetic microscopy have led to 20-fold enhancements in imaging contrast and resolution. DL techniques further refine magnetic image reconstruction, boosting biosensor imaging precision. Fluorescence-based biosensors benefit from on-chip data processing, noise suppression, and segmentation, enabling real-time detection of specific cells or tissues.

Furthermore, the integration of AI-driven processors with medical test strips enhances real-time biosensor monitoring, mitigating inaccuracies caused by environmental fluctuations or sample contamination. Advancements in biomedical engineering, biosensor miniaturization, and long-term implantable sensing technologies will be instrumental in the development of next-generation high-performance biosensor systems.

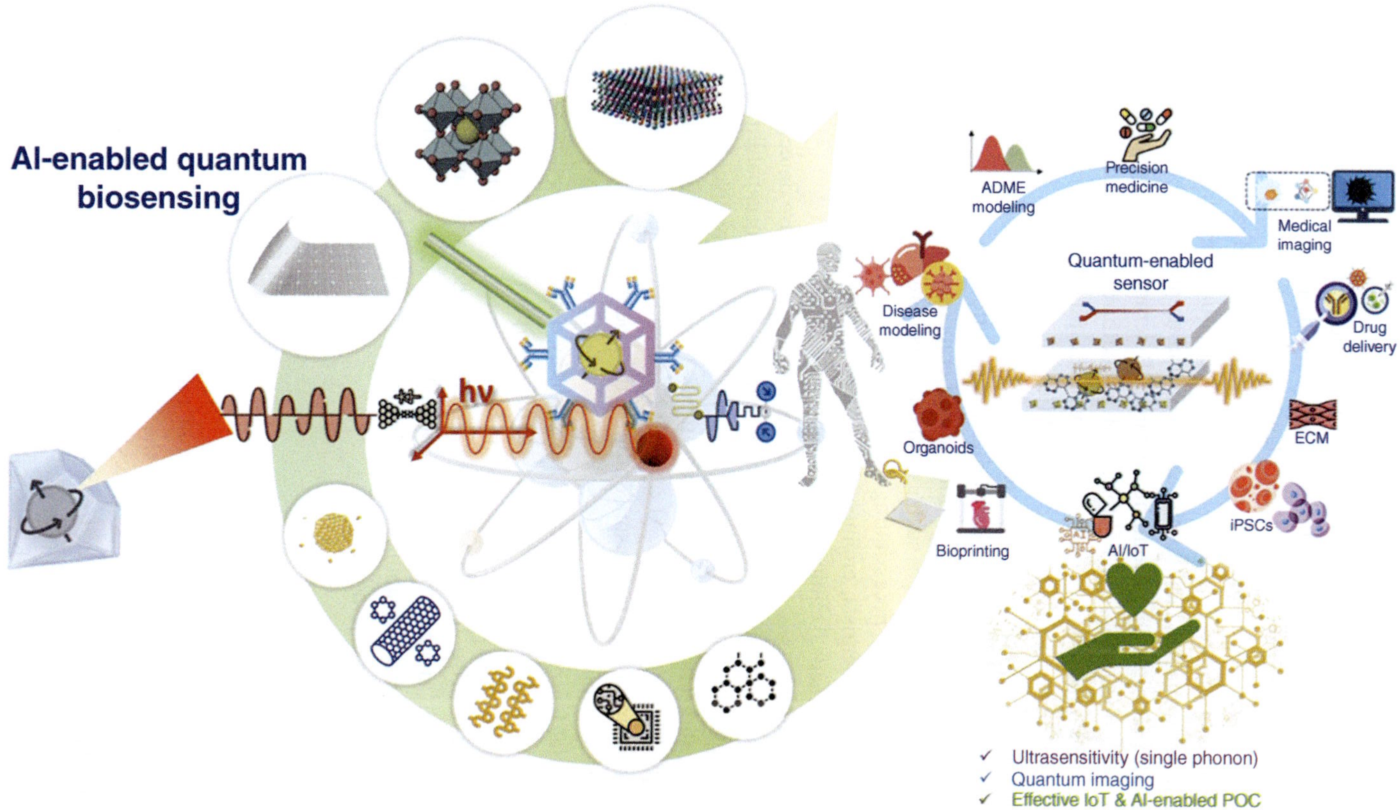

Figure 1.8 Futuristic plan of integration of AI into quantum biosensing to revolutionize medical diagnostics. Das et al. [28] / IOP Publishing / CC BY 4.0.

1.5 Conclusion

Quantum biosensing is revolutionizing modern diagnostics by offering incomparable sensitivity, specificity, and real-time biomolecular detection through fundamental QMs principles, including superposition, entanglement, and tunneling. These biosensors enable single-molecule detection, overcoming the limitations of conventional methods. The integration of QDs, NV centers in diamonds, and quantum photonics has further enhanced their performance, facilitating early precision medicine, disease detection, and continuous health monitoring. Beyond healthcare, QB demonstrates significant potential in environmental monitoring, food safety, and agriculture, highlighting its versatility across scientific disciplines. The convergence of AI, ML, and high-performance computing with QBs has expanded their diagnostic capabilities, enabling faster data analysis, noise reduction, and advanced pattern recognition in complex biological systems, as shown in Figure 1.8. These advancements support critical applications such as cancer biomarker identification, infectious disease detection, and therapeutic monitoring. Despite their transformative potential, challenges such as scalability, coherence preservation, specificity, and cost-efficiency remain, necessitating continuous research and innovation. Collaborative efforts among scientists, engineers, and medical professionals will be crucial in translating QBs into practical clinical applications.

As the field progresses, QBs are expected to redefine early disease detection, personalized treatment, and global healthcare outcomes. Their multiplexing capabilities, real-time monitoring, and precise biomarker detection enable rapid, non-invasive diagnostics, with significant applications in quantum-enhanced medical imaging, pathogen detection, and targeted therapy. The integration of AI and data analytics further improves diagnostic precision, while point-of-care biosensors optimize treatment regimens, reducing medication resistance risks. However, ethical, legal, and societal considerations-including privacy concerns, potential biases, and healthcare accessibility disparities-must be addressed to ensure equitable distribution and seamless clinical implementation. With continuous advancements, the synergy between nanotechnology, quantum computing, and biosensor miniaturization will drive the development of efficient, portable, and ultra-sensitive diagnostic platforms. Quantum biosensing is poised to redefine biomedical research and diagnostics, unlocking new possibilities for early disease intervention, precision medicine, and global healthcare improvements.

Acknowledgments

The authors are grateful to the Department of Biotechnology (DBT), New Delhi [BT/PR52345/BSA/33/37/2024 (CN19592)], and the Department of Atomic Energy (DAE)—Board of Research in Nuclear Sciences (BRNS), Mumbai (58/14/16/2024-BRNS/1135), for their research support. J.S. would like to acknowledge the University Grants Commission (UGC) Fellowship (221610016405) provided by the UGC, New Delhi. H V Muneshwar would like to acknowledge the CSIR fellowship (22D/23J02256) provided by the Council of Scientific and Industrial Research, New Delhi.

References

1 Haleem, A., Javaid, M., Singh, R.P. et al. (2021). Biosensors applications in medical field: a brief review. *Sensors International* 2: 100100.

2 Shrikrishna, N.S., Kolhe, P., and Gandhi, S. (2024). Sensitive detection of monocrotophos using a voltammetric immunosensor with randomly layered graphene oxide (GO) on fabricated electrode. *Journal of Environmental Chemical Engineering* 12 (2): 112059.

3 Bhat, G.P., Shrikrishna, N.S., Kumari, M., and Gandhi, S. (2025). Electrochemical biosensing of organophosphates. In: *Agricultural Electrochemistry* (ed. J.G. Manjunatha). American Chemical Society, 418 pages.

4 Malik, S., Singh, J., Goyat, R. et al. (2023). Nanomaterials-based biosensor and their applications: a review. *Heliyon* 9 (9): 19929.

5 Atanasov, P. and Wilkins, E. (1994). Biosensor for continuous glucose monitoring. *Biotechnology and Bioengineering* 43 (3): 262–266.

6 Scheller, F.W., Schubert, F., Neumann, B. et al. (1991). Second generation biosensors. *Biosensors & Bioelectronics* 6 (3): 245–253.

7 Taurino, I., Sanzò, G., Antiochia, R. et al. (2016). Recent advances in third generation biosensors based on Au and Pt nanostructured electrodes. *TrAC, Trends in Analytical Chemistry* 79: 151–159.

8 Tîlmaciu, C.M. and Morris, M.C. (2015). Carbon nanotube biosensors. *Frontiers in Chemistry* 3: 59.

9 Mokhtari, M., Khoshbakht, S., Ziyaei, K. et al. (2024). New classifications for quantum bioinformatics: Q-bioinformatics, QCt-bioinformatics, QCg-bioinformatics, and QCr-bioinformatics. *Briefings in Bioinformatics* 25 (2): 74.

10 Quantum Biology: The Hidden Code of Life | Twishmay's Blog | Curious about Life, The Universe, and Everything. https://twishmay.com/2024/09/03/quantum-biology-the-hidden-code-of-life

11 Shahdeo, D. and Gandhi, S. (2023). Aptasensors for point-of-care diagnostics of cancer. In: *Aptasensors for Point-of-Care Diagnostics of Cancer*, 7–19. IOP Publishing.

12 Shahdeo, D. and Gandhi, S. (2022). Next generation biosensors as a cancer diagnostic tool. In: *Biosensor-Based Advanced Cancer Diagnostics: From Lab to Clinics* (ed. R. Khan, A. Parihar, and S.K. Sanghi), 179–196. Elsevier.

13 Shahdeo, D., Kesarwani, V., Suhag, D. et al. (2021). Self-assembled chitosan polymer intercalating peptide functionalized gold nanoparticles as nanoprobe for efficient imaging of urokinase plasminogen activator receptor in cancer diagnostics. *Carbohydrate Polymers* 266: 118138.

14 Chen, B., Zou, L., Wu, Z., and Sun, M. (2016). The application of quantum dots in aquaculture pollution detection. *Toxicological and Environmental Chemistry* 98 (3–4): 385–394.

15 Safarpour, H., Safarnejad, M.R., Tabatabaei, M. et al. (2012). Development of a quantum dots FRET-based biosensor for efficient detection of *Polymyxa betae*. *Canadian Journal of Plant Pathology* 34 (4): 507–515.

16 Bhattarai, P. and Hameed, S. (2020). Basics of biosensors and nanobiosensors. In: *Nanobiosensors: From Design to Applications* (ed. A. Wu and W.S. Khan), 1–22. Wiley-VCH.

17 Sahoo, J., Sharma, R., and Gandhi, S. (2025). Review on biosensors for electroactive molecule detection. In: *Advancements in Voltammetry for Biosensing Applications* (ed. J.G. Manjunatha), 17–44. Springer Nature.

18 Sahoo, J., Sharma, R., Pachauri, V., and Gandhi, S. (2024). Biomimetic/bioderived nano engineered interfaces for biosensor applications: a review. *ACS Applied Nano Materials* 7 (17): 19854–19875.

19 Hareesha, N. and Manjunatha, J.G. (2021). Electro-oxidation of formoterol fumarate on the surface of novel poly(thiazole yellow-G) layered multi-walled carbon nanotube paste electrode. *Scientific Reports* 11 (1): 1–14.

20 Singh, A., Sharma, A., Ahmed, A. et al. (2021). Recent advances in electrochemical biosensors: applications, challenges, and future scope. *Biosensors* 11 (9): 336.

21 Raril, C., Manjunatha, J.G., Ravishankar, D.K. et al. (2020). Validated electrochemical method for simultaneous resolution of tyrosine, uric acid, and ascorbic acid at polymer modified nano-composite paste electrode. *Surface Engineering and Applied Electrochemistry* 56 (4): 415–426.

22 Pushpanjali, P.A., Manjunatha, J.G., Hareesha, N. et al. (2021). Voltammetric analysis of antihistamine drug cetirizine and paracetamol at poly(L-Leucine) layered carbon nanotube paste electrode. *Surfaces and Interfaces* 24: 101154.

23 Shrikrishna, N.S., Sharma, R., Sahoo, J. et al. (2024). Navigating the landscape of optical biosensors. *Chemical Engineering Journal* 490: 151661.

24 Skládal, P. (2024). Piezoelectric biosensors: shedding light on principles and applications. *Microchimica Acta* 191 (4): 184.

25 Jandas, P.J., Prabakaran, K., Luo, J., and Derry Holaday, M.G. (2021). Effective utilization of quartz crystal microbalance as a tool for biosensing applications. *Sensors and Actuators, A: Physical* 331: 113020.

26 Rundle, R.P. and Everitt, M.J. (2021). Overview of the phase space formulation of quantum mechanics with application to quantum technologies. *Advanced Quantum Technologies* 4 (6): 2100016.

27 Adams, B., Sinayskiy, I., van Grondelle, R., and Petruccione, F. (2022). Quantum tunnelling in the context of SARS-CoV-2 infection. *Scientific Reports* 12 (1): 1–14.

28 Das, S., Mazumdar, H., Khondakar, K.R. et al. (2024). Review—quantum biosensors: principles and applications in medical diagnostics. *ECS Sensors Plus* 3 (2): 025001.

29 Uppu, R., Midolo, L., Zhou, X. et al. (2021). Quantum-dot-based deterministic photon–emitter interfaces for scalable photonic quantum technology. *Nature Nanotechnology* 16: 1308–1317.

30 Mpofu, K.T., Mthunzi-Kufa, P., and Mpofu, K.T. (2024). Recent advances in quantum biosensing technologies. In: *Current Developments in Biosensor Applications and Smart Strategies* (ed. S. Karakuş), 1–20. Intechopen.

31 Yehia T. (2025). Time Evolution of Nodes in Quantum Superposition States. https://arxiv.org/abs/2502.00496v2 (accessed February 1, 2025).

32 Erhard, M., Krenn, M., and Zeilinger, A. (2020). Advances in high-dimensional quantum entanglement. *Nature Reviews Physics* 2: 365–381.

33 Moskal, P. (2021). Positronium and quantum entanglement imaging: a new Trend in positron emission tomography. Nuclear science symposium and medical imaging conference record.

34 Matarèse, B.F.E. and Purushotham, A. (2025). Quantum oncology. *Quantum Reports* 7 (1): 9.

35 Shrikrishna, N.S., Coudhury, S., and Gandhi, S. (2022). Magnetic properties-based biosensors for early detection of cancer. In: *Biosensor Based Advanced Cancer Diagnostics: From Lab to Clinics* (ed. R. Khan, A. Parihar, and S.K. Sanghi), 165–178. Elsevier/Academic Press.

36 Swan, M., Dos Santos, R.P., and Witte, F. (2022). Quantum neurobiology. *Quantum Reports* 4 (1): 107–126.

37 Tang, L., Yi, L., Jiang, T. et al. (2022). Measuring conductance switching in single proteins using quantum tunneling. *Science Advances* 8 (20): 8149.

38 Banerjee, A. and Mastrangelo, C.H. (2022). Resurgence of electron quantum tunneling sensors. *Micro* 2 (4): 679–69833.

39 Reddy, N.N. and Panda, D.K. (2021). A comprehensive review on tunnel field-effect transistor (TFET) based biosensors: recent advances and future prospects on device structure and sensitivity. *SILICON* 13: 3085–3100.

40 Ding, R., Chen, Y., Wang, Q. et al. (2022). Recent advances in quantum dots-based biosensors for antibiotics detection. *Journal of Pharmaceutical Analysis* 12 (3): 355–364.

41 García de Arquer, F.P., Talapin, D.V., Klimov, V.I. et al. (2021). Semiconductor quantum dots: technological progress and future challenges. *Science* 373 (6555): eaaz8541.

42 Quesada-González, D. and Merkoçi, A. (2025). Quantum dots for biosensing: classification and applications. *Biosensors & Bioelectronics* 273: 117180.

43 Tigari, G. and Manjunatha, J.G. (2020). Poly(glutamine) film-coated carbon nanotube paste electrode for the determination of curcumin with vanillin: an electroanalytical approach. *Monatshefte fuer Chemie* 151 (11): 1681–1688.

44 Pemmatte, P.A., Manjunatha, J.G., and Hareesha, N. (2021). An overview of recent developments of carbon-based sensors for the analysis of drug molecules. *Journal of Electrochemical Science and Engineering* 11 (3): 161–177.

45 Zhang, J., Zhang, S., Zhang, Y. et al. (2023). Colloidal quantum dots: synthesis, composition, structure, and emerging optoelectronic applications. *Laser & Photonics Reviews* 17 (3): 2200551.

46 Iannazzo, D., Espro, C., Celesti, C. et al. (2021). Smart biosensors for cancer diagnosis based on graphene quantum dots. *Cancers* 13 (13): 3194.

47 Majumdar, S., Thakur, D., and Chowdhury, D. (2020). DNA Carbon-nanodots based electrochemical biosensor for detection of mutagenic nitrosamines. *ACS Applied Bio Materials* 3 (3): 1796–1803.

48 Pourmadadi, M., Rahmani, E., Rajabzadeh-Khosroshahi, M. et al. (2023). Properties and application of carbon quantum dots (CQDs) in biosensors for disease detection: a comprehensive review. *Journal of Drug Delivery Science and Technology* 80: 104156.

49 Gorrini, F. and Bifone, A. (2023). Advances in stabilization and enrichment of shallow nitrogen-vacancy centers in diamond for biosensing and spin-polarization transfer. *Biosensors* 13 (7): 691.

50 Zhang, T., Pramanik, G., Zhang, K. et al. (2021). Toward quantitative bio-sensing with nitrogen-vacancy center in diamond. *ACS Sensors* 6 (6): 2077–2107.

51 Feng, Y., Zhao, Q., Shi, Y. et al. (2022). Recent applications of fluorescent nanodiamonds containing nitrogen-vacancy centers in biosensing. *Functional Diamond* 2 (1): 192–203.

52 The Nitrogen-Vacancy Center in Diamond | Pritzker School of Molecular Engineering | The University of Chicago. Available from: https://pme.uchicago.edu/awschalom-group/nitrogen-vacancy-center-diamond

53 Qureshi, S.A., Hsiao, W.W.W., Hussain, L. et al. (2022). Recent development of fluorescent nanodiamonds for optical biosensing and disease diagnosis. *Biosensors* 12 (12): 1181.

54 Xavier, J., Yu, D., Vollmer, F. et al. (2021). Quantum nanophotonic and nanoplasmonic sensing: towards quantum optical bioscience laboratories on chip. *Nano* 10 (5): 1387–1435.

55 Taha, B.A., Addie, A.J., Haider, A.J. et al. (2024). Exploring trends and opportunities in quantum-enhanced advanced photonic illumination technologies. *Advanced Quantum Technologies* 7 (3): 2300414.

56 Rivera, N. and Kaminer, I. (2020). Light–matter interactions with photonic quasiparticles. *Nature Reviews Physics* 2 (10): 538–561.

57 Wang, X., Ding, Q., Groleau, R.R. et al. (2024). Fluorescent probes for disease diagnosis. *Chemical Reviews* 124 (11): 7106–7164.

58 Ganganboina, A.B. and Doong, R.A. (2019). Graphene quantum dots decorated gold-polyaniline nanowire for impedimetric detection of carcinoembryonic antigen. *Scientific Reports* 9 (1): 1–11.

59 Hasanzadeh, M., Baghban, H.N., Shadjou, N., and Mokhtarzadeh, A. (2018). Ultrasensitive electrochemical immunosensing of tumor suppressor protein p53 in unprocessed human plasma and cell lysates using a novel nanocomposite based on poly-cysteine/graphene quantum dots/gold nanoparticle. *International Journal of Biological Macromolecules* 107 (Part A): 1348–1363.

60 Wang, P., Yu, L., Gong, J. et al. (2022). An activity-based fluorescent probe for imaging fluctuations of peroxynitrite (ONOO−) in the Alzheimer's disease brain. *Angewandte Chemie* 134 (36): 202206894.

61 Tabish, T.A., Hayat, H., Abbas, A., and Narayan, R.J. (2021). Graphene quantum dot–based electrochemical biosensing for early cancer detection. *Current Opinion in Electrochemistry* 30: 100786.

62 Memon, R., Shaheen, I., Qureshi, A., and Niazi, J.H. (2024). Enhanced detection of cardiac troponin-I using CdSe/CdS/ZnS core-shell quantum dot/TiO_2 heterostructure photoelectrochemical sensor. *Journal of Alloys and Compounds* 1008: 176592.

63 Fu, J., Wang, P., Wu, H. et al. (2025). Development of a graphene oxide multilayer quantum dot-based immunochromatographic strip for the ultrasensitive detection of H7 subtype avian influenza viruses. *Poultry Science* 104 (4): 104924.

64 Hatamluyi, B., Rezayi, M., Amel Jamehdar, S. et al. (2022). Sensitive and specific clinically diagnosis of SARS-CoV-2 employing a novel biosensor based on boron nitride quantum dots/flower-like gold nanostructures signal amplification. *Biosensors & Bioelectronics* 207: 114209.

65 Vasudevan, M., Perumal, V., Raja, P.B. et al. (2023). A quadruplet 3-D laser scribed graphene/MoS_2, functionalised N_2-doped graphene quantum dots and lignin-based Ag-nanoparticles for biosensing. *International Journal of Biological Macromolecules* 253 (2): 126620.

66 Feng, Y., Xie, J., Zhang, A. et al. (2024). Fabrication of core-shell quantum dots microbeads with high stability, enhanced sensitivity and potential application in C-reactive protein (CRP) and procalcitonin (PCT) immunochromatography in vitro diagnostic reagents. *Chinese Journal of Analytical Chemistry* 52 (10): 100447.

67 Akbari Nakhjavani, S., Mirzajani, H., Carrara, S., and Onbaşlı, M.C. (2024). Advances in biosensor technologies for infectious diseases detection. *TrAC, Trends in Analytical Chemistry* 180: 117979.

68 Shahdeo, D., Roberts, A., Archana, G.J. et al. (2022). Label free detection of SARS CoV-2 receptor binding domain (RBD) protein by fabrication of gold nanorods deposited on electrochemical immunosensor (GDEI). *Biosensors & Bioelectronics* 212: 114406.

69 Roberts, A., Chouhan, R.S., Shahdeo, D. et al. (2021). A recent update on advanced molecular diagnostic techniques for COVID-19 pandemic: an overview. *Frontiers in Immunology* 12: 732756.

70 Shrikrishna, N.S., Halder, S., Kesarwani, V. et al. (2024). Unveiling the potential: high-affinity aptamers for point of care detection of SARS-CoV-2 RBD protein and it's validation in clinical samples. *Journal of Chemical Engineering* 493: 152841.

71 Liang, L., Chen, M., Tong, Y. et al. (2021). Detection of mycobacterium tuberculosis IS6110 gene fragment by fluorescent biosensor based on FRET between two-dimensional metal-organic framework and quantum dots-labeled DNA probe. *Analytica Chimica Acta* 1186: 339090.

72 Hosseinchi Gharehaghaji, Z., Mohammad-Rezaei, R., and Khalilzadeh, B. (2024). Self-assembled prussian blue nanoparticles on MXene quantum dots modified AuNPs for sensitive and reproducible biosensing of ovarian specific antigen (CA-125). *Microchemical Journal* 206: 111633.

73 Prakashan, D., Shrikrishna, N.S., Byakodi, M. et al. (2023). Gold nanoparticle conjugate-based lateral flow immunoassay (LFIA) for rapid detection of RBD antigen of SARS-CoV-2 in clinical samples using a smartphone-based application. *Journal of Medical Virology* 95 (1): e28416.

74 Shrikrishna, N.S., Kaushik, A., and Gandhi, S. (2023). Smartphone-assisted detection of monocrotophos pesticide using a portable nano-enabled chromagrid-lightbox system towards point-of-care application. *Chemosphere* 330: 138704.

75 Shahdeo, D., Chauhan, N., Majumdar, A. et al. (2022). Graphene-based field-effect transistor for ultrasensitive immunosensing of SARS-CoV-2 spike S1 antigen. *ACS Applied Bio Materials* 5 (7): 3563–3572.

76 Loyez, M., Lobry, M., Hassan, E.M. et al. (2021). HER2 breast cancer biomarker detection using a sandwich optical fiber assay. *Talanta* 221: 121452.

77 Fan, X., Xing, Y., Wu, Z. et al. (2024). Electrochemical-enhanced charge state modulation of nitrogen-vacancy centers for ultrasensitive biodetection of MicroRNA-155. *ACS Applied Materials & Interfaces* 17 (1): 2592–2560.

78 Zhang, Y., Ding, L., Zhang, H. et al. (2022). A new optical fiber biosensor for acetylcholine detection based on pH sensitive fluorescent carbon quantum dots. *Sensors and Actuators B: Chemical* 369: 132268.

79 Maneesha, M., Preethi, P.C., Harisankar, A. et al. (2025). Engineering ZIF-67 with heteroatom-doped carbon dots for ultra-sensitive and selective detection of serum albumin in artificial body fluids. *Microchemical Journal* 210: 113046.

80 Putra, B.R., Nurwidayanti, E., Fadilah, S. et al. (2025). MWCNT-OH/graphene composite sensor for nonenzymatic detection of paraoxon-ethyl in agricultural samples. *Carbon Letters* 35: 1–20.

81 Shoshi, M.S., Huraiya, M.A., Vinoth Raj, R. et al. (2025). Revolutionary barium titanate-BlueP/TMDCs SPR sensor: ultra-sensitive detection of urine glucose levels. *Talanta Open* 11: 100401.

82 Zhang, H., Zhou, Y., Zhang, M. et al. (2025). Construction of $CdIn_2S_4$/MXene-TiO_2 Z-scheme heterojunction for high-gain organic photoelectrochemical transistor to achieve maximized transconductance at zero bias. *Small* 21 (5): 2408470.

83 Xu, D., Shan, L., Guo, B. et al. (2025). A fluorescent aptasensor for kanamycin detection in milk, seafood and water samples using DNA-AgNCs and exonuclease I-assisted recycling amplification strategy. *Food Chemistry* 478: 143291.

84 Wu, Y., Combs, C., Okosun, B.O. et al. (2025). Fe^{3+}-doped graphene quantum dots-based nanozyme for H_2O_2 detection in cellular metabolic distress. *ACS Applied Nano Materials* 8 (6): 2774–2784.

85 Zhang, Y., Lv, Y., Chen, X. et al. (2025). Ultrasensitive point-of-care multiplex diagnosis for influenza virus based robust quantum dot microsphere-lateral flow immunoassay. *Biosensors & Bioelectronics* 273: 117187.

86 Thulasinathan, B., Sujatha, D., Meurugan, S. et al. (2023). DNA-functionalized carbon quantum dots for electrochemical detection of pyocyanin: a quorum sensing molecule in *Pseudomonas aeruginosa*. *Biosensors & Bioelectronics* 227: 115156.

87 Hosseini, M., Ahmadi, E., Borghei, Y.S., and Ganjali, M.R. (2017). A new fluorescence turn-on nanobiosensor for the detection of micro-RNA-21 based on a DNA–gold nanocluster. *Methods and Applications in Fluorescence* 5: 015005.

88 Fan, L., Ji, X., Lin, G. et al. (2021). Green synthesis of stable platinum nanoclusters with enhanced peroxidase-like activity for sensitive detection of glucose and glutathione. *Microchemical Journal* 166: 106202.

89 Adel, R., Ebrahim, S., Shokry, A. et al. (2021). Nanocomposite of CuInS/ZnS and nitrogen-doped graphene quantum dots for cholesterol sensing. *ACS Omega* 6 (3): 2167–2176.

90 Abha, K., Nebu, J., AnjaliDevi, J.S. et al. (2019). Photoluminescence sensing of bilirubin in human serum using l-cysteine tailored manganese doped zinc sulphide quantum dots. *Sensors and Actuators B: Chemical* 282: 300–308.

91 Ghrera, A.S., Pandey, C.M., Ali, M.A., and Malhotra, B.D. (2015). Quantum dot-based microfluidic biosensor for cancer detection. *Applied Physics Letters* 106: 193703.

92 Zheng, G., Patolsky, F., Cui, Y. et al. (2005). Multiplexed electrical detection of cancer markers with nanowire sensor arrays. *Nature Biotechnology* 23: 1294–1301.

93 Zhang, G., Singh, R., Zhang, B. et al. (2024). ZnO-NWs and WS_2-QDs-functionalized multicore fiber-based W-shaped WaveFlex biosensor for rapid detection of hemoglobin

A1c in serum samples. *IEEE Transactions on Instrumentation and Measurement* 73 7007910.

94 Materon, E.M., Joshi, N., Shimizu, F.M. et al. (2021). Electrochemical sensors based on metal oxide-boron nitride nanocomposites in the detection of biomolecules and toxic chemicals. In: *Metal Oxides in Nanocomposite-Based Electrochemical Sensors for Toxic Chemicals* (ed. A. Pandikumar and P. Rameshkumar), 293–311. Elsevier.

95 Kukkar, M., Singh, S., Kumar, N. et al. (2017). Molybdenum disulfide quantum dot based highly sensitive impedimetric immunoassay for prostate specific antigen. *Microchimica Acta* 184: 4647–4654.

96 Ren, W., Fang, Y., and Wang, E. (2011). A binary functional substrate for enrichment and ultrasensitive SERS spectroscopic detection of folic acid using graphene oxide/Ag nanoparticle hybrids. *ACS Nano* 5 (8): 6425–6433.

97 Sahoo, J., Prakashan, D., Gandhi, S. (2025). 3D Printing of Biomaterials. Compendium of 3D Bioprinting Technology.

98 Trenfield, S.J., Awad, A., Goyanes, A. et al. (2018). 3D printing pharmaceuticals: drug development to frontline care. *Trends in Pharmacological Sciences* 39 (5): 440–451.

99 Sadée, W. and Dai, Z. (2005). Pharmacogenetics/genomics and personalized medicine. *Human Molecular Genetics* 14 (2): 207–214.

100 Pollard, T.D., Ong, J.J., Goyanes, A. et al. (2021). Electrochemical biosensors: a nexus for precision medicine. *Drug Discovery Today* 26 (1): 69–79.

101 Manickam, P., Mariappan, S.A., Murugesan, S.M. et al. (2022). Artificial intelligence (AI) and internet of medical things (IoMT) assisted biomedical systems for intelligent healthcare. *Biosensors* 12 (8): 562.

102 Qureshi, R., Irfan, M., Al, H. et al. (2023). Artificial intelligence and biosensors in healthcare and its clinical relevance: a review. *IEEE Access* 11: 61600–61620.

103 Cho, S.Y. and Jung, H.T. (2023). Artificial intelligence: a game changer in sensor research. *ACS Sensors* 8 (4): 1371–1372.

104 Cui, F., Yue, Y., Zhang, Y. et al. (2020). Advancing biosensors with machine learning. *ACS Sensors* 5 (11): 3346–3364.

2

Quantum Dots: Fluorescent Probes in Medical Diagnostics

Balamurugan Arumugam[1,2], Po-Ling Chang[1], Sathish Kumar Ponnaiah[3], and Sayee Kannan Ramaraj[2]

[1] *Department of Chemistry, National Sun Yat-sen University, Taiwan, Republic of China*
[2] *PG & Research Department of Chemistry, Thiagarajar College, Affiliated to Madurai Kamaraj University, Madurai, Tamil Nadu, India*
[3] *Department of Chemical Engineering, Pohang University of Science and Technology (POSTECH), Pohang 37673, Republic of Korea*

2.1 Introduction

Nanotechnology is evolving rapidly, changing science and technology, and it has opened a bright future for a range of novel nanomaterials with unique compositions and properties. The most distinctive feature for classifying these materials is their size [1]. For instance, three-dimensional nanomaterials, although larger than the nanoscale, consist of individual nanometer-sized building blocks, such as nanocomposites. Two-dimensional nanomaterials, including structures like nanoplates, nanofilms, and nanolayers, have one nanoscale dimension [2]. One-dimensional nanomaterials, such as nanotubes, nanorods, and nanowires, have two nanoscale dimensions [3]. Zero-dimensional nanomaterials, exemplified by nanopowders and nanoparticles, possess three nanoscale dimensions. Among them, nanoparticles have garnered the most courtesy due to their versatility in developing innovative and improved bioanalytical techniques for therapeutic applications. Quantum dots (QDs) are colloidal semiconductor nanocrystals that show good potential as photoluminescent probes and optical reagents in the biomedical environment [4]. QDs, in comparison to classic organic fluorophores, have better qualities when it comes to (i) narrow, intense emission (>100 nm), (ii) broad excitation, (iii) high quantum yields, (iv) long luminescence lifetimes, (v) photobleaching resistance, and (vi) large surface area to functionalize biology applications, including optosensing and bioimaging [5–7].

Tuning emission wavelengths, one of the advantages of multiplexed bioanalysis, comes with tunable QD emission wavelengths enabled by changing size or changing composition [8]. One light source is able to excite different-sized QDs with different emission peaks. Compared to conventional molecular probes, QDs have benefits due to (i) their high loading capacities to maintain a desired concentration of the agent, (ii) tunable surfaces that can be tailored to deliver to specific targets or circulate long term, and (iii)

Quantum Biosensing in Medical Diagnostics, First Edition. Edited by J. G. Manjunatha.

multimodal imaging. Although photoluminescent QDs have demonstrated significant effectiveness in clinical analysis, their use remains limited [9, 10]. New functionalization techniques are needed to address their reduced targeting efficiency. Furthermore, the inherent heavy metal composition of QDs raises concerns about toxicity, limiting their use in *in vivo* imaging and sensing [11]. Without molecular attachment, QDs often display a nonselective distribution, failing to meet essential imaging criteria. To ensure biocompatibility for *in vivo* applications, QD-based nanoprobes must undergo proper functionalization. The part of a probe to produce a signal is known as a fluorophore, and it is either an organic small-molecule dye or inorganic fluorescent substance [12]. Inorganic fluorophores frequently include d- or f-metal ions, such as QDs, like cadmium selenide (CdSe) and cadmium telluride (CdTe), or rare earth elements, like europium (Eu), terbium (Tb), samarium (Sm), erbium (Er), and neodymium (Nd) [13, 14]. Most organic fluorophores, such as cyanine dyes, rhodamines, coumarins, naphthalimides, BODIPYs, or aggregation-induced emission (AIE) materials such as tetraphenyl ethylene, can be added to these systems, providing a wide variety of structures and adjustable wavelengths to serve a variety of applications. One of the crucial facts of the probe design is a selection of fluorophores where their excitation and emission lie out of the tissue autofluorescence range (approximately 300–700 nm) [15]. As such, the majority of effective clinical imaging agents work in the 700–1700 nm near-infrared (NIR) regime that allows tissue penetration up to 5–10 mm. There are numerous such commercial agents, with longer wavelengths preferred because of their improved signal-to-noise ratio [16].

The chapter is a summary of latest developments in the application of QD fluorescent probes in the field of medicine. It is addressing the major forms of QDs in bioanalysis, their preparation, solubilizing, and stability in water. We take a cursory look into analytical approaches to characterize QDs and consider their clinical use, including molecular imaging and as biomarkers, and also consider recent advances in low-cost, handheld, easy-to-use QD-based biosensors that are useful in point of care. Finally, we look at the future prospects of QDs as diagnostic tools.

2.2 Properties of Quantum Dots (QDs)

QDs consist of a semiconductor core, typically made from a metalloid III–V or II–VI compound, which is enclosed by a semiconductor compound shell. A semiconductor element has electrical conducting properties that are neither too high nor too low, as those seen in a conductive or an insulating component. The III–V semiconductors can be found in gallium nitride (GaN) and gallium arsenide (GaAs) [17], indium arsenide (InAs) and indium phosphide (InP), and the group IIIs include cadmium telluride (CdTe), zinc selenide (ZnSe), zinc sulfide (ZnS), and cadmium selenide (CdSe [18]). A further study shows that through alloying of heavier elements, the properties of QDs may be engineered, including CdSe/ZnTe or CdTe/CdSe heterostructures. The behavior of QDs is also based on such factors as size, shape, composition, and surface chemistry. Important properties are emission wavelength, quantum yield, stability, toxicity, and biodegradability (see Table 2.1).

Functional groups added to the core or shell of QDs can generate new bioactive substances, making them suitable for biological applications. Various biological substances,

Table 2.1 Characteristics of diverse QD materials.

Material system	Emission wavelength	Quantum yield	Stability	Toxicity	Biodegradability
Metal chalcogenides	Visible to near-infrared	High (>80%)	Moderate to high	High	Low
Metal halides	Visible to near-infrared	High (>80%)	Low to moderate	Moderate to high	Low to moderate
C-dots	Ultraviolet to near-infrared	Low to high (<10–>80%)	High	Low	High
Si-dots	Visible to near-infrared	Low to high (<10–>80%)	Moderate to high	Low	High

such as lipids, proteins, and peptides, can adhere to the exterior of the QD shell. Thiol groups can be capped by creating covalent bonds with the superficial shell, which enhances the water solubility of QDs [19]. Polymers like polylactide coglycosides (PACG), polymethyl methacrylate (PMMA), polyvinyl alcohol (PVA), and polyoxometalates (POMs) can also be applied to QD surfaces. They are occasionally used solely as biosensors to detect and identify drugs and plasma components [20]. Surface-modified QDs can be used for the prevention, diagnosis, or treatment of certain diseases. Physical adsorption (physisorption), electrostatic attachment, covalent attachment, and multivalent chelation.

Furthermore, superficial adhesion can significantly impact the size of QDs [21]. Since the emission wavelength of QDs correlates with particle size, adjusting the particle size can result in a broad series of spectra (UV-IR). This characteristic can be leveraged to track and image multiple molecular targets with high resolution using QD signal multiplexing. On the other hand, organic dyes typically display comprehensive emission bands, making it more difficult to differentiate between signals. Organic dyes also generate substantial background signals, which significantly reduce detection sensitivity and, consequently, effectiveness of the probe [22]. Therefore, by being customized for a specific emission spectrum, QDs can reduce background signals.

2.2.1 Key Requirements for Applying QDs in Medicine

Most QDs have been associated with an intrinsic toxicity, which has been a major cause of concern for future applications in the biomedical field. Mostly, QDs based on telluride or cadmium selenide-based cores have always received much attention because they exhibit quantum confinement effects across the entire optical spectrum [23]. In a full investigation of all papers on the *in vivo* use of QDs (made by Walling et al.), the leakage of cadmium ions is reported to be the major cause of cytotoxicity. Such free ions have a strong affinity for thiol groups bound to key mitochondrial biomolecules, which cause oxidative stress and subsequent mitochondrial functional failures so pronounced that they cause significant cellular apoptosis [24]. However, additional surface coatings can reduce or even prevent the release of cadmium (Cd^{2+}). Irish scientists have used gelatin to produce CdTe QDs, helping to reduce the particles' toxicity [25]. On the other hand, Qian et al.

used the tripeptide glutathione as a stabilizer. Found in most organisms, glutathione has chelating properties and can be used in medicine to detoxify Cd^{2+} ions. According to Iyer et al., QDs surfaces were coated with peptides owing to their outstanding ligand properties and also with common factors: (i) peptides are stable at physiological pH and mimic the biological environment; (ii) the hydrophilic region of the peptide sequence is amenable to chemical modification through the incorporation of reactive moieties, including amine, carboxyl, and thiol groups, as well as peptide-based tags, to attach biomolecules of interest using conventional conjugation chemistry or enzymatic methods; (iii) molecular evolution techniques for peptide randomization can be improved to choose high-affinity binders for the QD surface; and (iv) peptide sequences can be combined in specific molar ratios to create multifunctional QDs for both *in vitro* and *in vivo* research (e.g. a QD containing biotin and polyethylene glycol (PEG), coupling a receptor ligand and PEG) [26].

In other words, the toxicity of nanoparticles is primarily influenced by their surface functionalization. Improved biocompatibility due to special surface coating could limit internalization of nanoparticles in cells and hence prevent the toxicity effects and consequently maintain cell integrity. According to Hardman's review, several factors, including intrinsic physicochemical properties and environmental conditions, affect the absorption, distribution, metabolism, excretion, and toxicity of QDs. The following factors have been identified as key determinants of QD toxicity: size, charge, concentration, oxidative, photolytic, and mechanical stability, as well as the bioactivity of the external surface modification [27]. In the future, if toxicity issue is addressed, QDs could be securely castoff fluorescent probes for biotic imaging, battered drug distribution tracking, and for modifying the physical and functional characteristics of cellular counterparts under controlled conditions.

In an intriguing study, Yong demonstrated that doping these CdTe QDs with manganese and functionalizing their exterior with lysine allows them uniform distribution in water environments [28]. No long-term adverse effects were detected. As a result, he proposed that these various average manganese-doped QDs could serve as probes in cell imaging and recognition of pancreatic cancer [29]. Chen and his associates detected the health, behavior, lifetime, and survival rates of rats after administering a three-month venous vaccination of thiol-capped CdHgTe QDs [30]. They found that the rats exhibited no abnormal behavior in any of the *in vivo* imaging studies, and the administration of 2 μg/g CdTeHg QDs showed no noticeable signs of toxicity. However, one critical question remains unresolved: can QDs be directly applied in clinical phase studies? Therefore, comprehensive *in vivo* toxicological evaluations and integrated cellular toxicity studies of QDs must be conducted before they can be utilized in human applications. Additional essential prerequisites for the medical application of QDs include effectively preventing the nonspecific adsorption of biomolecules on the superficial of nanoparticles and avoiding the aggregation of nanoparticles in biological environments. The surface of QDs can be functionalized with adsorbed peptides or hydrophilic polymers, such as PEG, to enhance their insufficient colloidal stability [31]. Storing QDs in a lyophilized form, away from direct sunlight, and at lower temperatures (e.g. 4 °C) can significantly enhance their long-term stability. The photophysical properties of QDs differ significantly from those of organic fluorophores. Their absorption spectrum is continuous for wavelengths shorter than their fluorescence emission wavelength. QDs can be engineered to achieve efficient monodispersity, which accounts for their symmetrical emission spectra, absence of red-tail effects, and relatively narrow bandwidth. This enables the spectrally distinct excitation of a broad range of colors using a single wavelength [32]. Perhaps, Liu et al. employed the conventional fluorescence resonance energy transfer (FRET) technique, using green and red QDs as FRET donors and acceptors, respectively, to simultaneously study the molecular communication among an antibody and immunoglobulin G on live cell casings without the need for organic dyes. Compared to conventional fluorophores, QDs are twenty times brighter and one hundred times more resistant to photobleaching. Nevertheless, there are two drawbacks when compared to traditional fluorescent dyes [33]. The first limitation of QDs is their poor photon polarization, which is primarily observed in elongated QDs, also referred to as nanorods. Their second limitation is their insolubility in water. The presence of hydrophobic molecules typically arises from the use of organic solvents, such as tri-n-octylphosphine oxide (TOPO), during their synthesis. This issue can be addressed by encapsulating QDs within a stable surfactant layer or a shell, such as silica [34].

Various influencing factors, are play significantly influence the photoactivation behavior of semiconductor nanoparticles. A thorough exploration of the effect of these parameters on the photon-sensitized increase of photoluminescence in polymer-stabilized CdSe/ZnS QDs and also in CdSe/ZnS QDs embedded in silica-based multicore colloidal particles was made by Dembski et al. at various experimental conditions [35]. The following parameters were examined: the dispersion medium, the density of the exterior silica shell of multi-core elements, the wavelength and intensity of the incident radiation, the local atmosphere of the QDs, and others. Zhang et al. also discovered that the photoluminescence lifespan of cellular thiol-covered CdTe QDs diverse significantly based on the intracellular atmosphere: it was noticeably shorter in acidic lysosomes (which have a pH of approximately 4.5–5.0) than it was in other cell regions [36]. However, the lifespan was lower than those in aquatic environments with the same pH values. Consequently, two potential causes for the reduction in intracellular QD photoluminescence lifetimes are the acidic environment and connections with biomolecules. Temperature also influences the photoluminescence of II–IV semiconductor QDs, causing the photoluminescence peak to redshift as the temperature increases. According to Zheng's group, thermally induced carrier detrapping causes the photoluminescence band of CdSe QDs to initially blue-shift and then red-shift as the temperature increases in the 120–300 K range [37].

2.3 Methods of Synthesis of Quantum Dots

QDs can be produced using various techniques. Some methods require expensive materials, labor-intensive processes, or extremely harsh reaction circumstances, while others are straightforward, reproducible, and inexpensive. Scientists have studied abundant approaches to synthesize ideal QDs for various applications. This section briefly covers the synthesis, solubilization, and biocoagulation of QDs.

2.3.1 Colloidal Chemistry

Colloidal QDs are widely used in most biological applications. High-boiling organic diluents, like hexadecylamine (HDA) and TOPO, are commonly used to synthesize high-quality

QDs at extremely high temperatures [38]. These long alkyl chain solvents can coordinate with the unsaturated metal atoms on the QDs, preventing the formation of bulky structures. Semiconductor materials are quickly inserted into the appropriate organic suspensions, which are heated and constantly stirred. These solvents comprise compounds capable of coordinating with the exteriors of the occasioned QDs. Because the process is simple, it is often referred to as the one-pot method. The synthesized QDs are stabilized by an organic ligand, which serves as a capping layer and is soluble only in nonpolar solvents like chloroform [39]. In addition to enhancing the stability of the QDs, the capping step enables their functionalization or interaction with other molecules. Common capping agents include compounds such as pyrophosphates and thiols (e.g. mercaptoacetic acid, dihydrolipoic acid, and trioctylphosphine [TOP]). The stability of QDs also depends on the length and surface thickness of the covering material. Major challenges, such as reduced brightness, general cellular attachment, defective coupling chemistry, and the development of greater hydrodynamic diameters (15–30 nm), are being addressed through advancements in QD research.

2.3.2 Organometallic Method

This is a mixing route, which is a mixture of the QD precursors, organic surfactants, and the solvents. In the presence of an elevated temperature, typically of about 300 °C, a hot coordinating solvent or hydrophilic inorganic ligands, e.g. TOPO, TOP, and HDA, are rapidly added to organometallic reagents (e.g. volatile metal alkyls such as dimethyl cadmium), metal oxide (such as CdO), and a chalcogen source (such as tri-n-octylphosphine selenide). The QDs are obtained with the QDs of the required size with a sharp photoluminescence (PL) spectrum due to precise control of the reaction parameters, such as temperature, pH, and nucleation time, including CdS, CdSe, and CdTe. The hydrophobic QDs produced in this fashion can be made water-soluble through direct aqueous synthesis, the ligand exchange (the exchange of hydrophobic surface ligands with ligands with hydrophilic bonds), or by enfolding them with a hydrophilic coating. Ligands that have two functional groups (bifunctional ligands) are frequently used in ligand-exchange processes, with their hydrophilic ends (e.g. carboxylate or amino- or multi-thiol ends, e.g. sodium thioglycolate) pointing outward, allowing the synthesis of species that are dispersible in water [40]. Alternatively, it is used to coat silica onto the QDs, where QDs are encapsulated through a process of alkoxysilanes (e.g. tetraethyl orthosilicate) and polycondensation, adding a further level of aqueous stability and biocompatibility.

2.3.3 Aqueous Phase Method

The reaction procedure is performed in water, giving QDs with high stability and good biocompatibility. Heavy metal salts (acetates, nitrates, or chlorides) are normally common precursors. Phosphine-free sources of chalcogen, such as NaHSe or Na_2SeO_3, are unstable under ambient conditions; synthesis is performed under an inert atmosphere. Nanoparticles are stabilized by various capping agents (mercapto acids, glutathione (GSH), thioglycolic acid (TGA), mercaptopropionic acid (MPA), mercaptosuccinic acid, and L-cysteine) to stave off aggregation by steric hindrance. By using mercaptoacetic acid as a capping ligand, Bao et al. [41] synthesized CdSe (QY 10%) and CdSe/CdS QDs (QY 30%)

and demonstrated their application in fluorescent imaging of blood capillaries and cells. In another study, Jiang et al. [42] synthesized CdSe QDs by combining a mixture of TGA and $CDCl_2$ in an alkaline solution with chalcogen precursors at room temperature. Recently, ultrasound-assisted techniques have been used for the aqueous synthesis of CdTe QDs and CdTe/CdS QDs. Labiadh et al. [43] formed an aqueous nucleation-doping method of synthesizing Mn-doped ZnS QDs (approximately 2.5 nm in diameter), then stabilized these QDs with 3-MPA and finally coated their shells with a ZnS shell to increase the photoluminescence by 10%–18%.

2.3.4 Epitaxial Growth

This technique involves the layer-by-layer growth or deposition of monocrystalline films [44]. It is used to form core-shell QD (CSQD) structures and QD films, where the luminescent semiconductor core (e.g. GaAs) is enclosed in a thin shell of a larger bandgap material (e.g. AlGaAs). The shell substantially enhances quantum yield (intensity) and photostability of the core by suppressing nonradiative recombination processes at the surface. Progress can be controlled by applying masking materials (e.g. SiO_2) and etching the mask into required shapes. Nevertheless, this approach's main drawback is reduced QD thickness caused by the cover designs. The primary application of the QDs produced by this method is in bioimaging.

2.3.5 Lithography

In the traditional top-down patterning technique called lithography, a concealed image pattern is formed on a solid infrastructure by applying a photosensitive resist layer. Then, the showing or invisible parts are removed to form the pattern [45]. The methods are classified as optical, X-ray, electron beam, or ion beam lithography, depending on the energy source used. During production, a suitable organic polymer film mask is placed over the quantum wells, followed by a thin metal layer as well as exposure to an electron or ion beam. This is achieved by merging high-resolution electron beam lithography with successive etching. Disadvantages of this technique include nonuniform size, poor interface quality, and irregular spatial resolution, which can lead to contamination, defect formation, and potential impairment to the whole crystal.

2.3.6 Eco-Friendly Synthesis

Different strategies for the preparation of QD comprise green techniques derived from natural bases. One example is a fast and cost-effective method to produce fluorescent carbon dots (CDs) through the leaves of *Ocimum sanctum* as a carbon basis. According to Kumar et al. [46], this method exhibited a high quantum product of 9.3% and excellent aqueous stability, making it suitable for lead (Pb^{2+}) ion detection. In a different research, apple juice was employed as a resource in a green hydrothermal method to synthesize luminous carbon dots [47]. Scientists have also explored using existing microbes, specifically bacteria and biomatrices, to drive synthesis processes. Cadmium sulfide (CdS) QDs with uniform size, a fluorescence emission peak at 470 nm, and blue-green fluorescence under UV light

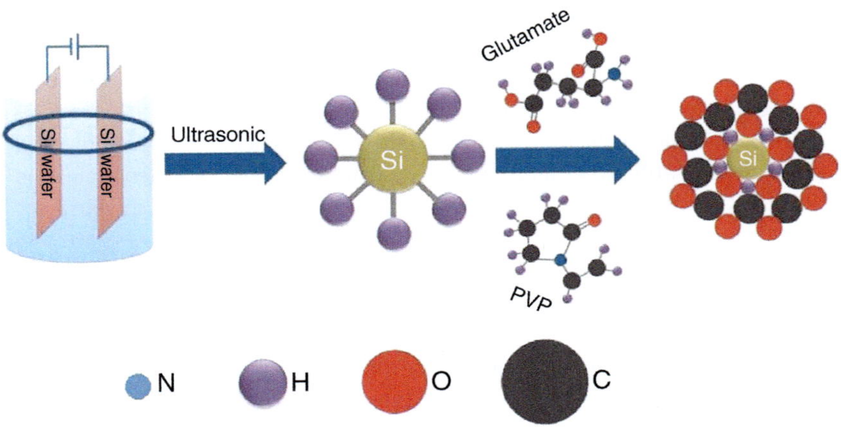

Figure 2.1 SiQDs modified with two components via electrochemical etching. Chen et al. [49] / with permission of Elsevier.

were synthesized by incubating an *Escherichia coli* culture medium with inorganic ions, such as a 1.03×10^{23} mol/L cadmium source, for two days [48].

2.3.7 Electrochemical Methods

The target material was electrochemically etched to produce QDs. This method is considered one of the best due to its low-cost synthesis and ability to generate QDs with unique chemical and physical properties. Based on anodic dissolution, this technique relies on combined chemical and electrical reactions. For instance, electrochemical etching allows the rapid production of silicon QDs (SiQDs) with a narrow size distribution and a broad range of fluorescence emissions. Two-component modifications of photostable SiQDs, using glutamate and polyvinyl pyrrolidone (PVP), have been developed. PVP improved the dispersion of SiQDs, while glutamate enhanced their fluorescence intensity. The complete synthesis procedure is illustrated in Figure 2.1 [49]. Furthermore, the electrochemical synthesis of graphene QDs measuring 3–5 nm at room temperature using $LiClO_4$ in propylene carbonate as the electrolyte has also been documented [50]. Continuous oxidation of graphene oxide (GO) at 1.05 V for varying durations under controlled reduction conditions produced graphene QD (GQD) particles exhibiting distinct characteristics.

2.4 Quantum Dots in Medical Diagnostics

QDs, due to their nanoscale dimensions, exhibit size-dependent optical and electronic features largely governed by quantum confinement effects. These semiconductor QDs are categorized by high quantum yields, incomparable photostability, and adjustable fluorescence emission, making them extremely beneficial for biomedical applications. In the framework of medical diagnostics, QDs are extensively used in both *in vitro* assays and *in vivo* imaging. Applying molecules known as targeting ligands gives us precise recognition of diseases and facilitates real-time visualization of pathological processes. Moreover, they can be incorporated with diverse biological molecules to recognize and find various disease biomarkers. The QDs now apply to the combination of fluorescence with magnetic resonance imaging (MRI) or positron emission tomography (PET), and computed tomography (CT), which improves analytical accuracy and provides complete molecular and functional information. While translational and clinical works need to be analyzed, QDs hold noteworthy potential to develop molecular diagnostics, biosensing, and personalized medicine.

2.4.1 Imaging Applications

QDs are currently widely used in bioimaging applications owing to advanced physicochemical properties, which make them an auspicious applicant in futuristic imaging technologies. Owing to their nanosized fluorescence emission through an extensive spectral range, high-resolution multiplexed imaging and extended visualization of energetic cellular actions are enabled, while revealing minimal exposure to photobleaching.

2.4.1.1 Real-Time Cellular Imaging and Intracellular Tracking

The use of QDs for real-time cellular imaging and intracellular tracking signifies a groundbreaking technique in life sciences that offers reflective insights into complex cellular functions and progressing the construction of diagnostics, therapeutics, and personalized medicine in biomedical applications. Among various properties of QDs, their nanosize and biocompatible nature produce significant opportunities as drug delivery vectors, with their PL enabling *in vivo* tracking. Compared to organic fluorophores, QDs, mainly carbon-based QDs provide outstanding brightness, comprehensive excitation spectra, more controlled influences, and low toxicity in biomedical applications [51].

The imaging of cellular compartments and organelles can provide a deeper insight into the early signs and development of various diseases, such as Alzheimer's, Parkinson's, diabetes, cancers, and others. QDs can be applied to investigate interactions with diverse proteins and monitor changes through bioimaging [52]. In addition to that, QDs have been employed for intracellular imaging in HeLa cells, Michigan Cancer Foundation (MCF)-7 cells, Caco-2 cells, HepG2 cells, PC12 cells, lung cancer cells, and pancreatic stem cells and their systematic biodistribution into mouse cells is used for *in vivo* imaging [53]. Tian et al. used carbon QDs (CQDs) as fluorescent tags in cells and zebrafish that have potential for human clinical trials as optical nanoprobes [54]. Meanwhile, hyaluronic acid-coated CQDs (HA-nCQDs) had an important effect, making it easier to detect tumor imaging by using the CD44 receptor of the cells. Its *in vivo* results reveal deeper involvement of drugs against tumor cells [55]. Highly photoluminescent BCQDs were prepared through a green hydrothermal approach, which remained steady at different pH, precursor quantities, and UV exposure that employed highly discriminating fluorescent sensors for vitamin B_{12} with a low detection limit [56]. High-quality GQDs by Narasimhan et al. [57] via nanosecond pulsed laser ablation with well-defined oxidation exhibited a strong fluorescent intensity with low cytotoxicity, efficient cellular uptake, and stability over 48 hours, suggesting useful biomarkers for cell and preclinical imaging.

2.4.1.2 Forster Resonance Energy Transfer (FRET) with QDs

Forster resonance energy transfer (FRET) is a nonradiative process in which energy is transported from an excited donor and ground-state acceptor fluorophores through long-range Coulombic dipole–dipole interactions. Quantum dots (QDs) can function as FRET donors, acceptors, or both simultaneously and offer size tunability with high extinction coefficients for efficient excitation. QD-based FRET systems boost up the accuracy of sensors by ratiometric, photostable statistics. Topical developments emphasize QD-QD FRET pairs to shortcut the restrictions of classical dye acceptors, including photobleaching and related noise. Improving hetero-structured QDs (e.g. cadmium-free InP cores with alloyed shells), scientists attained noticeable color shifts for instrument-free pH sensing, leveraging QDs' high intensity and tunable spectral overlap. Recently, Duong et al. [57] used CQDs for biosensing and food safety monitoring through binding-induced FRET of CQDs with aflatoxin B1 for its detection in food samples. Raju et al. developed N-doped CQDs using drumstick leaves as precursors developed for H_2O_2 and aspartic acid detection, and achieving good results. The synthesis of an "on–off–on" fluorophore (H_2O_2 quenches l-aspartic acid through fluorescence recovery), was carried out using the as-prepared N-CQDs, which do not require any additional treatments such as surface passivation, composite production, or incorporation of a photoreceptor [58].

2.4.2 Biomarker Detection

QDs have occurred as noteworthy tool for biomarker detection because of sensitivity and specificity that they offer in identification of biological targets. Their miniaturization and bi-functionalizable surface are good sources of functionalization using targeting molecules that effectively bind to the biomarkers such as proteins, DNA or cancer cells. Some of the major advantages include the fact that they are highly fluorescent and therefore the signals are clear and easier to detect in their complex biological samples. As well, QDs are multiplexable, agreeing concurrent monitoring of numerous biomarkers in a single assay, increasing both the exactness and efficacy of analysis.

2.4.2.1 Cancer Biomarkers and Tumor-Specific Antigen Detection

Cancer, a major public health threat in the twenty-first century, is characterized by uncontrolled growth of abnormal cells. Efficient treatment of cancer and favorable outcomes for patients depend on early cancer biomarker detection and tumor-specific antigen identification. Nanotechnology, particularly the use of QDs, provides an auspicious method for cancer diagnosis and treatment. Biomolecules, including cadmium-free antibodies, aptamers, nucleic acids, peptides, or small molecules attached with QD-based probes, lead to the unique and specific targeting ability of the probes to cancer cells owing to their coherent optical properties. These targeted probes enable accurate detection of cancer biomarkers, which are involved in cell interactions and regulate biological functions, are also being explored for cancer chemotherapy.

Fluorescence-based cell imaging is a well-established and highly effective method in biomedical applications, especially cancer biomarker detection. QDs possess bright, tunable, and enduring fluorescence, rendering them an appropriate possibility for bioimaging in complex biological environments. Recently, Zoghi et al. [59] explored curcumin-loaded chitosan CQDs, combined with Fe_2O_3, demonstrated controlled pH-dependent drug release, potent anticancer activity against MCF-7 cells, and enhanced stability that reveals QDs could be an auspicious tool for targeted cancer therapy. Samimi et al. [60] developed nitrogen-doped CQDs with quinic acid and gemcitabine for tumor bioimaging and targeted drug delivery. Their vivid blue luminescence allows tumor-specific fluorescence imaging, while quinic acid targets are triggered in the acidic tumor environment. Sparing healthy tissue. Fifteen minutes after injection, an optical signal was observed throughout the mouse's body, primarily in the intestinal region, and was eventually excreted through the urinary system, as the kidneys are the primary components responsible for removing overseas particles from the bloodstream. Over time, the N-CQD–quinic acid–Gem complex gradually distributed throughout the body and accumulated at the tumor site. This reflection confirms that N-CQD–quinic acid–Gem circulates systemically and preferentially accumulates at the target location, namely the tumor. Yao et al. [61] developed magneto-fluorescent CQDs from waste crab shells and doxorubicin (DOX)-loaded CQDs functionalized with folic acid. These CQDs exhibited remarkable anticancer activity against HeLa cells with minimal toxicity to normal cells. *in vivo* tests on zebrafish embryos confirmed high biocompatibility and promising hatching rates, promising their probability for future clinical applications. Oliveira et al., developed a biomarker, namely, CA19–9, for pancreatic cancer with the use of CdSe/ZnS QDs conjugated to anti-CA19–9 antibodies in a colorimetric fluorescent panel, which provides outstanding results of a linear range of 0.01–501.9 U/mL, LOD $\approx 1.7 \times 10^{-4}$ U/mL, and visual readability within 15 minutes [62]. These studies demonstrate the multi-functionality and efficacy of QDs as biomarkers of cancer, tumor-specific antigens, and others, thus suggesting that this discovery can transform the diagnosis and treatment approaches toward cancer significantly.

2.4.2.2 Role in Infectious Disease Diagnostics

Quantum dots (QDs) offer considerable prospects in the domain of infectious disease diagnostics owing to their unique optical properties. These characteristics enable QDs to surpass the traditional dyes and fluorophores, especially in sensitivity, multiplexing capabilities, and signal persistence. In infectious disease detection, it is in connection with biomolecules, including antibodies, aptamers, or nucleic acids, to specifically bind to pathogens or their biomarkers (e.g. viral proteins, bacterial toxins, nucleic acids), enabling early and sensitive detection at ultralow concentrations [63]. An outstanding example is QDs in COVID-19 diagnostics. A lateral flow immunoassay (LFIA) based on magnetic triple-shell QDs (MagTQDs) was found to detect SARS-CoV-2 nucleocapsid and spike antigens with limits of detection (LOD = 0.5 pg/mL) and a remarkable response time under 20 minutes [64]. Likewise, CdSe–ZnS QDs conjugated to SARS-CoV-2 B-cell epitope peptides showed detection of antibodies with a sensitivity of up to 98% that was significantly more than conventional ELISA [65].

QD-based electrochemiluminescence biosensors have been enormously more sensitive in detecting hepatitis viruses. For example, a hepatitis B virus (HBV) DNA and hepatitis C virus (HCV) RNA were sensed in serum of a CoE, with low sensitivity of 0.082 and 1.34 pmol/L, respectively, by using a CdTe QD-based biosensor [66]. In addition, QDs have been deployed in multiplex assays, which can simultaneously detect numerous viruses using a single test chip, including dengue, Zika, norovirus, and influenza detection, which is a great move in managing outbreaks in the resource-constrained environment. Bacterial diagnostics have also profited from the expertise of QDs. Perhaps *E. coli* O157:H7 was detected rapidly using streptavidin-coated CdSe/ZnS QDs conjugated with biotinylated antibodies with sensitivity up to 100-fold higher than conventional methods. Real-time detection of toxin profiling and bacterial loads can also be carried out through QDs, which is important in tracking areas of diseases such as cholera, tuberculosis (TB), and sepsis [67]. Moreover, QDs are applied in bioimaging to trace the interaction between a pathogen and host in live cells and life forms, as exhibited by live imaging of *Mycobacterium tuberculosis* in macrophages. These suitability in the process of immunological observation of infections and their response to immunity render them ideal in the development of vaccines and treatments [68]. In summary, QDs are transforming infectious disease diagnostics, providing more sensitive and rapid diagnostic tools. These are a particularly promising next generation of diagnostic technology due to their compatibility with point-of-care platforms, low limits of detection, and superior performance in complex biological matrices.

2.4.2.3 Use of QDs for Simultaneous Detection of Multiple Analytes

Carbon-based nanomaterials, such as CQDs, have gained interest due to their enlarged physical chemical properties, offering potential applications in various fields. These qualities denote a promising strategy that is suitable for biomedical and pharmaceutical applications, along with the detection of chemical ions and species for environmental monitoring objectives. Heavy metal ions and chemical sensing represent the significant applications in CQDs investigations. Metal ions, likewise, iron (Fe^{3+}), zinc (Zn^{2+}), and copper (Cu^{2+}), are essential for various biological processes, which makes the development of detection methods for these ions particularly significant. In contrast, noxious elements like mercury (Hg^{2+}) and lead (Pb^{2+}) can disrupt normal biological processes by chelating with biomolecules. The detection of these ions continues to be an intriguing research area for environmental sciences and monitoring. There is a growing interest in cost-effective, simple, and effective methods for detecting a variety of chemical analytes, and CQD-based detections have been widely researched in order to their outstanding features.

Huang et al. created a dual-emission ratiometric fluorescence probe for detecting an antibiotic and a metal ion (tetracycline and Fe^{3+}) in aqueous solutions using a boron-doped carbon QD (B-CQD)/CdTe–Eu^{3+} composite. The probe's fluorescence response is sensitive and selective to TC, making it suitable for ultrasensitive detection. The probe showed good linearity between fluorescence intensity ratio and TC and Fe^{3+} concentrations, with ultrasensitive limits of detection. The probe was successfully applied in environmental water samples, providing an innovative approach for environmental and food analysis [69]. In the same manner, N-doped carbon QDs-GO hybrid (NCQDs-GO) is developed for concurrent electrochemical anodic stripping voltammetry determination of cadmium and lead ions. The sensor has large electroactive surface areas and abundant oxygen-containing functional groups, improving sensor sensitivity. It has modified electrodes that offer greater analytical performance, a wide linear range, a low LOD (7.45 mg/L for Cd^{2+} and 1.17 mg/L for Pb^{2+}), respectable reproducibility, and selectivity [70]. In order to create an innovative fluorescence sensor array, a μPAD for simultaneous detection has been developed using molecular imprinting technology. The sensor array was grafted on glass-fiber paper using carbon QDs and molecularly imprinted polymers. The array showed discrete fluorescence quenching efficacy, allowing accurate differentiation of nitrophenol isomers (2-NP, 3-NP, 4-NP). The device achieved 100% accuracy in identifying blind samples, making it an ideal analytical tool for environmental monitoring, disease prognosis, and food safety [71].

2.4.3 Multiplexed Diagnostic Platforms

2.4.3.1 Use of QDs for Simultaneous Recognition of Various Analytes

The QDs have become an outstanding platform to detect multiple analytes simultaneously because of their physical–chemical characteristics. These characteristics enable QDs to emit light at various wavelengths depending on their size, enabling the detection of diverse analytes in parallel deprived of spectral overlap [72]. The surface chemistry of QDs can be adapted in such a way that these particles can be conjugated to an overall large number of biomolecules or chemical receptors so that they can bind selectively to a target, such as proteins, nucleic acids, or ecological contaminants. This has made it flexible, and as such, it has many uses in the form of medical diagnostics, environmental control, and food security.

QDs provide significant benefits to multiplexed-biomarker detection in medical diagnostics. This is evidenced by a notable number of cancer biomarkers being detected with QDs conjugated with various antibodies or aptamers. As an example, Chen et al. showed that QDs used in the simultaneous investigation of breast cancer tumor markers such as HER2 and AFP improve the sensitivity and specificity of the assay, as well as its overall accuracy relative to conventional methods [73]. Application of QDs in immunoassay has also led to the establishment of sensitive detection systems that can detect several biomarkers in one sample without much interference [74]. Also, QDs have been combined with other sensitive methods like FRET to enhance the sensitivity of multiplexed assays [57]. QDs are being applied in the development of environmental monitoring to investigate different pollutants such as heavy metals, pesticides, and industrial chemicals. They are more useful in water quality monitoring due to their selectivity to various ions. A recent study reported the use of QDs to detect both arsenic (As^{3+}) and mercury (Hg^{2+}) in drinking water with very high sensitivity and much better than the traditional method, atomic absorption spectroscopy [75]. By using sensor arrays, the use of QDs enables the detection of various contaminants simultaneously, and environmental pollution can be tracked over longer distances in real time [76]. Illustratively, a kit of heavy metals such as Pb^{2+}, Cd^{2+}, and copper (Cu^{2+}) has been identified in aquatic environments using QD-based sensors, therefore maintaining overall evaluations of the environment [77]. QDs have been used in food safety to monitor foodborne pathogens and contaminants in food protein, pesticides, etc. Their fluorescence can be manipulated in a manner that they react to certain analytes to give fast and accurate testing of food products. As an example, a QD-based biosensor detects multiple foodborne pathogens, such as Salmonella and *E. coli*, in fewer than 30 minutes [78]. The ability to detect rapidly and multiply makes the QD tool an attractive solution to on-site food testing that ensures safety of the food product as well as its quality within the provision chain.

The other QDs advantage is that, they are photostable. In contrast to organic dyes, QDs are far more resistant to photobleaching, and so the dyes can be monitored over a long time without re-excitation. This feature is particularly valuable in longitudinal studies, since there is a need to reliably detect the signals on long intervals [74]. Moreover, the QDs can be surface functionalized and thus directed to target certain structures or pathways in cells, which increases their applicability in biosensing and diagnostics even more. Nevertheless, regardless of all these benefits, some difficulties connected with the toxicity of QDs, particularly those based on heavy metals, such as Pb^{2+} or Cd^{2+}, might arise. Lately, research has been dedicated to finding fewer toxic options, including carbon-based QDs and silicon-based QDs, that retain all the optimal optical properties that are not associated with health risks [79]. These advancements provide an opportunity to use QD-based applications in clinical and environmental fields that are safer.

2.5 Recent Advancements and Emerging Trends

QDs have appeared as powerful nanomaterials in biomedical diagnostics owing to their exclusive optical properties, including high intensity, size-tunable emission, and excellent photostability. Recent advancements in QD-based biosensors are driving a new generation of extremely sensitive and selective diagnostic platforms.

2.5.1 Development of QD-Based Biosensors

QDs have emerged at the forefront of nanoscale sensing technologies due to their remarkable optical properties and versatility in biointerface engineering. These semiconductor nanocrystals offer a unique combination of features and exceptional photostability that make them exceptionally well-suited for biosensing applications [80]. Recent progress in nanobiotechnology has enabled the precise surface modification of QDs with a variety of biological ligands, allowing them to serve as highly effective platforms for molecular recognition [81]. Functionalization techniques, such as ligand exchange, silica encapsulation, and polymer coating, have facilitated the stable conjugation of QDs to target moieties like antibodies, DNA, and peptides [82]. This has expanded their utility in detecting a broad spectrum of clinically relevant analytes, ranging from nucleic acids and enzymes to cancer biomarkers and viral proteins, with detection limits reaching the femtomolar range [83]. Another significant advantage of QDs is their ability to support multiplexed detection. Thanks to their size-dependent emission properties, multiple QDs with distinct emission wavelengths can be simultaneously excited by a single light source, allowing for the parallel recognition of numerous targets in complex biological samples [84]. This capability is specifically beneficial for clinical diagnostics and on-site testing, where rapid and accurate screening of several biomarkers is essential. The incorporation of QDs into advanced sensing platforms, such as FRET-based sensors, electrochemiluminescent assays, and microfluidic-integrated devices, has further enhanced their analytical performance. These hybrid systems offer improved dynamic range, faster response times, and greater signal-to-noise ratios [85]. Notably, such systems have shown promise in minimally invasive diagnostics, including early cancer detection, infectious disease screening, and metabolic monitoring. A transformative trend in recent years is the convergence of QD biosensors with artificial intelligence (AI) and machine learning (ML). By employing algorithms capable of handling complex fluorescence datasets, AI enhances the interpretation of sensor output, automates decision-making, and enables pattern recognition beyond human capabilities [86]. This integration is pushing the boundaries of precision diagnostics, facilitating early disease identification, and enabling adaptive sensing platforms that can learn from biological variability in real-time. Together, these advances highlight the evolving role of QDs as not just passive labels but dynamic components in intelligent biosensing systems, pointing toward a future of personalized, high-throughput, and AI-enhanced diagnostics.

2.5.2 Integration with Other Nanomaterials

The functional integration of QDs with complementary nanomaterials has revolutionized the design of biosensing platforms, yielding hybrid systems that surpass the capabilities of individual components. By coupling QDs with entities such as graphene derivatives, plasmonic nanoparticles (e.g. gold and silver), carbon nanotubes (CNTs), and other nanoscale constructs, researchers have developed multifunctional biosensors that offer enhanced sensitivity, selectivity, and real-time detection capabilities in complex biological environments.

2.5.2.1 Graphene and Graphene Quantum Dots (GQDs)

Graphene and its derivatives, including GO, reduced graphene oxide (rGO), and GQDs, have emerged as highly compatible partners in QD-based biosensor development. Their exceptional electrical conductivity, vast surface area, and facile functionalization enable efficient electron transfer and signal amplification. In these hybrid constructs, QDs often serve as fluorescent probes, while graphene materials act as energy acceptors or electrochemical transducers, improving FRET and enhancing signal-to-noise ratios. Graphene QDs, owing to their quantum confinement and edge effects, exhibit size-tunable luminescence and high aqueous solubility. Their application in detecting a diverse biomolecule as well as hydrogen peroxide, micro RNA, DNA, and cardiac troponin has been widely reported, especially in FRET- and electrochemiluminescence-based platforms. The strategic interface between QDs and graphene substrates has enabled label-free detection of bacteria and other pathogens with remarkable accuracy using electrochemical impedance spectroscopy [87].

2.5.2.2 Gold–Graphene Nanocomposites

Further performance gains have been achieved by incorporating metallic nanoparticles into graphene-QD systems. For instance, the fusion of gold nanoparticles (AuNPs) with GQDs has produced biosensors capable of detecting uric acid with outstanding sensitivity and a detection limit as low as 0.2 mg/dL [88]. These gold graphene hybrids leverage the conductive network of graphene and the localized surface plasmon resonance (LSPR) of AuNPs to intensify signal transduction, enabling faster and more precise biosensing.

2.5.2.3 Plasmonic–Quantum Dot Hybrids

The integration of QDs with plasmonic nanostructures, particularly AuNPs and AgNPs, offers a highly effective route for enhancing optical signals. The LSPR effects of these metals dramatically amplify QD fluorescence, improving both excitation and emission efficiency. This leads to biosensors with ultralow detection limits, suitable for early-stage disease diagnostics, environmental monitoring, and pathogen detection [89]. Such hybrid systems can be precisely tuned by adjusting nanoparticle size, shape, and spacing, optimizing plasmonic coupling and emission intensity. Additionally, these constructs can be embedded into flexible substrates or wearable biosensors for portable, point-of-care diagnostics.

2.5.2.4 SERS and Multimodal Platforms

Surface-enhanced Raman scattering (SERS) biosensors based on QD nanomaterial hybrids are gaining attention for their ability to provide label-free molecular fingerprints with high specificity. When coupled with GQDs or plasmonic nanoparticles, SERS substrates can detect analytes at trace levels, offering a dual-mode sensing capability (fluorescence and Raman) ideal for high-throughput screening [90]. Altogether, the convergence of QDs with advanced nanomaterials enables the creation of intelligent, ultrasensitive, and multiplexed biosensing platforms. These hybrids harness the optoelectronic advantages of QDs and the structural or plasmonic benefits of complementary nanomaterials, paving the way for next-generation diagnostics that are portable, responsive, and applicable in clinical and environmental settings.

2.5.3 Role of AI and ML in QD-Based Diagnostics

The integration of AI and ML into QD-based diagnostic technologies is rapidly advancing the field of biosensing. QDs produce rich and complex fluorescence or spectral data due to their size-dependent optical properties and multiplexing capabilities. However, analyzing these data manually or through traditional algorithms can be challenging, especially when distinguishing subtle differences related to disease biomarkers. AI and ML algorithms address these challenges by automating data processing and enhancing the interpretation of QD sensor outputs. Supervised learning methods, such as support vector machines (SVM), arbitrary woods, and deep neural networks, can be qualified to recognize patterns in spectral data corresponding to specific biomolecules or pathological conditions. This allows for highly sensitive and specific detection, even in samples with low analyte concentrations or complex biological backgrounds [91]. Recent studies demonstrate that ML models can improve classification accuracy in cancer diagnostics by analyzing QD fluorescence signatures to differentiate between healthy and diseased states. For example, deep learning algorithms have been applied to fluorescence lifetime data from QD sensors to detect cancer biomarkers with higher precision than traditional threshold-based methods. Furthermore, unsupervised learning approaches help identify new patterns or biomarkers without prior labeling, opening avenues for novel diagnostic discoveries. In addition to data interpretation, AI optimizes sensor design and experimental parameters [92]. By simulating various sensor configurations and predicting their performance, ML can guide the fabrication of more efficient QD-based biosensors. Moreover, AI-driven analysis enables multiplexed detection, where multiple biomarkers are simultaneously identified from overlapping QD signals, enhancing diagnostic throughput and reducing testing times. The convergence of nanotechnology and AI thus holds significant promise for emerging point-of-care diagnostics that are swift, accurate, and adaptable. This synergy will play a significant role in personalized medicine, allowing earlier detection, better monitoring, and tailored treatment strategies for patients [93].

2.6 Conclusion and Future Perspectives

Quantum dots (QDs), representing the next generation of fluorescent probes, are widely used in biological and biomedical research. This chapter explored the fundamental mechanisms behind QD fluorescence and presented various strategies for tuning their emission wavelength and intensity to improve bioimaging applications. To move QDs closer to clinical use, researchers have extensively focused on surface modifications, bioconjugation techniques, and synthesis methods. Despite significant progress, several challenges remain, particularly in terms of sensitivity, toxicity, and specificity, which continue to hinder clinical adoption. One major concern is the prolonged retention of QDs in the body. Nevertheless, due to their unique photophysical properties, nontoxic QDs hold promise as alternatives to many current imaging and tracking technologies used in disease monitoring. Conventional diagnostic techniques are often expensive because they rely on sophisticated instrumentation, whereas QD-based approaches offer a more cost-effective solution, making them suitable for point-of-care and occasional use from a socioeconomic perspective. The ease of detecting and quantifying QDs further supports their potential for integration into portable diagnostic tools.

Beyond diagnostics, an innovative strategy could involve leveraging the inherent cytotoxicity of certain QDs to treat diseases through targeted action on specific tissues and cells. However, to realize this potential, comprehensive studies on the biological interactions of QDs are essential. If these challenges are overcome, QDs may provide a unified solution to a wide range of biological and medical problems.

References

1 Gao, Y., Lim, J., Teoh, S.-H., and Xu, C. (2015). *Emerging translational research on magnetic nanoparticles for regenerative medicine, Chemical Society Reviews* 44: 6306.

2 Byakodi, M., Shrikrishna, N.S., Sharma, R. et al. (2022). Emerging 0D, 1D, 2D, and 3D nanostructures for efficient point-of-care biosensing, *Biosensors and Bioelectronics: X* 12: 100284.

3 Schiavo, L., Cammarano, A., Carotenuto, G. et al. (2024). An overview of the advanced nanomaterials science, *Inorganica Chimica Acta* 559: 121802.

4 Vitshima, N.A., Silwana, B., Tsolekile, N., and Matoetoe, M.C. (2022). Effect of ZnS coating on the optoelectronic properties of aqueous glutathione capped AgInS quantum dots, *Journal of Alloys and Compounds* 900: 163386.

5 Yadav, S., Jena, S.R., Altaee, A. et al. (2021). Heterostructures of 2D materials-quantum dots (QDs) for optoelectronic devices: challenges and opportunities, *Emergent Materials* 4: 901.

6 Zhang, J., Zhang, S., Zhang, Y. et al. (2023). Colloidal Quantum Dots: Synthesis, Composition, Structure, and Emerging Optoelectronic Applications, *Laser & Photonics Reviews* 17: 2200551.

7 Manikandan, A., Chen, Y.-Z., Shen, C.-C. et al. (2019). *A critical review on two-dimensional quantum dots (2D QDs): From synthesis toward applications in energy and optoelectronics, Progress in Quantum Electronics* 68: 100226.

8 Sedlmeier, A. and Gorris, H.H. (2015). Surface modification and characterization of photon-upconverting nanoparticles for bioanalytical applications, *Chemical Society Reviews* 44: 1526.

9 Sharma, A., Tharuny, S., Pandey, T., and Pandey, V. (2025). Eco-Friendly Fabrication of FeS2 QD-Chitosan Biopolymer Composites: Green Synthetic Approach, *Biopolymers* 116: e70002.

10 Lu, M., Sun, J., Li, R. et al. (2025). Construction of 0-dimensional silver quantum dot/MoSe2@MXene/3-dimensional porous copper foam composite electrodes with heterogeneous interfaces for synergistic electrocatalytic degradation of antibiotics, *Journal of Colloid and Interface Science* 679: 503.

11 Zhang, M., Bishop, B.P., Thompson, N.L. et al. (2019). *Quantum dot cellular uptake and toxicity in the developing brain: implications for use as imaging probes, Nanoscale Advances* 1: 3424.

12 Wang, X., Ding, Q., Groleau, R.R. et al. (2024). Fluorescent Probes for Disease Diagnosis, *Chemical Reviews* 124: 7106.

13 Woidasky, J., Sander, I., Schau, A. et al. (2020). *Inorganic fluorescent marker materials for identification of post-consumer plastic packaging, Resources, Conservation and Recycling* 161: 104976.

14 Bradberry, S.J., D'Agostino, B., Caffrey, D.F. et al. (2023). Gunnlaugsson. Time-delayed Lanthanide Luminescent Sensors and Probes, in: Fluorescent Chemosensors, In: *Fluorescent Chemosensors*, 406–429. The Royal Society of Chemistry. https://doi.org/10.1039/9781839167324-00406.

15 Yang, M., Huang, J., Fan, J. et al. (2020). *Chemiluminescence for bioimaging and therapeutics: recent advances and challenges, Chemical Society Reviews* 49: 6800.

16 Huo, Y., Miao, J., Han, L. et al. (2017). Selective and sensitive visualization of endogenous nitric oxide in living cells and animals by a Si-rhodamine deoxylactam-based near-infrared fluorescent probe, *Chemical Science* 8: 6857.

17 Male, K.B., Lachance, B., Hrapovic, S. et al. (2008). Assessment of Cytotoxicity of Quantum Dots and Gold Nanoparticles Using Cell-Based Impedance Spectroscopy, *Analytical Chemistry* 80: 5487.

18 Taniguchi, S., Green, M., Rizvi, S.B., and Seifalian, A. (2011). The one-pot synthesis of core/shell/shell CdTe/CdSe/ZnSe quantum dots in aqueous media for in vivo deep tissue imaging, *Journal of Materials Chemistry* 21: 2877.

19 Mousavi, S.M., Hashemi, S.A., Gholami, A. et al. (2021). Precise Blood Glucose Sensing by Nitrogen-Doped Graphene Quantum Dots for Tight Control of Diabetes, *Journal of Sensors* 2021: 580203.

20 Mousavi, S.M., Hashemi, S.A., Kalashgrani, M.Y. et al. (2022). *Biomedical Applications of an Ultra-Sensitive Surface Plasmon Resonance Biosensor Based on Smart MXene Quantum Dots (SMQDs), Biosensors-Basel* 12: 743.

21 Mousavi, S.M., Hashemi, S.A., Yari Kalashgrani, M. et al. (2022). *Recent Advances in Inflammatory Diagnosis with Graphene Quantum Dots Enhanced SERS Detection, Biosensors-Basel* 12: 461.

22 Shu, Y., Lin, X., Qin, H. et al. (2020). *Quantum Dots for Display Applications, Angewandte Chemie* 132: 22496.

23 Elibol, E., Demirci, T., and Koç, D. (2025). High efficiency green and red emitting CdSeTe quantum dots: Synthesis and PLQY improvement for optical applications, *Journal of Alloys and Compounds* 1010: 178161.

24 Walling, M.A., Novak, J.A., and Shepard, J.R.E. (2009). Quantum Dots for Live Cell and In Vivo Imaging, *International Journal of Molecular Sciences* 10: 441.

25 Byrne, S.J., Williams, Y., Davies, A. et al. (2007). *"Jelly Dots": Synthesis and Cytotoxicity Studies of CdTe Quantum Dot–Gelatin Nanocomposites, Small* 3: 1152.

26 Qian, H., Dong, C., Weng, J., and Ren, J. (2006). *Facile One-Pot Synthesis of Luminescent, Water-Soluble, and Biocompatible Glutathione-Coated CdTe Nanocrystals, Small* 2: 747.

27 Hardman, R. (2006). *A Toxicologic Review of Quantum Dots: Toxicity Depends on Physicochemical and Environmental Factors, Environmental Health Perspectives* 114: 165.

28 Choi, A.O., Brown, S.E., Szyf, M., and Maysinger, D. (2008). Quantum dot-induced epigenetic and genotoxic changes in human breast cancer cells, *Journal of Molecular Medicine* 86: 291.

29 Li, H., Zhou, Q., Liu, W. et al. (2008). Progress in the toxicological researches for quantum dots, *Science in China, Series B Chemistry* 51: 393.

30 Chen, H., Wang, Y., Xu, J. et al. (2008). Non-invasive Near Infrared Fluorescence Imaging of CdHgTe Quantum Dots in Mouse Model, *Journal of Fluorescence* 18: 801.

31 Depalo, N., Mallardi, A., Comparelli, R. et al. (2008). Luminescent nanocrystals in phospholipid micelles for bioconjugation: An optical and structural investigation, *Journal of Colloid and Interface Science* 325: 558.

32 Lin, C.-A.J., Liedl, T., Sperling, R.A. et al. (2007). *Bioanalytics and biolabeling with semiconductor nanoparticles (quantum dots), Journal of Materials Chemistry* 17: 1343.

33 Liu, T.-C., Zhang, H.-L., Wang, J.-H. et al. (2008). *Study on molecular interactions between proteins on live cell membranes using quantum dot-based fluorescence resonance energy transfer, Analytical and Bioanalytical Chemistry* 391: 2819.

34 Tan, W., Wang, K., He, X. et al. (2004). Bionanotechnology based on silica nanoparticles, *Medicinal Research Reviews* 24: 621.

35 Dembski, S., Graf, C., Krüger, T. et al. (2008). Photoactivation of CdSe/ZnS Quantum Dots Embedded in Silica Colloids, *Small* 4: 1516.

36 Zhang, Y., Mi, L., Wang, P. et al. (2008). Photoluminescence Decay Dynamics of Thiol-Capped CdTe Quantum Dots in Living Cells under Microexcitation, *Small* 4: 777.

37 Zheng, J., Zheng, Z., Gong, W. et al. (2008). Abnormal temperature behavior of photoluminescence in CdSe/ZnSe self-assembled quantum dots, *Solid State Communications* 147: 429.

38 Xia, Y., Kapuria, N., He, M. et al. (2025). *Copper indium sulfide colloidal quantum dots: Advances in synthesis, structure-optoelectronic properties, and applications, Advanced Powder Materials* 4: 100283.

39 Mimona, M.A., Rimon, M.I.H., Zohura, F.T. et al. (2025). *Quantum dot nanomaterials: Empowering advances in optoelectronic devices, Chemical Engineering Journal Advances* 21: 100704.

40 Heuer-Jungemann, A., Feliu, N., Bakaimi, I. et al. (2019). *The Role of Ligands in the Chemical Synthesis and Applications of Inorganic Nanoparticles, Chemical Reviews* 119: 4819.

41 Bao, Y., Li, J., Wang, Y. et al. (2012). *Preparation of water soluble CdSe and CdSe/CdS quantum dots and their uses in imaging of cell and blood capillary, Optical Materials* 34: 1588.

42 Jiang, P., Zhu, D.-L., Zhu, C.-N. et al. (2015). *A highly reactive chalcogenide precursor for the synthesis of metal chalcogenide quantum dots, Nanoscale* 7: 19310.

43 Labiadh, H., Ben Chaabane, T., Piatkowski, D. et al. (2013). *Aqueous route to color-tunable Mn-doped ZnS quantum dots, Materials Chemistry and Physics* 140: 674.

44 Xin, K., Li, L., Zhou, Z. et al. (2024). *Epitaxial growth of quantum dots on van der Waals surfaces, Nature Synthesis* 3: 1176.

45 Myeong, S., Chon, B., Kumar, S. et al. (2022). *Quantum dot photolithography using a quantum dot photoresist composed of an organic–inorganic hybrid coating layer, Nanoscale Advances* 4: 1080.

46 Kumar, A., Chowdhuri, A.R., Laha, D. et al. (2017). *Green synthesis of carbon dots from Ocimum sanctum for effective fluorescent sensing of Pb2+ ions and live cell imaging, Sensors and Actuators B: Chemical* 242: 679.

47 Mehta, V.N., Jha, S., Basu, H. et al. (2015). *One-step hydrothermal approach to fabricate carbon dots from apple juice for imaging of mycobacterium and fungal cells, Sensors and Actuators B: Chemical* 213: 434.

48 Yan, Z.-Y., Du, Q.-Q., Qian, J. et al. (2017). *Eco-friendly intracellular biosynthesis of CdS quantum dots without changing Escherichia coli's antibiotic resistance, Enzyme and Microbial Technology* 96: 96.

49 Chen, Y., Sun, L., Liao, F. et al. (2019). Fluorescent-stable and water-soluble two-component-modified silicon quantum dots and their application for bioimaging, *Journal of Luminescence* 215: 116644.

50 Verma, A. and Stellacci, F. (2010). Effect of Surface Properties on Nanoparticle–Cell Interactions, *Small* 6: 12.

51 Sun, Y., Zhang, M., Bhandari, B., and Yang, C. (2022). *Recent Development of Carbon Quantum Dots: Biological Toxicity, Antibacterial Properties and Application in Foods, Food Reviews International* 38: 1513.

52 Wang, B., Cai, H., Waterhouse, G.I.N. et al. (2022). *Carbon Dots in Bioimaging, Biosensing and Therapeutics: A Comprehensive Review, Small Science* 2 (6): 2200012.

53 H. M. E. M. El-brolsy, N. A. N. Hanafy, M. A. El-Kemaryc. et al. (2022). Fighting Non-Small Lung Cancer Cells Using Optimal Functionalization of Targeted Carbon Quantum Dots Derived from Natural Sources Might Provide Potential Therapeutic and Cancer Bio Image Strategies, Int J Mol Sci 23: 13283.

54 Tian, X., Zeng, A., Liu, Z. et al. (2020). Carbon Quantum Dots: In vitro and in vivo Studies on Biocompatibility and Biointeractions for Optical Imaging, *International Journal of Nanomedicine* 15: 6519.

55 Karakoçak, B.B., Laradji, A., Primeau, T. et al. (2021). Hyaluronan-Conjugated Carbon Quantum Dots for Bioimaging Use, *ACS Applied Materials & Interfaces* 13: 277.

56 Wang, M., Liu, Y., Ren, G. et al. (2018). *Bioinspired carbon quantum dots for sensitive fluorescent detection of vitamin B12 in cell system, Analytica Chimica Acta* 1032: 154.

57 Van Duong, P., Thi, L.A., Hung, P.Q. et al. (2025). Probing Förster Resonance Energy Transfer in Carbon Quantum Dots for High-Sensitivity Aflatoxin B1 Detection, *Journal of Fluorescence* https://doi.org/10.1007/s10895-025-04246-6.

58 Raju, K.S., Das, G.S., and Tripathi, K.M. (2024). *Nitrogen-doped carbon quantum dots from biomass as a FRET-based sensing platform for the selective detection of H2O2 and aspartic acid, RSC Sustainability* 2: 223–232.

59 Zoghi, M., Pourmadadi, M., Yazdian, F. et al. (2023). Synthesis and characterization of chitosan/carbon quantum dots/Fe_2O_3 nanocomposite comprising curcumin for targeted drug delivery in breast cancer therapy, *International Journal of Biological Macromolecules* 249: 125788.

60 Samimi, S., Ardestani, M.S., and Dorkoosh, F.A. (2021). *Preparation of carbon quantum dots-quinic acid for drug delivery of gemcitabine to breast cancer cells, Journal of Drug Delivery Science and Technology* 61: 102287.

61 Yao, Y.-Y., Gedda, G., Girma, W.M. et al. (2017). Magnetofluorescent carbon dots derived from crab shell for targeted dual-modality bioimaging and drug delivery, *ACS Applied Materials & Interfaces* 9: 13887.

62 Oliveira, D., Oliveira, P., Xu, A. et al. (2025). *Optical immunosensor panel using quantum dot-antibody conjugates for highly sensitive detection of carbohydrate antigen 19–9 (CA19-9), Analytica Chimica Acta* 1333: 343399.

63 You, Q., Wang, P., Zhu, T. et al. (2025). Bifunctional MXene quantum dots-coated bimetallic Prussian blue analogues for sensitive sensing and accurate localization imaging of miRNAs in living cells, *Materials Today Bio* 32: 101747.

64 Zheng, Y., Song, K., Cai, K. et al. (2022). B-Cell-Epitope-Based Fluorescent Quantum Dot Biosensors for SARS-CoV-2 Enable Highly Sensitive COVID-19 Antibody Detection, *Viruses* 14: 1031.

65 Zhang, Z., Wang, X., Wei, X. et al. (2021). *Multiplex quantitative detection of SARS-CoV-2 specific IgG and IgM antibodies based on DNA-assisted nanopore sensing, Biosensors & Bioelectronics* 181: 113134.

66 Liu, L., Wang, X., Ma, Q. et al. (2016). *Multiplex electrochemiluminescence DNA sensor for determination of hepatitis B virus and hepatitis C virus based on multicolor quantum dots and Au nanoparticles, Analytica Chimica Acta* 916: 92.

67 Kuo, Y.-C., Wang, Q., Ruengruglikit, C. et al. (2008). *Antibody-Conjugated CdTe Quantum Dots for Escherichia coli Detection, The Journal of Physical Chemistry C* 112: 4818.

68 Nikolaev, V.V., Lepekhina, T.B., Alliluev, A.S. et al. (2024). Quantum Dot-Based Nanosensors for In Vitro Detection of Mycobacterium tuberculosis, *Nanomaterials* 14: 1553.

69 Huang, M. and Tong, C. (2022). A dual-emission ratiometric fluorescence probe for highly selective and simultaneous detection of tetracycline and ferric ions in environmental water samples based on a boron-doped carbon quantum dot/CdTe–Eu3^{+} composite. *Environmental Science. Nano* 9: 1712.

70 Li, L., Liu, D., Shi, A., and You, T. (2018). *Simultaneous stripping determination of cadmium and lead ions based on the N-doped carbon quantum dots-graphene oxide hybrid, Sensors and Actuators B: Chemical* 255: 1762.

71 Zhu, L., Mei, X., Peng, Z. et al. (2022). A paper-based microfluidic sensor array combining molecular imprinting technology and carbon quantum dots for the discrimination of nitrophenol isomers, *Journal of Hazardous Materials* 435: 129012.

72 Qin, T., Wang, J., Liu, Y., and Guo, S. (2022). *Carbon Quantum Dots based chemosensor array for monitoring multiple metal ions, Molecules* 27: 3843.

73 Chen, C., Peng, J., Xia, H. et al. (2010). *Quantum-dot-based immunofluorescent imaging of HER2 and ER provides new insights into breast cancer heterogeneity, Nanotechnology* 21: 095101.

74 Tran, L. and Park, S. (2021). *Highly sensitive detection of dengue biomarker using streptavidin-conjugated quantum dots, Scientific Reports* 11: 15196.

75 Latha, B.D., Soumya, K., More, N. et al. (2023). *Fluorescent carbon quantum dots for effective tumor diagnosis: A comprehensive review, Biomedical Engineering Advances* 5: 100072.

76 Ussia, M., Privitera, V., and Scalese, S. (2024). *Unlocking the Potential and Versatility of Quantum Dots: from Biomedical to Environmental Applications and Smart Micro/Nanorobots, Advanced Materials Interfaces* 11: 2300970.

77 Pudza, M.Y., Abidin, Z.Z., Abdul-Rashid, S. et al. (2020). *Selective and simultaneous detection of cadmium, lead and copper by tapioca-derived carbon dot-modified electrode, Environmental Science and Pollution Research* 27: 13315.

78 Wu, Y., Xu, X., Zhu, Y. et al. (2025). *Research progress on multiplexed pathogen detection using optical biosensors, Biosensors-Basel* 15: 378.

79 Mohammadi, R., Naderi-Manesh, H., Farzin, L. et al. (2022). *Fluorescence sensing and imaging with carbon-based quantum dots for early diagnosis of cancer: A review, Journal of Pharmaceutical and Biomedical Analysis* 212: 114628.

80 Michalet, X., Pinaud, F.F., Bentolila, L.A. et al. (2005). *Quantum Dots for Live Cells, in Vivo Imaging, and Diagnostics, Science* 307: 538.

81 Hanmante, P.S., Lohiya, R.T., Wankhede, A.G. et al. (2025). Quantum Dot Bioconjugates for Imaging, Labelling and Sensing, *Next Nanotechnology* 7: 100172.

82 Derfus, A.M., Chan, W.C.W., and Bhatia, S.N. (2004). Probing the Cytotoxicity of Semiconductor Quantum Dots, *Nano Letters* 4: 11.

83 Ding, L., Liu, H., Zhang, L. et al. (2018). *Label-free detection of microRNA based on the fluorescence quenching of silicon nanoparticles induced by catalyzed hairpin assembly coupled with hybridization chain reaction, Sensors and Actuators B: Chemical* 254: 370.

84 Castro, R.C., Saraiva, M.L.M.F.S., Santos, J.L.M., and Ribeiro, D.S.M. (2021). *Multiplexed detection using quantum dots as photoluminescent sensing elements or optical labels, Coordination Chemistry Reviews* 448: 214181.

85 Li, Y.-Q., Wang, J.-H., Zhang, H.-L. et al. (2010). *High-sensitivity quantum dot-based fluorescence resonance energy transfer bioanalysis by capillary electrophoresis, Biosensors & Bioelectronics* 25: 1283.

86 Ali, M.K., Javaid, S., Afzal, H. et al. (2023). *Exploring the multifunctional roles of quantum dots for unlocking the future of biology and medicine, Environmental Research* 232: 116290.

87 Rasheed, P.A., Ankitha, M., Pillai, V.K., and Alwarappan, S. (2024). *Graphene quantum dots for biosensing and bioimaging, RSC Advances* 14: 16001.

88 Onifade, O.A., Mohamad Azri, D.D., Abu Bakar, M.H. et al. (2025). Surface Functionalized Plasmonic Sensors for Uric Acid Detection With Gold-Graphene Stacked Nanocomposites, *Photonic Sensors* 15: 250132.

89 Li, X., Wang, Y., Fu, Q. et al. (2020). *Plasmon-Emitter Hybrid Nanostructures of Gold Nanorod-Quantum Dots with Regulated Energy Transfer as a Universal Nano-Sensor for One-step Biomarker Detection, Nanomaterials* 10: 444.

90 Kurniawan, D., Chen, Y.-Y., Sharma, N. et al. (2022). Graphene Quantum Dot-Enabled Nanocomposites as Luminescence- and Surface-Enhanced Raman Scattering Biosensors, *Chem* 10: 498.

91 Bocan, A., Siavash Moakhar, R., del Real Mata, C. et al. (2025). *Machine-Learning-Aided Advanced Electrochemical Biosensors, Advanced Materials* 2417520. https://doi.org/10.1002/adma.202417520.

92 Agarwal, A. and Chumbhale, N. (2025). Quantum Dots in Neuroimaging: Machine Learning for Brainwave Analysis and Toxicity Assessment. https://doi.org/10.2139/ssrn.5237091.

93 Goumas, G., Vlachothanasi, E.N., Fradelos, E.C., and Mouliou, D.S. (2025). *Biosensors, Artificial Intelligence Biosensors, False Results and Novel Future Perspectives, Diagnostics* 15: 1037.

3

Single-Molecule Detection with Quantum Biosensors

Jayeeta Chattopadhyay[1] *and Tara Sankar Pathak*[2]

[1] *Amity Institute of Applied Sciences, Amity University Jharkhand, Ranchi, Jharkhand, India*
[2] *Chemistry Department, Surendra Institute of Engineering and Management, Siliguri, West Bengal, India*

3.1 Introduction

Single-molecule detection (SMD) represents a transformative frontier in analytical science, enabling the observation and characterization of individual molecules rather than averaged signals from large ensembles. This capability is crucial for uncovering unique properties and behaviors of biological entities that are often obscured in bulk measurements [1]. The advent of quantum biosensors (QBs) has further revolutionized this field by leveraging the unique principles of quantum mechanics to achieve unprecedented levels of sensitivity, specificity, and precision in detecting biological molecules and events [2]. Quantum biosensors integrate biological recognition elements, such as enzymes, antibodies, or DNA, with quantum systems that serve as highly sensitive transducers. These devices convert biological interactions into measurable optical or electrical signals, often through discernible alterations in light emissions or electrical properties when interacting with target molecules. This approach allows for the detection and tracking of individual molecules or even single atoms within biological samples, providing invaluable insights into various biological processes. The importance of SMD with quantum biosensors stems from their ability to surpass the limitations of traditional analytical techniques, which often struggle with low concentrations of analytes, complex biological matrices, and the need for external light sources or labeling [3]. Quantum biosensors offer distinct advantages, including high sensitivity, real-time monitoring, and label-free detection capabilities, making them highly promising for applications ranging from disease monitoring and drug discovery to environmental surveillance [4]. This review article will explore the fundamental principles, diverse technological platforms, key advantages, recent applications, and persistent challenges associated with SMD using quantum biosensors, focusing on advancements made in recent years.

Quantum Biosensing in Medical Diagnostics, First Edition. Edited by J. G. Manjunatha.

3.2 Fundamental Principles of Quantum Biosensing for Single-Molecule Detection

Quantum biosensors harness specific quantum mechanical phenomena to achieve their remarkable detection capabilities. These principles enable the precise and sensitive measurement of biological interactions at the single-molecule level.

3.2.1 Quantum Confinement

Quantum confinement is a fundamental principle exploited in various quantum biosensors, particularly those utilizing nanomaterials like quantum dots (QDs). QDs are zero-dimensional (0D) nanomaterials, typically 2–10 nm in diameter, exhibiting unique optical and electronic properties due to the confinement of electrons and holes within their small dimensions [5]. When light interacts with QDs, QDs emit fluorescence based on their bandgap energy. This process involves an electron moving to a higher energy level upon light absorption and then releasing light as it returns to its original state [6]. The color of the emitted light is directly influenced by the QD's size, allowing for tunable emission spectra [7]. This size-dependent emission, coupled with broad absorption spectra and narrow emission profiles, makes QDs ideal candidates for optical sensing applications [8, 9]. Their high quantum yield, photostability, and resistance to photobleaching further enhance their utility compared to traditional fluorophores, enabling long-term tracking and multi-color imaging at the single-molecule level.

3.2.2 Quantum Properties for Enhanced Sensing

The superior performance of quantum biosensors is rooted in their ability to leverage unique quantum properties: superposition, entanglement, and quantum noise reduction [10].

3.2.2.1 Superposition and Quantum Coherence

Superposition allows a quantum system to exist in multiple states simultaneously, enabling quantum sensors to interact with external fields in ways classical systems cannot. Quantum coherence, the ability of a quantum system to maintain its superposition state over time, is critical for stable and precise measurements [10]. For instance, nitrogen-vacancy (NV) centers in diamond, a prominent quantum spin system, maintain long coherence times even at room temperature, allowing for optical initialization, readout, and coherent microwave control [10]. This property is essential for highly sensitive magnetometry and thermometry applications within biological systems. The ability to maintain coherence over longer durations directly translates to improved sensor stability and precision, as demonstrated in atomic clocks and advanced sensing protocols [11].

3.2.2.2 Entanglement

Quantum entanglement, where two or more particles become correlated regardless of distance, allows quantum sensors to surpass classical limitations in measurement resolution [10]. Entangled states, such as N00N states, enable enhanced phase measurements that exceed classical shot noise limits, leading to improved resolution and signal-to-noise ratio (SNR) [12]. This "quantum advantage" is particularly valuable in biological contexts where high-intensity illumination, common in classical methods, could damage delicate samples. By using entangled photons, quantum optical sensors can achieve high-precision measurements at low light levels, minimizing photodamage while maximizing data quality.

3.2.2.3 Quantum Noise Reduction

Classical sensors are inherently limited by thermal and shot noise, which introduce uncertainty into measurements. Quantum sensors, however, can utilize squeezed states of light or atoms to reduce measurement uncertainty beyond classical limits [10]. This technique, exemplified by squeezed light interferometry, allows for the detection of minute signals that would otherwise be obscured by noise [13]. The ability to operate at the quantum noise limit, as demonstrated in evanescent single-molecule biosensing, allows for a significant reduction in optical intensity while maintaining state-of-the-art sensitivity. This is crucial for noninvasive probing of biological samples, as it minimizes photodamage and enables extended observation periods.

3.2.3 Transduction Mechanisms

Quantum biosensors employ various transduction mechanisms to convert biological interactions into measurable signals.

3.2.3.1 Optical Transduction

Many quantum biosensors rely on optical transduction, where the binding of an analyte to a bioreceptor leads to a change in light properties such as fluorescence intensity, wavelength, or refractive index. Quantum dots, for instance, emit fluorescence that changes when they interact with target biomolecules, allowing for quantification of their presence and concentration [14]. Recent innovations include self-illuminating plasmonic biosensors that embed quantum light sources directly into chip-scale devices, leveraging quantum tunneling to detect biomolecules without an external light source. This label-free, real-time detection method modulates the spectrum and intensity of emitted light upon contact with biomolecules, offering a compact and scalable solution for high-performance sensing.

3.2.3.2 Electronic Transduction

Single-molecule electronic biosensors convert physiological changes of single biomolecules into measurable electrical signals. These devices, often employing field-effect biosensors, use a sensing nanochannel where biomolecule binding or interaction alters the electrical resistance or modulates electron fluxes, leading to detectable current changes. Scaling the sensing channel nanomaterials to the size of biological entities ensures that even a single captured molecule can significantly affect the device's conductivity. Examples include silicon nanowires and carbon nanotubes, which act as molecularly gated switches [15]. Despite their high sensitivity and real-time capabilities, electronic biosensors face challenges like the Debye screening effect in high ionic strength physiological solutions, which can hinder charge detection [1]. Strategies to mitigate this include modifying channel morphology or designing biomolecule decoration to alter the local ionic environment.

3.3 Types of Quantum Biosensors for Single-Molecule Detection

Recent advancements have seen the development of several distinct types of quantum biosensors, each leveraging specific quantum phenomena for SMD.

3.3.1 Quantum Dots (QDs)-Based Biosensors

Quantum dots (QDs) are widely utilized in biosensors due to their exceptional fluorescent properties and tunable characteristics.

3.3.1.1 Properties and Synthesis Methods

QDs are zero-dimensional nanomaterials, typically 2–10 nm in diameter, that emit fluorescence when excited by light. Their emission color is size-dependent, offering broad absorption and narrow, customizable emission spectra. Compared to traditional fluorophores, QDs exhibit decreased photobleaching, larger size, and increased sensitivity, making them highly advantageous for biosensing. While some QDs are made of heavy metals, posing toxicity concerns, the development of carbon and graphene QDs offers less toxic, more soluble, and more versatile alternatives [16]. Synthesis methods include top-down (e.g. laser ablation, chemical ablation) and bottom-up approaches (e.g. deposition, solvothermal method) [7]. Environmentally sustainable methods like hydrothermal and microwave-assisted synthesis are also emerging, enhancing scalability and reducing harmful byproducts.

3.3.1.2 Integration into Biosensors

For integration into biosensors, QDs are functionalized with bioreceptors such as molecularly imprinted polymers, aptamers, or antibodies, enabling specific analyte binding. QDs function as both bioreceptors and transducers, converting the analyte's interaction into a detectable light signal [6]. While traditional QD biosensors require external light sources, innovations like self-illuminating plasmonic biosensors embed quantum light sources directly into chip-scale devices, eliminating this need and enabling compact, real-time sensing.

3.3.1.3 Recent Breakthroughs in QD Sensors

Recent years have seen significant breakthroughs in QD sensor technology [17]. A Cardiff-led team developed a high-performance near-infrared QD sensor with record responsivity (up to 20,000 A/W) for low-light biomedical imaging and next-gen optical communications. This was achieved by modifying lead sulfide (PbS) QDs with long-chain ligands, which suppressed charge leakage and enhanced vertical charge injection, leading to industry-leading detectivity. These advancements promise applications in low-dose medical imaging, night-vision systems, and wearable health devices. Single-molecule fluorescence detection using QDs has also gained popularity for its ultrahigh sensitivity and low sample consumption, enabling precise quantification of targets by directly counting individual fluorescence bursts. Efforts are ongoing to find more suitable functional nanomaterials, such as graphene quantum dots, for constructing these highly sensitive biosensors.

3.3.2 Nitrogen-Vacancy (NV) Centers in Diamond

NV centers in diamond are leading platforms in quantum sensing due to their exceptional spin coherence time and optomagnetic properties.

3.3.2.1 Properties and Development for Magnetic Quantum Sensing

An NV center forms when a nitrogen atom replaces a carbon atom, and an adjacent carbon atom is missing, creating a negatively charged state (NV−). This defect acts as a quantum spin system that can be optically initialized, read out, and coherently controlled with microwave pulses, even at ambient conditions. The photoluminescence (PL) of the NV− center changes when excited with microwaves, allowing for the quantitative measurement of external magnetic fields through Zeeman splitting. NV centers are highly sensitive magnetometers, capable of measuring extremely small magnetic fields (as low as 5×10^{-14} T) and enabling device miniaturization down to the nanoscale [18]. Their biocompatibility and insensitivity to parameters like pH and viscosity make them ideal for biological sensing. Developing these sensors involves characterizing diamond samples, performing optically detected magnetic resonance (ODMR), and observing PL changes under external magnetic fields.

3.3.2.2 Breakthroughs in Integrating NV Centers into Living Cells

A significant breakthrough involves encasing diamond nanoparticles with specially engineered shells, making them ideal for use in living cells. Previous attempts to use nanodiamonds as biosensors showed reduced performance due to surface effects [19]. Inspired by QLED televisions, researchers developed a silicon-oxygen (siloxane) shell that not only protects the diamond surface but also fundamentally alters its quantum behavior, leading to up to a fourfold improvement in spin coherence, a 1.8-fold increase in fluorescence, and significant charge stability. This innovation solved a longstanding mystery in quantum sensing by revealing that the material interface drives electron transfer from the diamond into the shell, depleting electrons that normally reduce quantum coherence [19]. This development provides a new framework for engineering coherence and charge stability in quantum nanomaterials, promising hypersensitive sensors for early disease diagnosis, including cancers, within living cells.

3.3.2.3 New Techniques Using Nanodiamonds in Microdroplets

Another recent technique combines nanodiamonds in tiny moving microdroplets of liquid for quantum sensing. This method, utilizing a green laser, microwaves, and diamond dust suspended in water droplets, is precise, fast, sensitive, and requires only small amounts of material. This makes it particularly useful for studying trace chemicals or individual cells [20]. A potential future application is in creating self-driving bioreactors, where each nanodiamond droplet could function as a microscopic "beaker" holding a single cell, enabling real-time intracellular data for fine-tuning bioreactor processes in challenging environments.

3.3.3 Superconducting Qubits

Superconducting qubits, a leading approach in quantum information science, are also being explored for their potential in single-molecule sensing.

3.3.3.1 Principles of Superconducting Qubits

Superconducting qubits are based on superconducting electrical circuits, particularly Josephson tunnel junctions, which act as nonlinear, nondissipative elements [21]. Unlike conductors, where charge carriers are single electrons, superconductors utilize Cooper pairs, which are bosons that can attain a single quantum energy level (Bose-Einstein condensate). These systems exhibit quantum properties like superposition, interference, and entanglement. The nonlinearity of the Josephson inductance breaks the degeneracy of energy levels, allowing the system's dynamics to be restricted to two qubit states. This inherent low dissipation in superconductors theoretically allows for long coherence times, which are critical for quantum computing and sensing applications [22].

3.3.3.2 Recent Advances in Single-Molecule Sensing with Superconducting Qubits

Recent research has demonstrated the coherent control of a gatemon based on a single molecule: a one-dimensional carbon nanotube integrated into a circuit quantum electrodynamics (cQED) architecture. This nanotube-based gatemon qubit exhibits voltage tunability over more than 4 GHz and achieved a record coherence time of 200 ns, with potential for further improvement. This breakthrough positions carbon nanotubes as contenders for future quantum technologies and provides a sensitive platform for probing the microscopic behavior of fermions in quantum materials. The cQED architecture allows for probing the nanotube qubit by observing frequency shifts in the readout resonator depending on the qubit's state. While large-scale quantum computing with superconducting qubits still faces challenges related to tunability, size, and the need for dilution refrigerators, these advancements in single-molecule integration are a significant step toward harnessing individual molecules for quantum sensing applications. Furthermore, new techniques have been developed to enhance quantum sensing with superconducting qubits by counteracting decoherence, the unpredictable behavior caused by environmental noise. A coherence-stabilized protocol has been shown to temporarily counteract this decay, leading to larger sensing signals and improved sensitivity, surpassing conventional methods by up to 1.65 times [23].

3.4 Advantages of Quantum Biosensors for Single-Molecule Detection

Quantum biosensors offer profound advantages over conventional techniques, fundamentally transforming the landscape of detection and analysis.

3.4.1 Unprecedented Sensitivity and Resolution

Quantum biosensors achieve levels of sensitivity and resolution previously unattainable, surpassing classical limits.

3.4.1.1 Beyond Classical Limits

Quantum light sources, such as squeezed states and N00N states, enable enhanced phase measurements that move beyond the classical shot noise limits, leading to improved resolution and sensitivity. This allows for the detection of extremely weak signals, even single-photon interactions, and minute ripples in space-time in other quantum sensing applications. Quantum magnetometers, like NV center-based diamond sensors, can detect weak magnetic fields at the nanoscale, with applications in noninvasive brain activity mapping and early disease detection [24]. The ability to reduce measurement uncertainty beyond classical limits, often by leveraging quantum noise reduction techniques, allows for a four-order-of-magnitude reduction in optical intensity while maintaining state-of-the-art sensitivity.

3.4.1.2 Digital Readout and Molecular Counting

Single-molecule sensors provide digital signals, enabling the direct counting of molecules and interactions. This offers a distinct advantage over ensemble-averaged sensing, which typically provides no signal at low receptor occupancy. Digital detection allows for the analysis of characteristics of each SMD event, such as signal amplitude or bound-state lifetime, providing distributions of molecular properties rather than just averages. This molecular counting approach allows for absolute quantification without the need for calibration and can expand the dynamic range of detection significantly.

3.4.1.3 Real-Time and Continuous Monitoring

Quantum biosensors facilitate real-time and continuous monitoring of biomolecules, which is crucial for understanding dynamic natural processes and for applications like point-of-care diagnostics and personalized medicine [2]. Devices such as self-illuminating plasmonic biosensors offer label-free, real-time tracking of biomarkers in diseases like infections, cancer, and metabolic disorders [5]. Continuous monitoring sensors for a wide range of biomarkers are considered the next landmark in point-of-care diagnostics, building on the success of continuous glucose monitoring sensors.

3.4.2 Noninvasiveness and Biocompatibility

A critical advantage of quantum biosensors, particularly in biological applications, is their noninvasive nature and increasing biocompatibility.

3.4.2.1 Low Light Levels and Minimal Sample Damage

Quantum optical sensors enable high-precision measurements at low light levels, which is vital for probing delicate, photosensitive biological specimens without causing damage or altering their behavior. This is in contrast to traditional high-intensity illumination methods that can lead to photodamage, affecting enzyme activity or cellular viability

[12]. Techniques like "imaging with undetected photons" further enhance noninvasiveness by decoupling sensing and detection wavelengths, allowing imaging in regions like the mid-infrared without direct exposure of the sample to those wavelengths. This ensures the preservation of the biological sample's integrity during prolonged observation [24].

3.4.2.2 Integration into Biological Systems

The development of biocompatible quantum sensing platforms is a key focus. For instance, NV centers in diamonds, known for their unique sensing capabilities, are also biocompatible, making them ideal for intracellular measurements like temperature detection and magnetic field sensing within living cells. Recent breakthroughs, such as encasing nanodiamonds in specially engineered siloxane shells, have made these quantum biosensors ideal for integration into living cells without triggering an immune response [19]. This allows for the tracking of cell growth and early diagnosis of diseases within their natural environment.

3.4.3 Ability to Uncover Hidden Molecular Properties

Quantum biosensors provide a unique window into the complexities of biological systems by revealing properties often obscured by ensemble averaging.

3.4.3.1 Heterogeneity and Stochastic Variability

Single-molecule analysis, enabled by quantum biosensors, can reveal the heterogeneity and stochastic variability within a sample, providing a more complete picture than ensemble-averaged measurements [2]. This is particularly important in affinity-based assays, where real-time observation of molecular association and dissociation events provides digital signals that capture the distribution of molecular properties rather than just their average [6].

3.4.3.2 Rare Event Detection

One of the most compelling benefits of single-molecule analysis is its ability to detect rare events that would be lost or masked in ensemble-average analyses [3]. This capability is crucial for applications such as early diagnosis of diseases where biomarkers are present at extremely low concentrations. For example, single-molecule sensitivity in diagnostics provides a digital readout of the number of detected biomarkers, even in complex matrices like blood, where traditional methods might fail due to low concentration or nonspecific interactions [6].

3.5 Applications of Single-Molecule Detection with Quantum Biosensors

The unparalleled capabilities of quantum biosensors for SMD are driving transformative applications across diverse fields.

3.5.1 Biomedical and Clinical Diagnostics

Quantum biosensors hold immense promise for revolutionizing healthcare, from early disease detection to personalized medicine.

3.5.1.1 Early Disease Diagnosis and Personalized Medicine

Single-molecule biosensors address the unmet need for real time, highly specific, and accurate detection of individual biological molecules, which is critical for early diagnosis of diseases like cancer and for monitoring medical treatments. Their exceptional sensitivity and minimal sample volume requirements make them practical for analyzing patient-derived samples such as blood, saliva, and urine [2]. The ability to detect biomarkers at picogram to femtomolar concentrations rivals the most advanced sensors available today, enabling earlier and more reliable detection of pathologies [5]. This precision supports the development of future point-of-care devices and personalized medicine, tailoring treatments to individual patient profiles.

3.5.1.2 Drug Discovery and Neuroscience Research

Quantum biosensors offer cutting-edge tools for drug discovery and neuroscience. They enable the design and monitoring of complex drug delivery systems, the elucidation of pharmacological mechanisms, and the study of pharmacokinetics [25]. In neuroscience, these sensors can provide insights into neuronal signaling and synaptic function at the single-molecule level. The ultrasensitive detection of molecular signatures and precise analysis of drug-receptor interactions and drug metabolism are facilitated by the high sensitivity of quantum biosensors at low exposure levels.

3.5.1.3 Detection of Pathogens and Biomarkers in Bodily Fluids

Quantum biosensors are utilized for noninvasive detection of microorganisms and viral markers in bodily fluids like sweat, saliva, urine, and blood. Their label-free nature and real-time monitoring capabilities make them ideal for tracking biomarkers in infections, cancer, and metabolic disorders [5]. Miniaturization and integration with microfluidic systems enable rapid and point-of-care detection of bacterial infections, leveraging nanomaterials like carbon nanotubes, quantum dots, and graphene for ultrasensitive detection of antigens and nucleic acids. This capability is critical for faster, more accurate diagnosis and timely management of infectious diseases.

3.5.2 Environmental Monitoring and Food Safety

Beyond healthcare, quantum biosensors are extending their reach into environmental and food safety applications.

3.5.2.1 Detection of Contaminants

The technology offers potential for detecting environmental contaminants, contributing to improved water testing and air quality control [2]. Biosensors can monitor pollutants and nutrients, providing information on the status of plants and ecosystems [2]. In food safety, they address the challenge of rapid and cost-effective detection of contaminants such as toxins and pathogens, ensuring safer food resources [26]. Quantum-enhanced imaging and spectroscopy can provide superior imaging capabilities for environmental monitoring, even in low-light or obscured conditions.

3.5.3 Materials Science and Fundamental Research

Quantum biosensors are also making significant contributions to materials science and fundamental research by enabling the characterization of materials at an unprecedented scale.

3.5.3.1 Characterization of Quantum Materials

Single-qubit sensing, particularly using NV centers in diamond, allows for the measurement of high-speed fluctuations in magnetic materials at nanoscale dimensions [27]. This capability provides new ways to understand materials' behavior near critical phase transitions, which is key to developing new technologies, including advancements in spintronics and quantum computing. The ability to probe local changes in spin fluctuations offers deeper insights into wide classes of quantum materials and their potential for information-processing technologies. Quantum sensors constructed from polarized spin-triplet electrons in photoexcited organic chromophores, such as pentacene-doped para-terphenyl, also show potential for quantum sensing in chemical systems and for sensing pressure and temperature. Single-molecule magnets (SMMs) are also being explored for quantum sensing in materials science, functioning as quantum thermometers and local probes, and potentially for detecting dark matter particles [28].

3.5.3.2 Probing Molecular Dynamics and Interactions

Quantum biosensors enable the direct measurement of dynamic motion of biomolecules using quantum probes like X-rays, electron beams, and neutrons, revealing functional mechanisms on a real-time axis. This allows for the precise determination of distances between labeled sites in biological macromolecules, capturing their diverse conformational spectrum. Single-molecule scale magnetic resonance using NV centers offers prospects for sensing and imaging technology at the single-molecule level, providing insights into protein structures and dynamics that are lost in ensemble measurements [17]. The ability to track single biomolecules as small as 3.5 nm with quantum noise-limited precision allows for monitoring surface-molecule interactions over extended periods, opening new avenues to explore single-molecule biophysics. Atomic defects in diamond, such as NV centers, are also used to probe nearby spins at the nanometer scale, enabling nuclear magnetic resonance (NMR) studies of a few thousand nuclear spins and revealing underlying physics of new designer materials.

3.6 Challenges and Limitations

Despite their immense potential, the widespread adoption and full realization of SMD with quantum biosensors face several significant challenges.

3.6.1 Fabrication and Integration Complexities

The creation and integration of quantum biosensors involve intricate processes that present substantial hurdles.

3.6.1.1 Material Immobilization and Contamination

A primary difficulty in fabricating optical biosensors is the immobilization of biomaterials on transducers. Material losses during this process can be significant, and ensuring the strong attachment of biomolecules to the transducer is crucial to avoid contamination and leakage [2]. The behavior of biomolecules when adsorbed on surfaces is not yet fully understood, complicating efforts to achieve uniformity and reproducibility in sensor fabrication.

3.6.1.2 Miniaturization and Cost

While significant strides have been made in miniaturizing quantum sensors, many existing setups remain prohibitively expensive and large, limiting their practicality for mass use and point-of-care applications. For instance, traditional NV-based quantum sensors can be the size of a kitchen table [18]. Developing compact, scalable, and cost-effective devices that maintain high sensitivity is an ongoing challenge. The integration of complex components onto chip-scale devices, while promising, requires overcoming technical hurdles in precise placement and control of quantum elements.

3.6.1.3 Debye Screening Effect in Physiological Solutions

A critical factor affecting the performance of single-molecule electronic biosensors is the Debye screening effect [1]. In ionic solutions, an ionic double layer forms around the conductive channel, screening the charge of target molecules. This effect significantly hinders the detection of biomolecules in physiological solutions, which typically have high ionic strength (above 100 mM), whereas electrical detection often requires low ionic strength conditions (around 10 mM). Overcoming this barrier is crucial for practical applications in biological environments, requiring strategies such as modifying channel materials, modulating devices with electrical fields, or designing specific biomolecule decorations [1].

3.6.2 Signal Processing and Data Analysis

The nature of quantum signals and single-molecule events introduces complexities in data acquisition and interpretation.

3.6.2.1 Low Signal-to-Noise Ratio in Complex Samples

Detecting single molecules, especially in complex biological matrices like blood or serum, is challenging due to background noise and non-specific signals [6]. While quantum sensors offer improved SNRs compared to classical methods, achieving quantum noise-limited performance for unlabeled biomolecules is still difficult due to very low scattered power (femtowatts) and technical noise sources like laser fluctuations, electronic noise, and acoustic vibrations [13]. These technical noises are particularly problematic in the hertz to kilohertz frequency band relevant for biomolecule dynamics.

3.6.2.2 Quantum State Reconstruction Challenges

Reconstructing quantum states from measurement data is a formidable challenge, particularly as system sizes grow. Many quantum machine learning approaches for quantum state tomography rely on idealized assumptions incompatible with current hardware constraints. While protocols operating on classical measurement data and executable on noisy intermediate-scale quantum (NISQ) devices are being developed, achieving high-fidelity reconstructions remains an area of active research [29]. The inherently short coherence times of molecular vibrations also pose significant challenges for practical quantum optomechanical applications.

3.6.2.3 Decoherence and Environmental Interference

Quantum coherence, essential for sensor stability and precision, is fragile and susceptible to environmental noise [10]. Decoherence, the loss of quantum coherence due to interaction with the environment, is a long-standing problem [25]. While new protocols are being developed to temporarily counteract decoherence and stabilize quantum states, this remains a critical limitation for achieving sustained high-precision measurements [25]. The unpredictable behavior of nanoscale entities and interference from the target medium also pose challenges for stochastic detection methods [27].

3.6.3 Toxicity and Biocompatibility (For Certain QDs)

While quantum dots offer many advantages, their potential toxicity remains a concern for broader biological and medical applications. QDs, especially those containing heavy metals, can harm organisms if they enter the environment. Although carbon and graphene QDs are less toxic and more versatile, ensuring long-term biocompatibility, controlling biodistribution, and preventing accumulation in the body are ongoing challenges for all QD types. Efforts to encapsulate QDs in hydrophilic, soluble materials have been made to address water insolubility, but issues like the release of toxic ligands during functionalization and the need for stringent purification persist.

3.7 Conclusion

SMD with quantum biosensors represents a revolutionary advancement in analytical science, offering unprecedented sensitivity, resolution, and the ability to probe molecular heterogeneity that is obscured in traditional ensemble measurements. Recent years have witnessed significant breakthroughs across various quantum platforms, including self-illuminating plasmonic biosensors, highly sensitive NV centers in diamond, and single-molecule-based superconducting qubits. These innovations are poised to transform diverse fields, from early disease diagnosis and personalized medicine to environmental monitoring and fundamental materials science. The unique advantages conferred by quantum phenomena—superposition, entanglement, and quantum noise reduction—enable these biosensors to operate at low light levels, minimize sample damage, and provide digital, real-time readouts of individual molecular events [6]. This capability is critical for detecting trace biomarkers, understanding complex biological dynamics, and advancing drug discovery. However, the field is not without its challenges. Complexities in fabrication, particularly concerning material immobilization, miniaturization, and cost, continue to be significant hurdles. Overcoming the Debye screening effect in physiological solutions remains crucial for the practical application of electronic biosensors [1]. Furthermore, managing signal processing challenges, such as low SNRs in complex samples and the inherent fragility of quantum coherence, requires sophisticated engineering and data analysis techniques. While progress in developing less toxic quantum dots is promising, ensuring long-term biocompatibility and controlled biodistribution for *in vivo* applications is an ongoing area of research. The trajectory of quantum biosensors suggests a future where highly precise, noninvasive, and compact diagnostic tools are commonplace, enabling a deeper understanding of biological processes at their most fundamental level. Continued interdisciplinary research and technological innovation are essential to overcome existing limitations and unlock the full potential of SMD with quantum biosensors, paving the way for a new era of scientific discovery and practical applications.

References

1 Gao, C., Gao, Q., Zhao, C. et al. (2024). Technologies for investigating single-molecule chemical reactions. *National Science Review* 11: https://doi.org/10.1093/nsr/nwae236.

2 Das, S., Mazumdar, H., Khondakar, K.R. et al. (2024). Review—Quantum biosensors: principles and applications in medical diagnostics. *ECS Sensors Plus* 3: 025001. https://doi.org/10.1149/2754-2726/ad47e2.

3 Gooding, J.J. and Gaus, K. (2016). Single-molecule sensors: challenges and opportunities for quantitative analysis. *Angewandte Chemie* 55: 11354–11366. https://doi.org/10.1002/anie.201600495.

4 Mousavi, S.M., Kalashgrani, M.Y., Gholami, A. et al. (2023). Recent advances in quantum dot-based lateral flow immunoassays for the rapid, point-of-care diagnosis of COVID-19. *Biosensors-Basel* 13 (8): 786. https://doi.org/10.3390/bios13080786.

5 Chattopadhyay, J. and Bag, B. Carbon quantum dot electrocatalysts in hydrogen evolution reaction. *Energy Technology* 12: 2300913. https://doi.org/10.1002/ente.202300913.

6 Quesada-González, D. and Merkoçi, A. (2025). Quantum dots for biosensing: classification and applications. *Biosensors and Bioelectronics* 273: 117180. https://doi.org/10.1016/j.bios.2025.117180.

7 Osypiw, A.R.C., Lee, S., Jung, S.-M. et al. (2022). Solution-processed colloidal quantum dots for light emission. *Materials Advances* 3: 6773–6790. https://doi.org/10.1039/D2MA00375A.

8 Muhammad, H., Masab, M., Ullah, N. et al. (2025). A comprehensive review of current trends and future perspective of quantum dots regarding their optical and photoswitchable properties. *Renewable and Sustainable Energy Reviews* 223: 115984. https://doi.org/10.1016/j.rser.2025.115984.

9 Qureshi, Z.A., Dabash, H., Ponnamma, D., and Abbas, M.K.G. (2024). Carbon dots as versatile nanomaterials in sensing and imaging: efficiency and beyond. *Heliyon* 10: e31634. https://doi.org/10.1016/j.heliyon.2024.e31634.

10 Khan, M.A., Ghafoor, S., Zaidi, S.M.H. et al. (2024). From quantum communication fundamentals to decoherence mitigation strategies: addressing global quantum network challenges and projected applications. *Heliyon* 10: e34331. https://doi.org/10.1016/j.heliyon.2024.e34331.

11 Crawford, S.E., Shugayev, R.A., Paudel, H.P. et al. (2021). Quantum sensing for energy applications: review and perspective. *Advanced Quantum Technologies* 4: 2100049. https://doi.org/10.1002/qute.202100049.

12 Huang, J., Zhuang, M., and Lee, C. (2024). Entanglement-enhanced quantum metrology: from standard quantum limit to Heisenberg limit. *Applied Physics Reviews* 11 (3): 031302. https://doi.org/10.1063/5.0204102.

13 Montenegro, V., Mukhopadhyay, C., Yousefjani, R. et al. (2025). Review: quantum metrology and sensing with many-body systems. *Physics Reports* 1134: 1–62. https://doi.org/10.1016/j.physrep.2025.05.005.

14 Naresh, V. and Lee, N. (2021). A review on biosensors and recent development of nanostructured materials-enabled biosensors. *Sensors* 21 (4): 1109. https://doi.org/10.3390/s21041109.

15 Chang, X., Huo, Y., Zhao, C. et al. (2023). Single-molecule electronic biosensors: principles and applications. *Advanced Sensor Research* 2: 2200084. https://doi.org/10.1002/adsr.202200084.

16 Mimona, M.A., Rimon, M.I.H., Zohura, F.T. et al. (2025). Quantum dot nanomaterials: empowering advances in optoelectronic devices. *Chemical Engineering Journal Advances* 21: 100704. https://doi.org/10.1016/j.ceja.2025.100704.

17 Chattopadhyay, J. and Srivastava, N. (2021). *Application of Nanomaterials in Chemical Sensors and Biosensors*. CRC Press https://doi.org/10.1201/9781003009085.

18 Sánchez Toural, J.L., Marzoa, V., Bernardo-Gavito, R. et al. (2023). Hands-on quantum sensing with NV−Centers in diamonds. *C - Journal of Carbon Research.* 9 (1): 16. https://doi.org/10.3390/c9010016.

19 Chauhan, S., Jain, N., and Nagaich, U. (2020). Nanodiamonds with powerful ability for drug delivery and biomedical applications: recent updates on in vivo study and patents. *Journal of Pharmaceutical Analysis* 10 (1): 1–12. https://doi.org/10.1016/j.jpha.2019.09.003.

20 Sarkar, A., Jones, Z.R., Parashar, M. et al. (2024). High-precision chemical quantum sensing in flowing monodisperse microdroplets. *Science Advances* 10 (50): eadp4033. https://doi.org/10.1126/sciadv.adp4033.

21 Ezratty, O. (2023). Perspective on superconducting qubit quantum computing. *European Physical Journal A: Hadrons and Nuclei* 59: 94. https://doi.org/10.1140/epja/s10050-023-01006-7.

22 Vigneau, F., Fedele, F., Chatterjee, A. et al. (2023). Probing quantum devices with radio-frequency reflectometry. *Applied Physics Reviews* 10 (2): 021305. https://doi.org/10.1063/5.0088229.

23 de Leon, N.P., Itoh, K.M., Kim, D. et al. (2021). Materials challenges and opportunities for quantum computing hardware. *Science* 372: eabb2823. https://doi.org/10.1126/science.abb2823.

24 Moreva, E., Cimini, V., Gianani, I. et al. (2025). Quantum photonics sensing in biosystems. *APL Photonics* 10 (1): 010902. https://doi.org/10.1063/5.0232183.

25 Yan, J., Cheng, L., Li, Y. et al. (2025). Advancements in single-molecule fluorescence detection techniques and their expansive applications in drug discovery and neuroscience. *Biosensors-Basel* 15 (5): 283. https://doi.org/10.3390/bios15050283.

26 Popenda, A., Wiśniowska, E., and Manuel, C. (2024). Biosensors in environmental analysis of microplastics and heavy metal compounds – a review on current status and challenges. *Desalination and Water Treatment* 319: 100456. https://doi.org/10.1016/j.dwt.2024.100456.

27 Boretti, A., Rosa, L., Blackledge, J., and Castelletto, S. (2019). Nitrogen-vacancy centers in diamond for nanoscale magnetic resonance imaging applications. *Beilstein Journal of Nanotechnology* 10: 2128–2151. https://doi.org/10.3762/bjnano.10.207.

28 Datta, S.N., Pal, A.K., and Panda, A. (2023). Design of magnetic organic molecules and organic magnets: experiment, theory and computation with application and recent advances. *Chemical Physics Impact* 7: 100379. https://doi.org/10.1016/j.chphi.2023.100379.

29 Kurmapu, M.K., Tiunova, V.V., Tiunov, E.S. et al. (2023). Reconstructing complex states of a 20-qubit quantum simulator. *PRX Quantum* 4: https://doi.org/10.1103/PRXQuantum.4.040345.

4

Photonic Quantum Sensors: Light-Based Diagnostic Tools

Monima Sarma and Tanmay Chatterjee

Department of Chemistry (Science and Humanities), PES University, Bengaluru, Karnataka, India

4.1 Introduction

Quantum technologies are proceeding at a rapid pace. The diverse and promising technological applications stemming from quantum information are becoming increasingly apparent. Among these, quantum sensing represents a leading frontier. This broad field includes quantum protocols for estimation and discrimination that demonstrably outperform classical approaches. The photonic regime provides the most advantageous environment, largely due to the ease with which quantum features can be generated, manipulated, and detected. Photonic quantum sensing, in particular, quantum biosensing, leverages quantum mechanics for biological sensing, offering a potent means to advance our knowledge of life at the molecular level. Quantum biosensing (QB) offers significant advantages over classical biosensing by employing quantum mechanics for detection and analysis. This approach allows for highly sensitive and precise detection of biological molecules, pathogens, and other biological entities. QB's ability to detect and track individual molecules or even single atoms provides unique insights into biological processes. This pinpoint accuracy stems from detecting subtle changes in magnetic and electrical fields at the single-atom level, a stark contrast to classical methods that rely on large atomic ensembles. By integrating quantum phenomena, QB enhances the sensitivity, accuracy, and selectivity of traditional biosensors. In this context, light-driven diagnostic tools, i.e. photonic quantum sensors, have played a crucial role in providing valuable information for their advancements toward biological applications. In our understanding of light's wave-like behavior, the phenomenon of interference plays a fundamental role. Interferometers, designed to exploit this very phenomenon, allow us to observe and analyze interference patterns, thereby providing significant information about light's properties. Classical interferometers, including the well-known Michelson, Mach-Zehnder, and double-slit configurations, have been essential tools in experiments that have led to significant discoveries in optics. These devices are not just limited to fundamental research; they are also crucial in diverse industrial applications. Their ability to precisely measure optical properties such

Quantum Biosensing in Medical Diagnostics, First Edition. Edited by J. G. Manjunatha.

as wavelength, amplitude, coherence, phase, and polarization makes them indispensable for precision measurements, quality control, and the advancement of optical systems. The field has further evolved with the development of quantum technologies, giving rise to quantum interferometers, quantum imaging, quantum magnetometry, quantum lidar, etc. [1–5] This chapter focuses on the fundamental tools and techniques of photonic quantum sensing, emphasizing its promising uses in biomedical diagnostics and imaging.

4.2 Various Light-Based Diagnostic Tools and Technologies

4.2.1 Quantum Interferometers

Quantum interferometers, which employ quantum resources to investigate interference, are vital experimental tools in quantum physics. Their applications are extensive within quantum information technology, impacting fields like quantum computing, communication, and measurement. Quantum interferometers find use in a variety of areas [6–11], i.e.

- In quantum metrology, they enable the extraction of information regarding the structure and dynamics of biological and chemical molecules.
- For precision measurements, these devices offer high accuracy in determining optical properties, including wavelength, phase, amplitude, coherence, and polarization.
- Quality control processes can also benefit from the use of interferometers.
- Furthermore, interferometers play a crucial role in the development of advanced optical systems.

In this section, we will discuss three prominent quantum interferometers, namely, the Hong–Ou–Mandel (HOM) interferometer, the N00N state interferometer, and the Franson interferometer, and summarize their unique features. For a quick understanding of the key aspects of each interferometer, Table 4.1 provides a summary of their setups, theoretical models (single- and multi-mode), principles, advantages, and applications.

4.2.1.1 HOM Interferometers

The HOM effect, demonstrated in 1987 by Hong, Ou, and Mandel at the University of Rochester, is a quantum optical phenomenon involving two-photon interference. When two identical photons, one entering each input port of a 50:50 beam splitter, arrive simultaneously, they always exit together through the same output port. This means there's no chance of detecting one photon in each output (a coincidence event). They have an equal probability of existing together in either of the two output ports. In short, perfectly indistinguishable, simultaneously arriving photons always emerge together from the same output port, and this results in a suppressed coincidence count, known as the HOM dip. However, if the photons become distinguishable (e.g. by arriving at different times or having different wavelengths), the likelihood of detecting them separately in each output increases. This interference effect allows precise measurement of bandwidth, path lengths, and timing. Because the HOM effect relies on the quantum nature of light (second quantization), it cannot be explained by classical optics [12–15].

To aid understanding of the HOM interferometer, we present its theoretical models progressively, from the simplest to the most complex one.

Table 4.1 Summary of HOM, NOON state, and Franson interferometers.

	HOM interferometer	NOON state interferometer	Franson interferometer
Setup	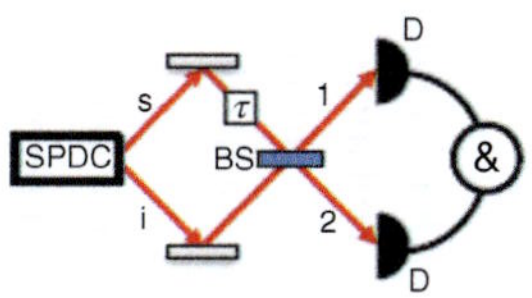	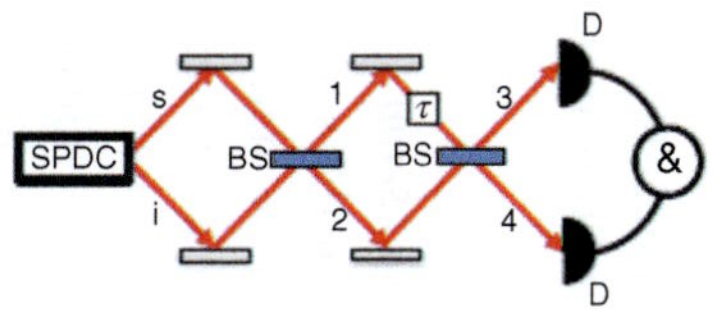	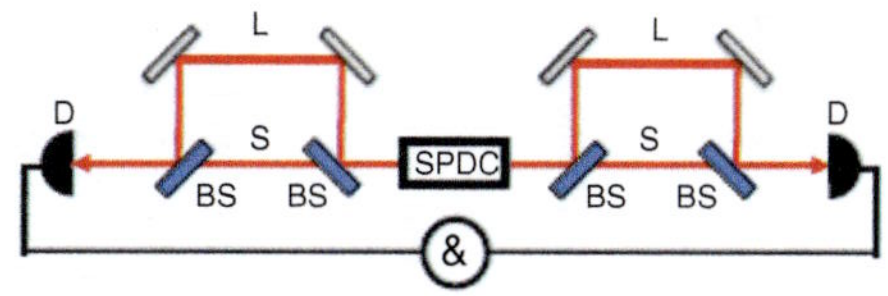
Principle	Accounts for the antibunching effect of bosons	Accounts for multi-photon wave packet interference	Accounts for the interference of time-frequency entanglement
Advantages	Insensitive to phase fluctuation	Super sensitive	Non-local interference
Applications	Quantum computation, imaging, communication, etc.	Quantum sensing and measurement, etc.	Quantum imaging and communication, etc.

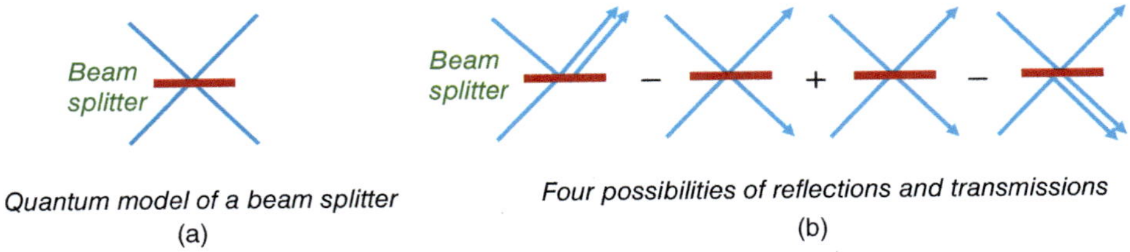

Figure 4.1 The different possibilities of reflections and transmissions. Jin et al. [15] / arXiv / CC BY 4.0.

Quantum optical description of a beam splitter: Let us first consider a photon entering a beam splitter. This gives rise to two possibilities, i.e. the photon is either reflected or transmitted, the probabilities of which depend on the beam splitter's reflectivity. For a 50:50 beam splitter, the reflection and transmission are equally likely [16]. Now, let us consider two photons entering separate input ports of the beam splitter. Here, four combinations of transmission and reflection are possible, namely, (i) reflection of the top photon and transmission of the bottom; (ii) transmission of both; (iii) reflection of both; and (iv) transmission of the top photon and reflection of the bottom (see Figure 4.1).

4.2.1.1.1 HOM Interferometers in Different Modes of Setup

a) *Single-mode theory of HOM interferometer:* In the single-mode treatment of the HOM interferometer, photons are restricted to a single mode in all degrees of freedom. These input photons can be generated via spontaneous parametric down-conversion (SPDC), where they emerge as signal and idler pairs (Figure 4.2a).
b) *Multi-frequency theory of HOM interferometer for biphotons with no spectral correlation:* This theory addresses the interference of multi-mode frequency-entangled states in an HOM interferometer. This approach simplifies the treatment of multi-photon interference (Figure 4.2b). However, a key limitation is its assumption of uncorrelated signal and idler photons, rendering it unsuitable for biphotons exhibiting arbitrary correlations [17].
c) *Multi-frequency theory of HOM interferometer for biphotons with arbitrary distribution:* It is a method to characterize the spectral correlation of biphotons. It can be used to measure the phase drifts and displacements of light fields. However, this treatment is only valid in the frequency domain. It is also essential to model the HOM interference in the time domain (Figure 4.2c).
d) *Multi-frequency theory of fourfold hom interferometer between two independent sources:* It describes the interference of two independent photons. It assumes that the signal and idler photons emanating from the first and second SPDC sources have joint spectral amplitudes (Figure 4.2d). It is used in photonic quantum information processing, such as quantum key distribution, teleportation, and quantum repeaters [16]. Though the initial demonstration of the HOM interferometer employed photons generated via SPDC in a nonlinear crystal; however, HOM interference is not restricted to SPDC. Indistinguishable photons can also be produced from various other physical systems such as quantum dots, nitrogen-vacancy (NV) centers in diamonds, silicon-vacancy, and microwaves [16].

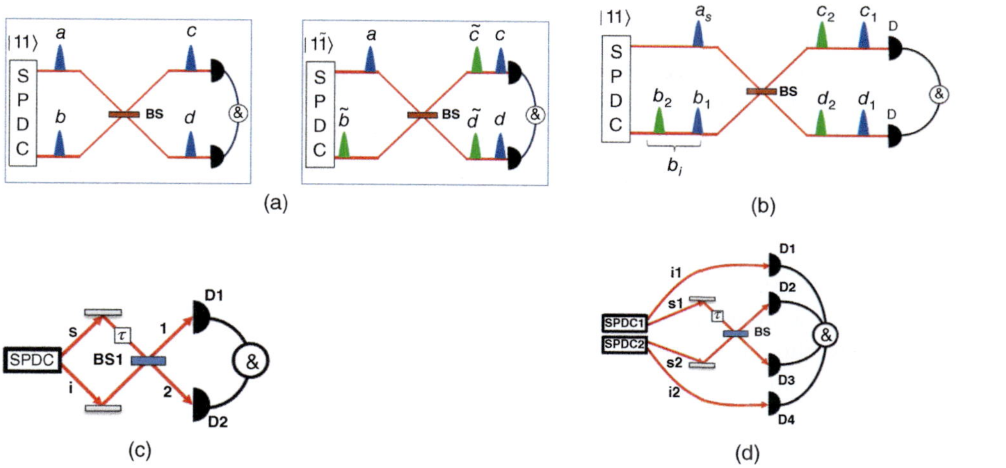

Figure 4.2 (a) Single-mode HOM interferometer when two photons are completely overlapped; two photons are not overlapped. (b) Multi-frequency HOM interference (no spectral correlation). (c) Multi-frequency HOM interference (with arbitrary distribution). (d) Multi-frequency fourfold HOM interference (with independent sources). Jin et al. [15] / arXiv / CC BY 4.0.

Furthermore, some typical HOM interferometers are also known, such as HOM interferometers using independent photon sources (HOMI-IPS), detection-dependent multi-photon HOM interferometers, and spectrally resolved techniques [17–24].

Moreover, many quantum imaging techniques can be considered as forms of quantum biosensing, especially when they are used to detect specific biological markers or processes within the image. Two typical important examples of quantum imaging are quantum optical coherence tomography (QOCT) and HOM interferometer-based microscopy, respectively. Quantum optical coherence tomography (QOCT) merges optical coherence tomography (OCT) with quantum optics to provide high-resolution imaging and depth profiling of biological samples. Optical coherence tomography (OCT) is a well-established, noninvasive imaging technique used extensively in medical and biological research for tissue imaging and analysis [25–27]. A key limitation of conventional OCT is that increasing resolution by broadening the light source bandwidth is often counteracted by dispersion within the sample, which degrades resolution. This presents a major obstacle in OCT systems. Quantum OCT (QOCT) offers a solution to this dispersion-induced resolution loss. QOCT employs HOM interference of frequency-entangled photon pairs [28]. A primary benefit of QOCT is its ability to maintain resolution (defined by the HOM dip width) even with group velocity dispersion (GVD) present in the sample. This advantage stems from the frequency correlation of the entangled photons, which allows for "dispersion cancellation." Hong-Ou-Mandel (HOM) interferometer-based microscopy is another significant example of quantum imaging. In 2022, Ndagano et al. developed a scan-free quantum imaging method using HOM interference to reform the surface depth profile of transparent samples [29]. They successfully imaged micrometer-scale depth features with a remarkably minimal photon flux of merely seven photon pairs per frame.

Using a single-photon avalanche diode (SPAD) camera, they determined both bunched and anti-bunched photon-pair distributions at the HOM interferometer output. This work highlights the potential of HOM microscopy for label-free imaging of transparent samples even with extremely low light levels [29].

The HOM interferometers discussed above offer several key advantages for photonic quantum biosensing, primarily stemming from their exquisite sensitivity to changes in the quantum state of light, such as:

a) *Extreme sensitivity to refractive index changes:* The HOM dip is highly sensitive to any factors that affect the indistinguishability of the photons. In a biosensing context, the presence of a target biomolecule can alter the refractive index of the medium through which the photons propagate. Even minute changes in refractive index, caused by the binding of a single molecule, can shift the HOM dip, providing a highly sensitive detection mechanism [30].
b) *Label-free detection:* Unlike many traditional biosensing methods that require fluorescent labels, HOM interferometry can enable label-free detection. This is a significant advantage, as labeling can be complex, potentially alter the biomolecule's behavior, and increase background noise [31].
c) *Single-molecule detection potential:* The extreme sensitivity of HOM interferometers opens the possibility of detecting single biomolecules. By carefully designing the sensor and optimizing the experimental conditions, it may be possible to detect the binding of a single target molecule through its effect on the HOM dip [32].
d) *Background-free measurement:* In ideal conditions, the HOM dip can reach near-perfect extinction, meaning almost zero coincidence counts. This provides a very low background for measurements, enhancing the signal-to-noise ratio and improving detection sensitivity [14].
e) *Compatibility with integrated photonics:* HOM interferometers can be implemented on integrated photonic chips, enabling miniaturization and mass production of compact and robust biosensors. This is crucial for developing portable and point-of-care diagnostic devices [33].

In short, HOM interferometers offer a powerful platform for photonic quantum biosensing due to their extreme sensitivity, label-free detection capability, potential for single-molecule detection, low background noise, and compatibility with integrated photonics. These features make them promising candidates for future applications in diagnostics, environmental monitoring, and fundamental biological research.

4.2.1.2 NOON State Interferometers

A NOON state, crucial in quantum optics, is a highly entangled quantum state involving multiple particles. It is a superposition where N particles are either all in one mode (a) and none in another (b), or vice versa. While these particles are typically photons, any bosonic field could theoretically form NOON states. These states are significant in quantum sensing, quantum imaging, quantum spectroscopy, etc. [7, 34, 35], particularly for achieving precise phase measurements within optical interferometers. Just like HOM interferometers, the NOON state interferometers also function in different modes.

4.2.1.2.1 NOON State Interferometers in Different Modes of Setup

i) *Single-mode theory of NOON state interferometers:* The term "single-mode" in this context highlights that the analysis concentrates on the interference between the two modes comprising the NOON state, despite the NOON state itself being a two-mode entity. The focus is on how the phase difference between these two modes influences the overall state and the resulting interference pattern, as illustrated in Figure 4.3a [7]. However, this single-mode model simplifies the situation by assuming no spectral correlation between the generated biphotons from SPDC. This simplification limits the model's applicability, as spectral correlation is a typical feature of quantum light sources. Therefore, the next section will explore a more comprehensive "multi-frequency" theory of NOON state interferometers that account for biphotons with arbitrary spectral distributions.

ii) *The multi-frequency theory of NOON state interferometer for biphotons with no spectral correlation*: When biphotons have no spectral correlation, it means the frequency of one photon provides no information about the frequency of the other (see Figure 4.3b). This simplifies the multi-frequency [36] analysis to some extent. In the absence of spectral correlation, the interference pattern will typically be "washed out" or have reduced visibility compared to the case with strong spectral correlation.

iii) *The multi-frequency theory of NOON state interferometer for biphotons with arbitrary distribution:* Here, the biphotons can have any possible distribution of frequencies, including Gaussian, non-Gaussian, and even more complex shapes. This is crucial for a realistic description of experimental scenarios. The spectral properties of biphotons are, in general, fully described by their joint spectral amplitude (JSA) function

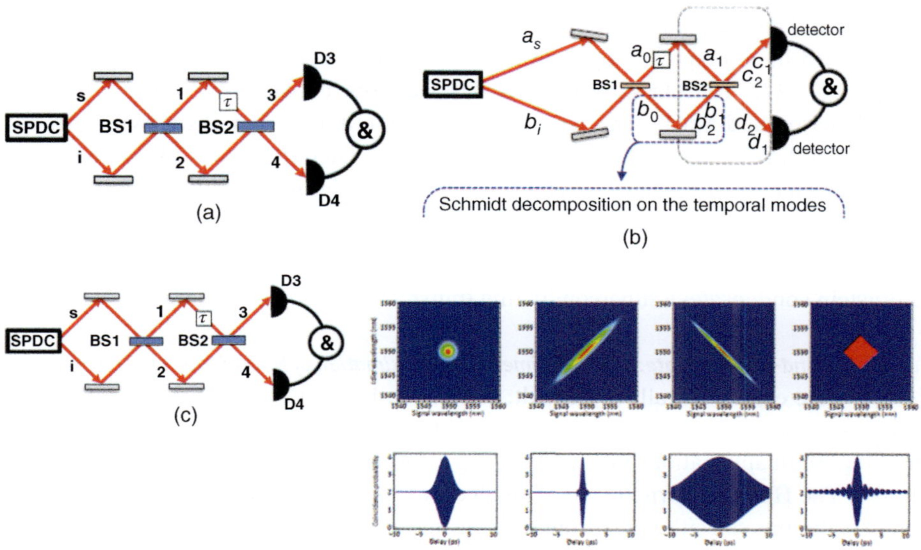

Figure 4.3 (a) Single-mode NOON state interferometer setup. Jin and Shimizu [7] / Optica Publishing Group / CC BY 4.0. (b) Multi-frequency NOON state interferometer setup. Adapted from Jin et al. [36]. (c) Joint Spectral Amplitudes of the NOON state setup and the corresponding Interference Patterns. With permission from Jin and Shimizu [7] / Optica Publishing Group.

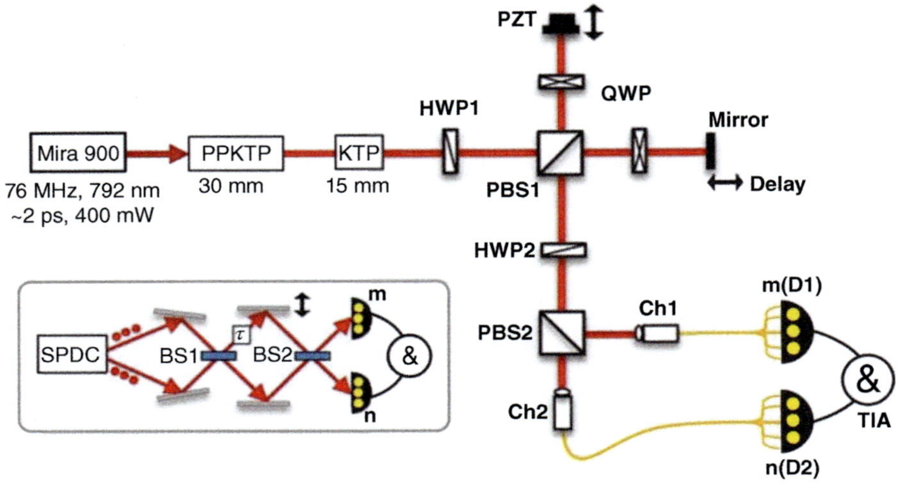

Figure 4.4 The experimental setup, where the inset depicts a standard configuration of the HB state interference using the path mode. Jin et al. [36] / Springer Nature / CC BY 4.0.

(see Figure 4.3c). This function encodes the probability amplitude for detecting one photon at a particular frequency and its partner at another frequency. The JSA can have any functional form, reflecting the arbitrary spectral distribution of the biphotons. This finally relates to the quantum analog of the Wiener-Khinchin theorem (WKT). The WKT establishes a relationship between a signal's power spectrum and its autocorrelation function via Fourier transformation. In classical optics, this fundamental theorem underpins Fourier transform infrared spectroscopy, a technique widely used in commercial applications like chemical, polymer, and pharmaceutical analyses [7, 37]. The classical WKT is derived from classical interferometers like the Michelson and Mach-Zehnder. Similarly, a quantum version of the WKT can be formulated, linking the spectral information of correlated biphotons using quantum interferometers such as the HOM and N00N state interferometers [7].

N00N states can also be generated from different systems, such as photonic systems (bulk nonlinear, waveguides, integrated chips), quantum dots, plasmons, trapped ions, and atoms (Bose-Einstein condensates), superconducting transmons, and optomechanics.

4.2.1.2.2 Holland-Burnett State: The Experimental Approximation of N00N State Preparing the ideal N00N state experimentally is difficult; consequently, the Holland-Burnett (HB) state is often employed as an approximation of the N00N state. It is a quantum state of light that has important applications in phase sensing measurements with enhanced sensitivity [36, 38–40]. The Holland-Burnett state is created by sending two photons through a beam splitter. The photons can either go through the beam splitter or be reflected (see Figure 4.4). This creates a superposition of states, where the photons are in both places at once. The Holland-Burnett (HB) state, in simpler terms, also has the potential to enhance biosensing due to its increased sensitivity to phase changes.

4.2.1.2.3 Multi-Mode HOM Interferometer and NOON State Interferometer Conventional HOM and NOON state interferometers typically utilize biphotons correlated within a single, discrete spectral mode. However, biphotons can also exhibit correlations across multiple discrete spectral modes, creating a two-body, high-dimensional entangled state known as entangled qudits [41–43]. A multi-mode HOM or NOON state interferometer is thus defined as one employing these frequency-entangled qudits. A key feature of multi-mode HOM interference is the significantly narrower interference fringes compared to single-mode HOM interference. Both multi-mode HOM and NOON state interference patterns are governed by two components: an "envelope factor" and a "details factor." The envelope factor is determined by the characteristics of a single spectral mode, while the details factor is governed by the mode number, N [44].

Some applications of the NOON state interferometers [38, 45]:

1) *Enhanced Sensitivity Because of Refractive Index Sensing as Well as Absorption and Dispersion Measurements*
   Because the refractive index of a substance changes with the presence and concentration of biomolecules, it serves as a useful indicator. NOON state interferometers offer superior precision in detecting even tiny variations in refractive index compared to traditional interferometers, allowing for the detection of minute quantities of target substances. Also, biological samples exhibit unique absorption and dispersion spectra. NOON state interferometers can measure these spectral characteristics with greater sensitivity, providing valuable data for biomolecule identification and characterization. The high sensitivity of NOON state interferometers could be exploited to detect early disease diagnosis and monitoring as well as analyze DNA and RNA molecules, with applications in genetic testing and diagnostics. In addition, real-time monitoring of biological processes at the single-molecule level is also possible with the NOON state interferometers.
2) *Biological Imaging and Microscopy*
   Integrating NOON state interferometers into imaging systems can improve both resolution and contrast, leading to enhanced visualization of cells, tissues, and other biological structures. In addition, NOON state-based microscopy has the potential to enable nondestructive imaging of sensitive biological samples by reducing the required light exposure and thereby minimizing photodamage.

However, even with so much progress, there remain some challenges, such as generating high-quality NOON states and integrating them into practical biosensing devices, and developing compact and portable NOON state-based biosensors. NOON state interferometers have the potential to significantly advance quantum biosensing by providing enhanced sensitivity, resolution, and precision. While there are challenges to overcome, ongoing research in this field promises to open up new possibilities for biological and medical applications.

4.2.1.3 Franson Interferometers

Used in quantum optics, the Franson interferometer allows the observation of nonlocal correlations between entangled photons by analyzing their temporal relationships. J.D. Franson introduced this setup in 1989 to experimentally probe Bell's inequalities.

4.2.1.3.1 Franson Interferometers in Different Modes of Setup

i) *Single-mode model of Franson interferometer:* A single-mode model of a Franson interferometer simplifies the analysis by considering only one spatial mode for the photons propagating through the interferometer. This means that the photons are assumed to travel along well-defined paths within the interferometer, without considering the effects of multiple spatial modes or diffraction. The setup and interference pattern are shown in Figure 4.5a. The single-mode model is a useful approximation for many Franson interferometer experiments, especially those where the spatial mode is well-controlled and the effects of diffraction are minimal. It provides a good starting point for understanding the basic principles of Franson interferometry and analyzing experimental results.
   While the single-mode model is a valuable tool, it is important to recognize its limitations. In real experiments, there may be deviations from the ideal single-mode behavior due to factors such as imperfect optical components, misalignments, and multi-mode fibers. In such cases where these effects are significant, a more complex multi-mode model may be necessary to accurately describe the behavior of the Franson interferometer [47].
ii) *Multi-frequency model of Franson interferometer*
   A multi-frequency model of a Franson interferometer goes beyond the simplified single-mode model by considering the frequency spectrum of the photons involved. This is crucial when dealing with realistic light sources that emit photons with a range of frequencies, such as those produced in SPDC. The Franson interference with different joint spectral distributions is shown in Figure 4.5b [46]. The multi-frequency model is crucial for the precise characterization of entangled photon sources, optimization of experimental parameters, advanced quantum information protocols, etc.

In summary, the multi-frequency model of a Franson interferometer provides a more complete and accurate description of the experimental reality compared to the single-mode model, especially when dealing with broadband sources and the intricacies of frequency entanglement [48].

4.2.1.3.2 *Original Version* The Franson interferometer, which is a quantum optics device, was first proposed by J. D. Franson in 1989 [5]. A continuous-wave (CW) laser was initially used to pump a nonlinear crystal, generating photon pairs subsequently analyzed for time-bin entanglement. These photons passed through two unbalanced Mach-Zehnder interferometers (MZIs) with differing path lengths. The resulting time delay within each MZI exceeded the photons' coherence time, preventing individual MZI interference. However, coincident detection of the photons revealed interference fringes upon scanning the phase of one MZI, demonstrating nonlocal quantum entanglement inexplicable by classical physics.

4.2.1.3.3 *Revised Versions* A later iteration of the Franson interferometer employed a pulsed laser for pumping, directly creating time-bin entanglement within discrete time slots. This pulsed approach offers advantages like increased entanglement rates and facilitates novel experimental possibilities.

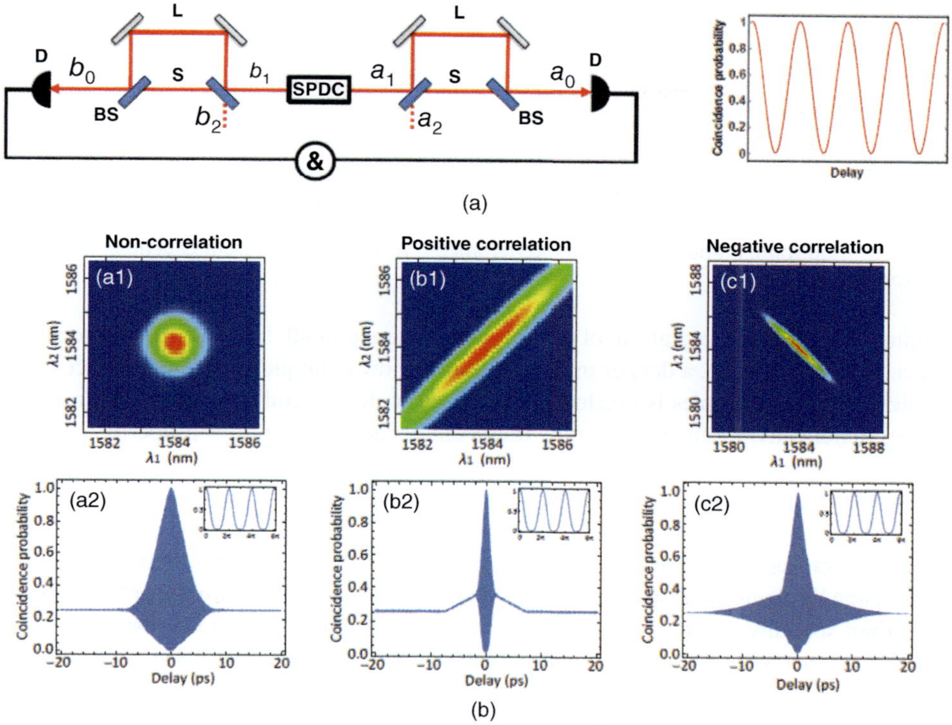

Figure 4.5 (a) Setup of the Franson interferometer and interference pattern. Jin et al. [15] / arXiv / CC BY 4.0. (b) The Franson interference uses biphotons with different spectral correlations. With permission from Jin et al. [46]. Science China Press.

Table 4.2 Table showing major differences between original and revised Franson interferometers.

Feature	Original version	Revised version
Laser type	Continuous wave (CW) laser	Pulsed laser
Entanglement type	Polarization entanglement	Time-bin entanglement
Interference type	Nonlocal Interference	Nonlocal Interference

The major differences between the original [5] and revised Franson interferometers [49–53] are summarized in Table 4.2.

While Franson interferometers are renowned for their role in quantum communication and fundamental physics tests, their potential applications in biosensing are recently emerging and promising, namely, for ultrasensitive detection of biomolecules, enhanced imaging techniques, drug discovery and development, environmental monitoring, etc. However, all these applications require precise control over experimental parameters.

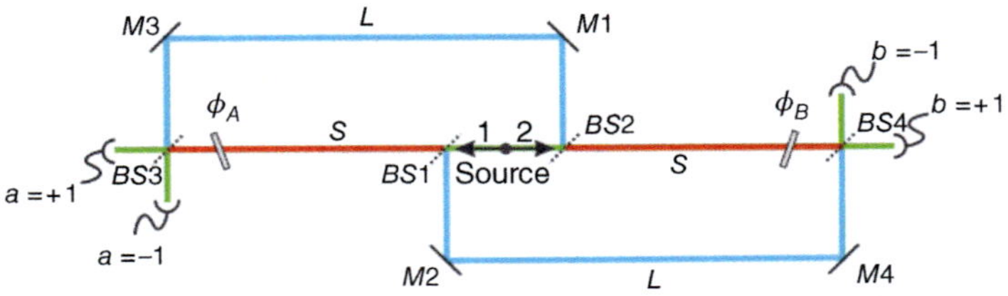

Figure 4.6 Hug configuration of Franson interferometer. With permission from Cabello et al. [49] / American Physical Society.

Miniaturization and integration of these systems are crucial for practical biosensing applications. In addition, a deeper understanding of how entangled photons interact with complex biological samples is needed to optimize the design and application of Franson interferometers in biosensing. Despite these challenges, the unique properties of the Franson interferometer offer exciting possibilities for developing highly sensitive and specific biosensing platforms.

a) The Hug Franson Interferometer The "hug" Franson interferometer modifies the traditional design by placing the two unbalanced MZIs in close physical proximity, or "hugging," rather than spatially separating them. This compact configuration enhances stability and simplifies experimental setups by mitigating the impact of environmental disturbances like temperature fluctuations and vibrations that can disrupt delicate interference patterns. The Hug Franson interferometer (see Figure 4.6) offers some key features, such as compactness, enhanced stability, and simplified implementation [49].

Overall, the Franson interferometer provides a practical and robust platform for quantum optics experiments using time-bin entangled photons. Its compact design and enhanced stability make it a valuable asset for applications ranging from quantum communication and metrology to fundamental tests of quantum mechanics. Despite its advantages, the "hug" Franson interferometer presents some challenges. The proximity of the MZIs can introduce unwanted interactions, potentially affecting measurement accuracy. Furthermore, this design may not be suitable for experiments requiring significant separation between the interferometers. Even with these challenges, the "hug" Franson interferometer's balance of stability, simplicity, and performance makes it a popular and effective tool in quantum optics research [54].

b) The Conjugate-Franson Interferometer To fully understand biphotons, we need to know how their nonclassical correlations behave in both frequency and time. The traditional Franson interferometer only gives us information about frequency correlations, leaving out the crucial time–domain picture [51]. This is a significant drawback, especially when working with quantum frequency combs, as it prevents us from distinguishing combs that differ only in their spectral phase [55]. As Chang et al. [56] noted, the standard Franson interferometer's interference patterns only reveal the joint spectral intensity (JSI). To access the corresponding joint temporal intensity (JTI), we need conjugate-Franson interferometry. This technique, first proposed by Zhang et al. [51], was initially conceived to provide robust security against collective attacks in quantum key distribution (QKD) systems that

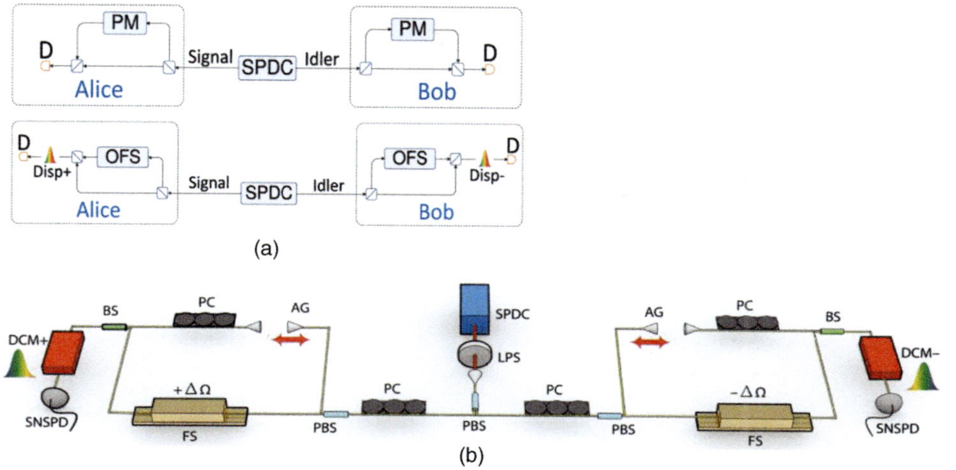

Figure 4.7 (a) The original Franson interferometer. With permission from Zhang et al. [51] / American Physical Society. (b) Its conjugate version is shown here. With permission from Chen et al. [52] / American Physical Society.

Table 4.3 Key differences between the Franson interferometer and its conjugate version.

Feature	Traditional Franson interferometer	Conjugate-Franson interferometer (CFI)
Primary focus	Frequency–domain correlations	Time–domain correlations
Method	Uses time delays in Mach-Zehnder interferometers	Uses frequency shifts in Mach-Zehnder interferometers
Input	Entangled photon pairs	Entangled photon pairs
Output	Interference pattern related to joint spectral intensity (JSI)	Interference pattern related to joint temporal intensity (JTI)
Sensitivity	Sensitive to spectral correlations	Sensitive to temporal correlations

use energy-time entanglement. Figure 4.7 shows the original and conjugate versions [52] of the Franson interferometer (upper panel). Furthermore, some key differences [52] between the original version and its conjugate are also summarized in Table 4.3.

c) The Three-Photon Franson Interferometer Multi-particle quantum interference is crucial for both understanding and utilizing quantum information as well as for conducting fundamental tests of quantum mechanics [20, 57]. Greenberger et al. first proposed extending the original Franson interferometer to three photons to explore the Einstein-Podolsky-Rosen paradox using three-photon correlations [58]. This three-photon interferometer has since been explored in several theoretical studies [59, 60], while experimental creation and measurement of three-photon polarization entanglement has been achieved using standard polarization analyzers [61, 62]. A typical three-photon Franson interferometer, illustrated in Figure 4.8, involves sending three correlated photons individually into three

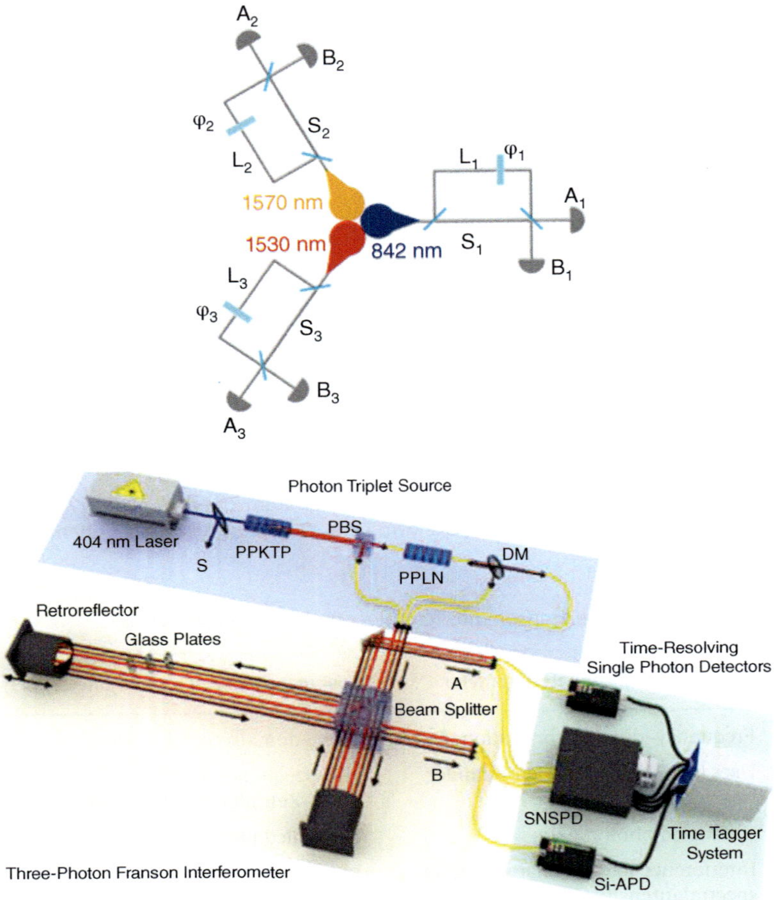

Figure 4.8 Schematic representation and experimental setup of the three-photon Franson interferometer. With permission from Agne et al. [53] / American Physical Society.

unbalanced Mach-Zehnder interferometers (UMZIs) and measuring coincidences between any two output ports from two different UMZIs.

While the three-photon Franson interferometer offers a valuable tool for investigating fundamental quantum phenomena like Bell inequalities and entanglement in higher dimensions, it presents significant experimental hurdles. Generating high-quality three-photon entangled states is difficult, and the setup and alignment of three separate interferometers, along with the requirement for precise coincidence detection, adds considerable complexity. Nevertheless, this interferometer marks important progress in exploring complex quantum effects and leveraging multi-photon entanglement for future quantum technologies. The three-photon Franson interferometer's reliance on genuine three-photon interference grants it superior sensitivity compared to single- or two-photon interference, making it potentially ideal for detecting subtle refractive index variations in biological samples. Moreover, using three-photon excitation offers advantages like higher spatial resolution and reduced photodamage compared to single-photon methods, which is beneficial for live cell and tissue imaging. This heightened sensitivity also suggests applications in detecting trace biomarkers in biological fluids, potentially enabling earlier disease diagnosis. Although still in its nascent stages, the use of three-photon Franson interferometry in biosensing holds substantial promise. Continued progress in quantum technology and theoretical research could lead to the development of innovative and highly sensitive biosensing platforms [53].

Thus, the Franson interference arises from energy-time or time-bin entanglement, which can also be generated through various physical processes in different media. These include spontaneous parametric down-conversion (SPDC) in bulk crystals or waveguides; spontaneous four-wave mixing (SFWM) in silica fiber, integrated waveguides, or micro resonators; SFWM in warm atomic ensembles; and two-photon emission from quantum dots (QDs).

Continued research and technological advancements are paving the way for the widespread use of Franson interferometers in various biomedical and environmental applications [47]. Transitioning quantum interferometers from laboratory demonstrations to practical applications is a crucial step in realizing their full potential. Quantum interferometers can surpass the sensitivity limits of classical devices by exploiting quantum entanglement and squeezing. Exploring new strategies for generating and utilizing entangled states and squeezed states can lead to further improvements in measurement precision.

4.2.2 Squeezed Light Interferometers

Interferometer sensitivity is fundamentally limited by quantum noise, stemming from the inherent uncertainty in light's phase and amplitude. Squeezed light, by reducing this quantum noise in one quadrature (such as phase), enables significantly more precise interferometric measurements. Injecting squeezed light into the interferometer effectively lowers the phase measurement noise below the standard quantum limit, allowing for the detection of much smaller phase shifts. Mach-Zehnder and Michelson interferometers are commonly employed in biosensing applications utilizing this squeezed light enhancement. These squeezed light interferometers (Figure 4.9) offer substantial improvements in sensitivity, precision, and the advantage of label-free detection. Consequently, they hold great promise for diverse biosensing applications, including biomolecule detection, cellular and tissue imaging, and monitoring physiological processes [10].

Advancements using squeezed light have achieved up to 15 dB of shot-noise reduction, demonstrating its value in high-precision applications like gravitational wave detection. Squeezing techniques can reduce light intensity fluctuations, often relative to a reference beam, a process known as relative intensity squeezing. Current squeezing devices are relatively compact, utilizing second-order nonlinearities in monolithic cavities or microstructures, or third-order nonlinearities in atomic media or optical fibers. Having proven its worth in astronomical observations by minimizing noise in LIGO and GEO600 experiments, squeezed light offers similar promise for quantum microscopy. Recent demonstrations highlight its potential to enhance imaging resolution and sensitivity in biological applications, marking a significant step in quantum-enhanced microscopy. For instance, a quantum microscope employing squeezed light has been developed

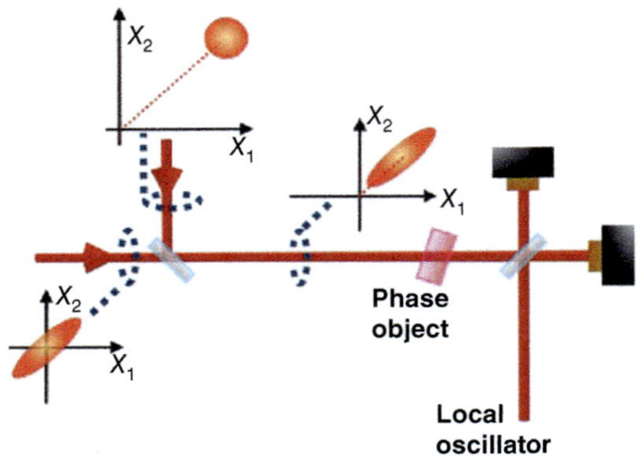

Figure 4.9 Measurement schemes adopting quantum light based on squeezed states. With permission from Moreva et al. [63] / AIP Publishing.

for improved 2D cell imaging. This system uses squeezing to lower the noise floor in coherent Raman microscopy, leading to an enhanced signal-to-noise ratio in imaging lipid concentrations within yeast cells, surpassing limitations imposed by photodamage [63].

Additionally, quantum-correlated light, particularly amplitude-squeezed states, has been used to improve the information gained per photon when tracking live yeast cells. This demonstrates the feasibility of using squeezed light within frequency ranges relevant to biological applications, while also addressing challenges associated with classical noise and spatial mode distortion [64].

4.2.3 Quantum Magnetometers

Quantum magnetometry is emerging as a powerful tool for biosensing, offering unprecedented sensitivity and precision in detecting biological entities and processes. By leveraging the principles of quantum mechanics, these sensors can detect minute magnetic fields generated by biological samples, providing new avenues for diagnostics, monitoring, and fundamental research. Quantum magnetometers for biosensing typically rely on detecting the magnetic fields produced by magnetic nanoparticles, which make use of several quantum devices, such as optically pumped magnetometers, nitrogen-vacancy (NV) centers in diamond, and superconducting quantum interference devices (SQUIDs) [10] Quantum magnetometry spans a wide range of potential functions in biosensing, viz., early diagnosis of diseases, point-of-care testing, drug discovery and development, and monitoring biological processes.

Quantum nanoprobes, nanoscale quantum systems robust enough for biological environments, are employed to label single molecules and biosamples. A prominent example is nanodiamond probes. In contrast to quantum dots, they are much more eco-friendly and exhibit surface chemistry enabling the attachment of various biomolecules. Nanodiamonds can be devised through various methods, involving detonation, laser ablation, and high-energy ball milling of high-pressure and high-temperature diamond microcrystals. Their quantum properties originate from the electron spin of nitrogen-vacancy (NV) centers, created via irradiation with high-energy electrons or helium atoms. Negatively charged NV-centers exhibit long spin coherence times (up to milliseconds) at room

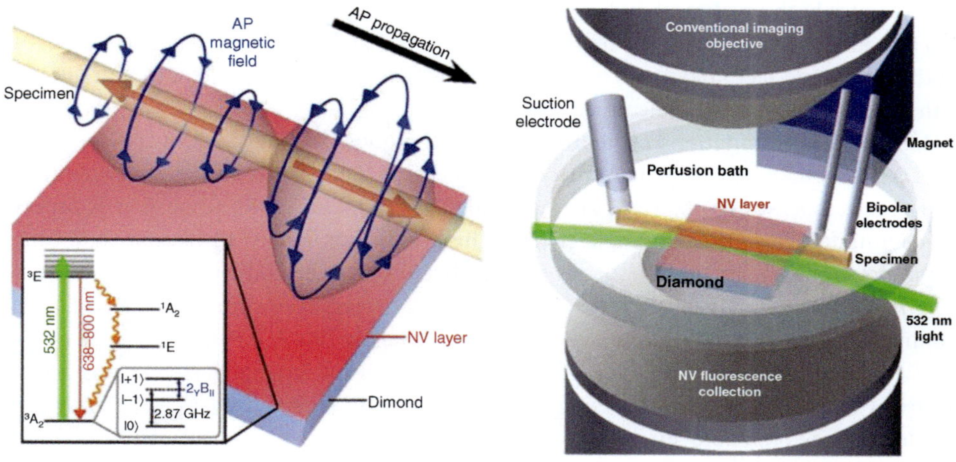

Figure 4.10 Detection of action potential using NV-centers inside the neuron of a living worm. (a) Experimental setup showing a diamond slab in blue with a layer of NV-center in red on which the axon of the specimen rests (orange). (b) The NV-centers are excited with green light, and their red fluorescence is collected by the bottom objective, while the top objective is used for imaging. With permission from Barry et al. [65] / National Academy of Sciences.

temperature. Their electron spin energy levels facilitate remote optical initialization and optical/microwave readout [20, 52, 57–60].

The quantum properties of NV-centers also enable highly precise extraction of other information from nanoscale samples. NV-centers exhibit exceptional fluorescence stability compared to other labels and are resistant to photobleaching. This makes them ideal candidates for fluorescence imaging. Notably, the fluorescence rate of NV-centers is influenced by their spin states, enabling spin state detection through optically detected magnetic resonance [66–68]. Single NV-centers in diamond plates have been employed to perform nanometer-cubed NMR-based measurements (nanotesla) in solution [69, 70], while ensembles of NV-centers have led to the finding of picotesla fields [71, 72]. This sensitivity to local magnetic fields has been harnessed to evaluate the spin density of biomolecules and rebuild 3D structures [73] and to identify the action potential of a single neuron in a living worm [65] (see Figure 4.10).

4.2.4 Quantum-Enabled Imaging Tools and Techniques

Advanced imaging technologies are essential for advancements in biotechnology. These tools are regularly employed in medical diagnoses, such as for vascular diseases and brain conditions, as well as for monitoring living tissues. Furthermore, they are critical for investigating and manipulating neural network structure and activity, both in laboratory settings and living organisms, enabling crucial insights into brain function, brain diseases, and complex cancers. Real-time monitoring of drug delivery, another key application, is vital for pharmaceutical development, highlighting the importance of imaging in this field [74–77]. Molecular imaging, which uses various imaging techniques to visualize and study molecular processes within living organisms, has become a valuable tool in drug development.

It accelerates both preclinical and clinical stages, resulting in cost and time savings, while also providing stronger evidence of drug effectiveness. Existing imaging systems face both fundamental and technological limitations [64], which constrain the breadth and quality of their biotechnological applications.

The resolution limitations of neuroimaging techniques like magnetoencephalography (MEG) [78], which detects neural currents via their magnetic fields and typically requires cryogenic temperatures, restrict our understanding of the brain. Its resolution is orders of magnitude larger than a single neuron, hindering our ability to observe real-time action potentials at the neuronal level—a significant obstacle. Thus, it is very important to enhance the sensitivity and resolution of optical imaging techniques [79], such as bioluminescence, fluorescence, and photoacoustic imaging [80–83]. Recent advancements have enabled femtocell-scale magnetic field measurements, crucial for capturing weak electrical signals from the brain, using SQUIDs. This technology produces real-time brain activity images, valuable for medical diagnosis and neuroscience research. However, the requirement for cryogenic cooling of SQUIDs creates several limitations: it confines MEG systems to specialized facilities, increases the sensor-to-subject distance, makes the technology susceptible to movement artifacts, and ultimately raises the overall cost of MEG systems [84, 85].

Apart from the above, stochastic optical reconstruction microscopy (STORM) is another optical microscopy technique that overcomes the diffraction limit in fluorescence microscopy, achieving roughly ten times higher resolution. It does this by sequentially activating and localizing individual fluorophores. Because only a small subset of fluorophores is active at any moment, their emissions are spatially distinct, allowing precise localization via centroid estimation. By repeating this process with different fluorophore subsets, a high-resolution image is constructed. Despite its wide use in bioimaging, STORM's performance is limited by fluorophore properties and lengthy acquisition times, which restrict its ultimate resolution and applicability [86]. Ongoing research focuses on improving these aspects. Notably, quantum effects offer potential avenues to further enhance the resolution of STORM and other super-resolution techniques.

In another significant advancement, it has been observed that encompassing principles of quantum mechanics in lidar technology has shown much-improved performance. Quantum lidar, using single-photon detection and quantum interference, achieves greater sensitivity for detecting subtle changes in biological and environmental samples. Researchers are particularly focused on applying quantum effects like Hong-Ou-Mandel interference and quantum erasure to achieve higher precision [87].

Thus, quantum imaging offers two key advantages: enhancing current imaging capabilities and enabling entirely new types of measurements and applications. By leveraging unique quantum effects at the nanoscale, these technologies can boost measurement sensitivity, push imaging resolution to unprecedented levels, and allow existing technologies to maintain performance at room temperature. This latter point is particularly significant, as it facilitates wider use of these technologies in point-of-care settings. Consequently, integrating such technologies with imaging has the potential to revolutionize biotechnology, offering significant advancements in disease diagnostics, drug pathway analysis, and diverse applications in our day-to-day lives [88–95].

4.3 Conclusion

We have discussed the application of quantum interferometers, specifically the HOM, N00N, and Franson interferometers, in light-based diagnostic tools for quantum biosensing. The theoretical background of these interferometers, including single- and multi-mode theories, and their ability to surpass classical sensitivity limits through entanglement and squeezing, was also discussed. The emerging field of quantum biology and the recent advancements at the intersection of quantum and biological science paint a very exciting future. The potential of these quantum technologies to advance biotechnology through improved disease diagnostics, drug pathway analysis, and fundamental investigations into quantum effects in living systems is immense. The inclusion of entanglement and squeezing phenomena in photonic quantum sensors represents a significant leap forward and opens exciting possibilities in diverse applications. These quantum sensors can detect extremely weak signals and subtle changes, expanding the range of measurable physical quantities. Ongoing research focuses on the miniaturization of such quantum light sensors and their use to make them more practical and widely applicable. However, their development faces substantial hurdles, such as scalability and integration, source and detection limitations, environmental sensitivity, cost, stability, and complexity. Overcoming these challenges is crucial for realizing the widespread application of photonic quantum sensors. However, ongoing research and development are steadily overcoming these challenges, paving the way for a future where photonic quantum sensors play a transformative role in various fields. In a nutshell, photonic quantum sensors represent a cutting-edge area of research with the potential to revolutionize sensing technologies.

References

1 Born, M. and Wolf, E. (1980). *Principles of Optics*, 6e. Cambridge University Press.

2 Hariharan, P. (2003). *Optical Interferometry*, 2e. Elsevier Inc https://doi.org/10.1016/B978-0-12-311630-7.X5000-X.

3 Moreau, P.-A., Toninelli, E., Gregory, T., and Padgett, M.J. (2019). Imaging with quantum states of light. *Nature Reviews Physics* 1: 367–380. https://doi.org/10.1038/s42254-019-0056-0.

4 Lawrie, B.J., Lett, P.D., Marino, A.M., and Pooser, R.C. (2019). Quantum sensing with squeezed light. *ACS Photonics* 6: 1307. https://doi.org/10.1021/acsphotonics.9b00250.

5 Franson, J.D. (1989). Bell inequality for position and time. *Physical Review Letters* 62: 2205. https://doi.org/10.1103/PhysRevLett.62.2205.

6 Zhang, Z., You, C., Magaña-Loaiza, O.S. et al. (2024). Entanglement-based quantum information technology: a tutorial. *Advances in Optics and Photonics* 16: 60–162. https://doi.org/10.1364/AOP.497143.

7 Jin, R.-B. and Shimizu, R. (2018). Extended Wiener–Khinchin theorem for quantum spectral analysis. *Optica* 5: 93. https://doi.org/10.1364/OPTICA.5.000093.

8 Bongs, K., Holynski, M., Vovrosh, J. et al. (2019). Taking atom interferometric quantum sensors from the laboratory to real-world applications. *Nature Reviews Physics* 1: 731–739. https://doi.org/10.1038/s42254-019-0117-4.

9 Dowling, J.P. and Milburn, G.J. (2003). Quantum technology: the second quantum revolution. *Philosophical Transactions of the Royal Society A* 361: 1655–1674. https://doi.org/10.1098/rsta.2003.1227.

10 Mauranyapin, N.P., Terrasson, A., and Bowen, W.P. (2022). Quantum biotechnology. *Advanced Quantum Technologies* 5: 2100139. https://doi.org/10.1002/qute.202100139.

11 Luo, K.-H., Santandrea, M., Stefszky, M. et al. (2021). Quantum optical coherence: from linear to nonlinear interferometers. *Physical Review A* 104: 043707. https://doi.org/10.1103/PhysRevA.104.043707.

12 Lewis-Swan, R. and Kheruntsyan, K. (2014). Proposal for demonstrating the Hong–Ou–Mandel effect with matter waves. *Nature Communications* 5: 3752. https://doi.org/10.1038/ncomms4752.

13 Makarov, D.N. (2020). Fluctuations in the detection of the HOM effect. *Scientific Reports* 10: 20124. https://doi.org/10.1038/s41598-020-77189-6.

14 Hong, C.K., Ou, Z.Y., and Mandel, L. (1987). Measurement of subpicosecond time intervals between two photons by interference. *Physical Review Letters* 59: 2044. https://doi.org/10.1103/PhysRevLett.59.2044.

15 Jin, R.-B., Zeng, Z.-Q., You, C., and Yuan, C. (2024). Quantum interferometers: principles and applications. *Progress in Quantum Electronics* 96: 100519. https://doi.org/10.48550/arXiv.2310.16378.

16 Ou, Z.Y. and Li, X. (2020). Quantum SU (1,1) interferometers: basic principles and applications. *APL Photonics* 5: 080902. https://doi.org/10.1063/5.0004873.

17 Ra, Y.-S., Tichy, M.C., Lim, H.-T. et al. (2013). Observation of detection-dependent multi-photon coherence times. *Nature Communications* 4: 2451. https://doi.org/10.1038/ncomms3451.

18 Valivarthi, R., Zhou, Q., John, C. et al. (2017). Measurement-device-independent quantum key distribution coexisting with classical communication. *Quantum Science and Technology* 4: 045002. https://doi.org/10.1088/2058-9565/ab2e62.

19 Lo, H.-K., Curty, M., and Qi, B. (2012). Measurement-device-independent quantum key distribution. *Physical Review Letters* 108: 130503. https://doi.org/10.1103/PhysRevLett.108.130503.

20 Menssen, A.J., Jones, A.E., Metcalf, B.J. et al. (2017). Distinguishability and many-particle interference. *Physical Review Letters* 118: 153603. https://doi.org/10.1103/PhysRevLett.118.153603.

21 You, C., Miller, A., León-Montiel, R.d.J., and Magaña-Loaiza, O.S. (2023). Multiphoton quantum van Cittert-Zernike theorem. *npj Quantum Information* 9: 50. https://doi.org/10.1038/s41534-023-00720-w.

22 Li, B., Yuan, B., Chen, C. et al. (2023). Spectrally resolved two-photon interference in a modified Hong-Ou-Mandel interferometer. *Optics and Laser Technology* 159: 109039. https://doi.org/10.48550/arXiv.2205.11135.

23 Zhai, Y., Dong, R., Li, B. et al. (2017). Frequency entanglement characterization of short-pulse pumped SPDC biphoton source with a Mach-Zehnder interferometer. *Journal of Physics B: Atomic, Molecular and Optical Physics* 50: 125502. https://doi.org/10.1088/1361-6455/aa5a1c.

24 Quan, R., Wang, M., Hou, F. et al. (2015). Characterization of frequency entanglement under extended phase-matching conditions. *Applied Physics B: Lasers and Optics* 118: 431–437. https://doi.org/10.1007/s00340-015-6006-3.

25 Huang, D., Swanson, E.A., Lin, C.P. et al. (1991). Optical coherence tomography. *Science* 254: 1178–1181. https://doi.org/10.1126/science.1957169.

26 Fercher, A.F., Drexler, W., Hitzenberger, C.K., and Lasser, T. (2003). Optical coherence tomography – principles and applications. *Reports on Progress in Physics* 66: 239. https://doi.org/10.1088/0034-4885/66/2/204.

27 Brezinski, M. (2006). *Optical Coherence Tomography: Principles and Applications*. Elsevier Science 1: 632.

28 Okano, M., Lim, H.H., Okamoto, R. et al. (2015). 0.54 μm resolution two-photon interference with dispersion cancellation for quantum optical coherence tomography. *Scientific Reports* 5: 18042. https://doi.org/10.1038/srep18042.

29 Ndagano, B., Defienne, H., Branford, D. et al. (2022). Quantum microscopy based on Hong-Ou-Mandel interference. *Nature Photonics* 16: 384–389. https://doi.org/10.1038/s41566-022-00980-6.

30 Ono, T., Okamoto, R., and Takeuchi, S. (2010). Quantum two-photon interference with attosecond-scale time delay. *Nature Communications* 1 (1): 1–6. https://doi.org/10.1007/s00340-003-1337-x.

31 Lounis, B. and Orrit, M. (2005). Single-photon sources. *Reports on Progress in Physics* 68 (5): 1129. https://doi.org/10.1088/0034-4885/68/5/R04.

32 Wu, J., Zhang, Y., Yang, M. et al. (2025). Observation of quantum interference in single–molecule junctions via electrochemical gating. *Current Opinion in Electrochemistry* 51: 101688. https://doi.org/10.1016/j.coelec.2025.101688.

33 Politi, A., Cryan, J., Rarity, J.G. et al. (2008). Silica-on-silicon waveguide quantum circuits. *Science* 320 (5876): 646–649. https://doi.org/10.48550/arXiv.0802.0136.

34 Mitchell, M.W., Lundeen, J.S., and Steinberg, A.M. (2004). Super-resolving phase measurements with a multiphoton entangled state. *Nature* 429: 161–164. https://doi.org/10.1038/nature02493.

35 Ono, T., Okamoto, R., and Takeuchi, S. (2013). An entanglement-enhanced microscope. *Nature Communications* 4: 2426. https://doi.org/10.1038/ncomms3426.

36 Jin, R.-B., Fujiwara, M., Shimizu, R. et al. (2016). Detection-dependent six-photon Holland-Burnett state interference. *Scientific Reports* 6: 36914. https://doi.org/10.1038/srep36914.

37 Griffiths, P.R. and De Haseth, J.A. (2007). *Fourier Transforms*. In: *Fourier Transform Infrared Spectrometry*, 2e (ed. J.D. Winefordner). Wiley https://doi.org/10.1002/047010631X.

38 Xiang, G.Y., Hofmann, H.F., and Pryde, G.J. (2013). Optimal multi-photon phase sensing with a single interference fringe. *Scientific Reports* 3: 2684. https://doi.org/10.1038/srep02684.

39 Holland, M.J. and Burnett, K. (1993). Interferometric detection of optical phase shifts at the Heisenberg limit. *Physical Review Letters* 71: 1355. https://doi.org/10.1103/PhysRevLett.71.1355.

40 Li, F., Sun, Y., and Zhang, X.D. (2022). Deep-learning-based quantum imaging using NOON states. *Journal of Physics Communications* 6: 035005. https://doi.org/10.1088/2399-6528/ac5e25.

41 Cozzolino, D., Lio, B.D., Bacco, D., and Oxenløwe, L.K. (2019). High-dimensional quantum communication: benefits, progress, and future challenges. *Advanced Quantum Technologies* 2: 1900038. https://doi.org/10.1002/qute.201900038.

42 Erhard, M., Krenn, M., and Zeilinger, A. (2020). Advances in high-dimensional quantum entanglement. *Nature Reviews Physics* 2: 365–381. https://doi.org/10.1038/s42254-020-0193-5.

43 Yang, Z.-X., Zeng, Z.-Q., Tian, Y. et al. (2023). Spatial–spectral mapping to prepare frequency entangled qudits. *Optics Letters* 48: 2361–2364. https://doi.org/10.1364/OL.487300.

44 Guo, Y., Yang, Z.-X., Zeng, Z.-Q. et al. (2023). Comparison of multi-mode Hong-Ou-Mandel interference and multi-slit interference. *Optics Express* 31: 32849–32864. https://doi.org/10.1364/OE.501645.

45 Nielsen, J.A.H., Neergaard-Nielsen, J.S., Gehring, T., and Andersen, U.L. (2023). Deterministic quantum phase estimation beyond N00N states. *Physical Review Letters* 130: 123603. https://doi.org/10.1103/PhysRevLett.130.123603.

46 Jin, R.-B., Zeng, Z.-Q., Xu, D. et al. (2024). Spectrally resolved Franson interference. *Science China – Physics Mechanics & Astronomy* 67: 250312. https://doi.org/10.1007/s11433-023-2326-7.

47 Cameron, P., Courme, B., Vernière, C. et al. (2024). Adaptive optical imaging with entangled photons. *Science* 383: 1142–1148. https://doi.org/10.1126/science.adk7825.

48 Zhao, Y., Chen, L., Hu, H. et al. (2020). Numerical investigation of parametric frequency dependence in the modeling of octave-spanning Kerr frequency combs. *IEEE Photonics* 12 (2): 1–9. https://doi.org/10.1109/JPHOT.2020.2976558.

49 Cabello, A., Rossi, A., Vallone, G. et al. (2009). Proposed Bell experiment with genuine energy-time entanglement. *Physical Review Letters* 102: 040401. https://doi.org/10.1103/PhysRevLett.102.040401.

50 Lima, G., Vallone, G., Chiuri, A. et al. (2010). Experimental Bell-inequality violation without the post selection loophole. *Physical Review A* 81: 040101. https://doi.org/10.1103/PhysRevA.81.040101.

51 Zhang, Z., Mower, J., Englund, D. et al. (2014). Unconditional security of time-energy entanglement quantum key distribution using dual-basis interferometry. *Physical Review Letters* 112: 120506. https://doi.org/10.1103/PhysRevLett.112.120506.

52 Chen, C., Shapiro, J.H., and Wong, F.N. (2021). Experimental demonstration of Conjugate-Franson interferometry. *Physical Review Letters* 127: 093603. https://doi.org/10.1103/PhysRevLett.127.093603.

53 Agne, S., Kauten, T., Jin, J. et al. (2017). Observation of genuine three-photon interference. *Physical Review Letters* 118: 153602. https://doi.org/10.1103/PhysRevLett.118.153602.

54 Jogenfors, J. and Larsson, J.-Å. (2014). Energy-time entanglement, elements of reality, and local realism. *Journal of Physics A: Mathematical and Theoretical* 47: 424032. https://doi.org/10.1088/1751-8113/47/42/424032.

55 Lingaraju, N.B., Lu, H.-H., Seshadri, S. et al. (2019). Quantum frequency combs and Hong–Ou–Mandel interferometry: the role of spectral phase coherence. *Optics Express* 27: 38683–38697. https://doi.org/10.1364/OE.379749.

56 Chang, K.-C., Cheng, X., Sarihan, M.C. et al. (2021). 648 Hilbert-space dimensionality in a biphoton frequency comb: entanglement of formation and Schmidt mode decomposition. *npj Quantum Information* 7: 48. https://doi.org/10.1038/s41534-021-00388-0.

57 Pan, J.-W., Chen, Z.-B., Lu, C.-Y. et al. (2012). Multiphoton entanglement and interferometry. *Reviews of Modern Physics* 84: 777. https://doi.org/10.1103/RevModPhys.84.777.

58 Greenberger, D.M., Horne, M.A., Shimony, A., and Zeilinger, A. (1990). Bell's theorem without inequalities. *American Journal of Physics* 58: 1131–1143. https://doi.org/10.1119/1.16243.

59 Barnett, S.M. and Imoto, and N., Huttner, B. (1998). Photonic de broglie wave interferometers. *Journal of Modern Optics* 45: 2217. https://doi.org/10.1080/09500349808231234.

60 Greenberger, D.M., Horne, M.A., and Zeilinger, A. (1993). Multiparticle interferometry and the superposition principle. *Physics Today* 46: 22. https://doi.org/10.1063/1.881360.

61 Bouwmeester, D., Pan, J.-W., Daniell, M. et al. (1999). Observation of three-photon Greenberger-Horne-Zeilinger entanglement. *Physical Review Letters* 82: 1345. https://doi.org/10.1103/PhysRevLett.82.1345.

62 Hamel, D.R., Shalm, L.K., Hübel, H. et al. (2014). Direct generation of three-photon polarization entanglement. *Nature Photonics* 8: 801–817. https://doi.org/10.1038/nphoton.2014.218.

63 Moreva, E., Cimini, V., Gianani, I. et al. (2025). Quantum photonics sensing in biosystems. *APL Photonics* 10 (1): 010902. https://doi.org/10.1063/5.0232183.

64 Taylor, M.A., Janousek, J., Daria, V. et al. (2013). Biological measurement beyond the quantum limit. *Nature Photonics* 7: 229. https://doi.org/10.1038/nphoton.2012.346.

65 Barry, J.F., Turner, M.J., Schloss, J.M. et al. (2016). Optical magnetic detection of single-neuron action potentials using quantum defects in diamond. *Proceedings of the National Academy of Sciences of the United States of America* 113: 14133–14138. https://doi.org/10.1073/pnas.1601513113.

66 Danilenko, V.V. (2004). On the history of the discovery of nanodiamond synthesis. *Physics of the Solid State* 46: 595–599. https://doi.org/10.1134/1.1711431.

67 Boudou, J.-P., Curmi, P.A., Jelezko, F. et al. (2009). High yield fabrication of fluorescent nanodiamonds. *Nanotechnology* 20: 235602. https://doi.org/10.1088/0957-4484/20/23/235602.

68 Balasubramanian, G., Lazariev, A., Arumugam, S.R., and Duan, D.-W. (2014). Nitrogen-Vacancy color center in diamond – emerging nanoscale applications in bioimaging and biosensing. *Current Opinion in Chemical Biology* 20: 69–77. https://doi.org/10.1016/j.cbpa.2014.04.014.

69 Staudacher, T., Shi, F., Pezzagna, S. et al. (2013). Nuclear magnetic resonance spectroscopy on a (5-nanometer)3 sample volume. *Science* 339: 561. https://doi.org/10.1126/science.1231675.

70 Mamin, H.J., Kim, M., Sherwood, M.H. et al. (2013). Nanoscale nuclear magnetic resonance with a nitrogen-vacancy spin sensor. *Science* 339: 557–560. https://doi.org/10.1126/science.1231540.

71 Wolf, T., Neumann, P., Nakamura, K. et al. (2015). Subpicotesla diamond magnetometry. *Physical Review X* 5: 41001. https://doi.org/10.1103/PhysRevX.5.041001.

72 Bennett, J.S., Vyhnalek, B.E., Greenall, H. et al. Precision magnetometers for aerospace applications: a review. *Sensors* 21: 5568. https://doi.org/10.3390/s21165568.

73 Balasubramanian, G., Chan, I.Y., Kolesov, R. et al. (2008). Nanoscale imaging magnetometry with diamond spins under ambient conditions. *Nature* 455: 648–651. https://doi.org/10.3390/s21165568.

74 Lao, Y., Xing, D., Yang, S., and Xiang, L. (2008). Noninvasive photoacoustic imaging of the developing vasculature during early tumor growth. *Physics in Medicine and Biology* 53: 4203–4212. https://doi.org/10.1088/0031-9155/53/15/013.

75 Pfefferbaum, A., Sullivan, E.V., Mathalon, D.H. et al. (1995). Longitudinal changes in magnetic resonance imaging brain volumes in abstinent and relapsed alcoholics. *Alcoholism: Clinical and Experimental Research* 19: 1177–1191. https://doi.org/10.1111/j.1530-0277.1995.tb01598.x.

76 Chan, V. and Perlas, A. (2010). *Atlas of Ultrasound-Guided Procedures in Interventional Pain Management*, 13. Springer.

77 Aaronson, S. (2009). Why quantum chemistry is hard. *Nature Physics* 5: 707–708. https://doi.org/10.1038/nphys1415.

78 Cohen, D. (1972). Magnetoencephalography: detection of the Brain's electrical activity with a superconducting magnetometer. *Science* 175: 664–666. https://doi.org/10.1126/science.175.4022.664.

79 Luker, G.D. and Luker, K.E. (2007). Optical imaging: current applications and future directions. *Journal of Nuclear Medicine* 49: 1–4. https://doi.org/10.2967/jnumed.107.045799.

80 Contag, C.H. and Bachmann, M.H. (2002). Advances in in vivo bioluminescence imaging of gene expression. *Annual Review of Biomedical Engineering* 4: 235–260. https://doi.org/10.1146/annurev.bioeng.4.111901.093336.

81 Sarma, M. and Chatterjee, T. (2023). Graphene-based sensors. *IOP Publishing* 2: https://doi.org/10.1088/978-0-7503-5578-0ch2.

82 Wöll, D. and Flors, C. (2017). Super-resolution fluorescence imaging for materials science. *Small Methods* 1: 1700191. https://doi.org/10.1002/smtd.201700191.

83 Attia, A.B.E., Balasundaram, G., Moothanchery, M. et al. (2019). A review of clinical photoacoustic imaging: current and future trends. *Photoacoustics* 16: 100144. https://doi.org/10.1016/j.pacs.2019.100144.

84 Boto, E., Hill, R.M., Rea, M. et al. (2021). Measuring functional connectivity with wearable MEG. *NeuroImage* 230: 117815. https://doi.org/10.1016/j.neuroimage.2021.117815.

85 Baillet, S. (2017). Magnetoencephalography for brain electrophysiology and imaging. *Nature Neuroscience* 20: 327–339. https://doi.org/10.1038/nn.4504.

86 Bates, M., Huang, B., Dempsey, G.T., and Zhuang, X. (2007). Multicolor super-resolution imaging with photo-switchable fluorescent probes. *Science* 317: 1749–1753. https://doi.org/10.1126/science.1146598.

87 Das, S., Mazumdar, H., Khondakar, K.R. et al. (2024). Review—quantum biosensors: principles and applications in medical diagnostics. *ECS Sensors Plus* 3: 025001. https://doi.org/10.1149/2754-2726/ad47e2.

88 Broadway, D.A., Tetienne, J.-P., Stacey, A. et al. (2018). Quantum probe hyperpolarisation of molecular nuclear spins. *Nature Communications* 9: 1246. https://doi.org/10.1038/s41467-018-03578-1.

89 Sarma, M. and Chatterjee, T. (2024). *The Voltammetry for Forensic Materials Sensing*, 25–46. American Chemical Society Publishers. https://doi.org/10.1021/bk-2024-1481.ch002.

90 Golman, K., Olsson, L.E., Axelsson, O. et al. (2003). Molecular imaging using hyperpolarized 13C. *Journal de Radiologie* 76: S118–S127. https://doi.org/10.1259/bjr/26631666.

91 Chen, H., Larson, P.E.Z., Gordon, J.W. et al. (2018). Technique development of 3D dynamic CS-EPSI for hyperpolarized ^{13}C pyruvate MR molecular imaging of human prostate cancer. *Magnetic Resonance in Medicine* 80: 2062–2072. https://doi.org/10.1002/mrm.27179.

92 Sarma, M. (2022). Carbon nanomaterial-based electrochemical sensors for biomedical applications. *IOP Publishing* 15: https://doi.org/10.1088/978-0-7503-5127-0ch15.

93 Tigari, G., Manjunatha, J.G., Nagarajappa, H., and Prinith, N.S. (2021). Research developments in carbon materials based sensors for determination of hormones: review. *Journal of Electrochemical Science and Engineering* 12 (1): 3–23. https://doi.org/10.5599/jese.1094.

94 Hareesha, N. and Manjunatha, J.G. (2021). Electro-oxidation of formoterol fumarate on the surface of novel poly(thiazole yellow-G) layered multi-walled carbon nanotube paste electrode. *Scientific Reports* 11: 12797. https://doi.org/10.1038/s41598-021-92099-x.

95 Monnappa, A.B., Manjunatha, J.G., and Bhatt, A.S. (2020). Design of a sensitive and selective voltammetric sensor based on a cationic surfactant-modified carbon paste electrode for the determination of alloxan. *ACS Omega* 5 (36): 23481–23490. https://doi.org/10.1021/acsomega.0c03517.

5

Quantum Resonance Imaging: A New Era in Medical Imaging

Dhivya Antony[1], Arasan Saroja Anakath[2], Pandurangan Anandan[3], and Chinnapiyan Vedhi[4]

[1]*Department of Chemistry, St. Joseph's College of Engineering, Chennai, Tamil Nadu, India*
[2]*Department of Computer Science and Engineering, Saveetha School of Engineering, Saveetha Institute of Medical and Technical Sciences, Chennai, Tamil Nadu, India*
[3]*Department of Physics, Thiru Kolanjiappar Government Arts College, Virudhachalam, Tamil Nadu, India*
[4]*Department of Chemistry, V.O. Chidambaram, Thoothukudi, Tamil Nadu, India*

5.1 Introduction

Over the last few decades, the science of medical imaging has advanced dramatically, moving from traditional X-rays to sophisticated techniques like magnetic resonance imaging (MRI) and positron emission tomography (PET). However, the emergence of quantum resonance imaging (QRI) promises to usher in a paradigm shift. This innovative technology uses quantum mechanics principles to improve resolution, sensitivity, and diagnostic accuracy beyond what regular imaging technologies can achieve [1]. The quantum principles are highly advanced in imaging performance and discuss the practical implications and ongoing efforts to address these challenges. The image processing using computers for image acquisition has proven invaluable for generating images of remarkable resemblance to the patient's actual internal anatomy; in particular, the high imaging fidelities afforded by contemporary X-ray CT and MRI rivals in many cases the realism of photographs of *ex-vivo* frozen sections. However, the use of computers has not been limited to the image acquisition-reconstruction front end of medical imaging, and these uniquely adaptable machines find numerous post-acquisitions use. The field of medical image processing (IP) is becoming a science in its own right with well-defined principles and theory [2].

IP improves image quality with respect to the four objective measures like signal-to-noise ratio (SNR), spatial resolution, contrast-to-noise ratio, and artifact reduction. Other well-known IP methods include methods for quantifying the sizes and shapes of organs, which include segmentation and volumetric, and methods for co-registering spatially different images from different scans. Quantum correlation enhanced optical imaging represents a transformative approach to optical microscopy and imaging technologies, leveraging quantum mechanical phenomena such as entanglement and squeezed states to surpass the limitations of classical imaging techniques. By exploiting quantum correlations

Quantum Biosensing in Medical Diagnostics, First Edition. Edited by J. G. Manjunatha.

between photons, this method allows for enhanced resolution, sensitivity, and contrast, enabling the detection of faint signals and minute structures that would otherwise remain undetectable. Quantum-enhanced imaging techniques, such as quantum interferometry and quantum squeezing, reduce the standard quantum limit, offering new capabilities in fields like biological imaging, medical diagnostics, astronomy, and quantum metrology [3]. Despite the promise of quantum-enhanced optical imaging, challenges such as photon loss, decoherence, and system complexity remain significant hurdles. This chapter will delve deeper into the quantum mechanics principles underpinning QRI, providing a more technical understanding of its operation and potential advancements in medical imaging.

5.1.1 Background of Quantum Resonance Imaging with Traditional MRI

QRI builds upon MRI principles but extends its capabilities using quantum technology, potentially revolutionizing medical imaging and diagnostics. MRI originated in 1975, when Peter Mansfield and Andrew Maudsley developed a line scan technique that resulted in the first *in vivo* human image in 1977. That year, Hinshaw, Bottomley, and Holland scanned a wrist, while Damadian's team created a cross-section of the human chest. Hugh Clow and Ian R. Young obtained the first transverse NMR image of a human head in 1978, followed by coronal and sagittal images in 1980 [4]. Edelstein's team at Aberdeen University increased scanning efficiency, cutting scan time from five minutes in 1980 to five seconds in 1986. Paul Lauterbur and Peter Mansfield established MRI in 1974 by using magnetic field gradients for spatial localization, allowing for 2D and 3D imaging. Lauterbur, utilizing CAT-scan principles, released the first NMR images in 1973, while Mansfield concentrated on high-resolution scanning. Their accomplishments earned them the 2003 Nobel Prize. Modern MRI is based on Fourier Transform NMR, in which RF pulses excite numerous frequencies, and the free induction decay (FID) gives spectrum information. Image quality is improved using techniques like wavelet transforms and spin-echo sequences [5]. Magnetic field gradients are used for spatial localization, which allows for comprehensive anatomical imaging and revolutionizes medical diagnostics.

5.1.2 Technological Review on Magnetic Resonance Imaging (MRI)

Noninvasive, high-resolution glutamate detection in the brain, with enhanced sensitivity at ultrahigh fields (7T+) by Whole-Body MRI. It detects glutamate changes despite small contamination, and future 3D imaging advances may aid in Central Nervous System (CNS) issue diagnosis and stroke monitoring. MRI investigations employed 7T (human) and 9.4T (animal) scanners with specialized RF coils for high-resolution metabolic imaging [6]. Molecular MRI improves diagnoses with its multiparametric capabilities. Advances such as frequency-encoding contrast agents, dual-modality MRI-PET, and nanotechnology-based MRI-guided therapies are revolutionizing personalized medicine and real-time monitoring [7]. Metal alloy nanoparticles (MNPs) such as FeCo and FePt improve MRI sensitivity, with coated FeCo MNPs surpassing conventional contrast agents in cell labeling and T1-weighted imaging [8]. 19F MRI detects fluorine atoms in tracers, enhancing sensitivity with fluorinated probes and lanthanide contrast agents (CAs) [9]. Optimized trifluoromethyl positioning increases sensitivity by up to 25 times, whereas improper spacing lowers signal quality. Medium-weight probes linked to biocompatible vectors enhance imaging at clinical doses. Perfluorocarbons (PFCs) utilized in ^{19}F MRI are chemically stable, biologically inert, and removed via the lungs, which reduces toxicity [9]. Paramagnetic contrast compounds improve MRI; however, typical coordination complexes present stability and pharmacokinetic issues. Nanoparticles coupled with paramagnetic ions enhance stability, biodistribution, and water exchange kinetics, allowing for multimodal imaging and personalized therapy in cancer, cardiovascular, and neurological imaging, as well as drug administration [10]. Liposomes, made up of lipid bilayers, can transport both hydrophilic and lipophilic molecules, making them excellent for medication administration. Their pharmacokinetics are determined by size, charge, lipid composition, and distribution technique. Polyethylene glycol (PEG) increases circulation time, but cholesterol and phospholipids improve stability, so optimizing their medical applications [11]. Figure 5.1 groups all technology beneficial in MRI in different fields.

Molecular MRI allows for cancer monitoring, which improves early identification, diagnosis, and treatment planning, especially in prostate, ovarian, and lung cancers. Integrated PET/MRI improves diagnosis accuracy through simultaneous imaging [12]. MRI-based gene expression imaging utilizes a gadolinium-based contrast agent triggered by β-galactosidase [13]. Silicon-based photodetectors increase PET signals, although attenuation correction remains difficult. The study focuses on dual-mode contrast agents and biodistribution studies for applications such as plaque imaging. MRFM and NV-center scanning probes are used in atomic-scale MRI to achieve high-resolution biomolecular imaging. Scanning Tunneling Microscopy (STM) with Electron Spin Resonance (ESR)

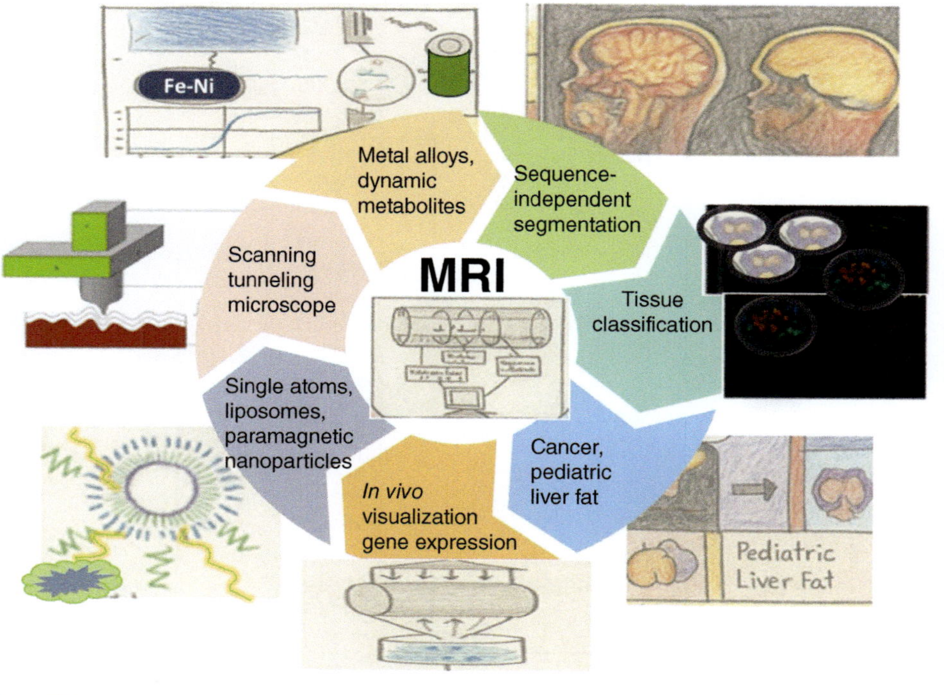

Figure 5.1 Technological benefits in MRI.

achieves sub-Ångström MRI resolution, providing insight into atomic-scale magnetism. Dynamic NMR imaging tracks metabolite transport, with gradient-enhanced HMQC increasing 13C sensitivity and outperforming PET for metabolite tracking. MRI segmentation improvements, such as modified FLASH sequences, improve T2 estimation and pathology identification [14]. MRI-based liver fat measurement with MRI-C and MRI-M assesses hepatic steatosis more accurately than biopsies do. Functional BOLD-MRI in pediatrics helps neurological research, with continual developments in imaging technology and software [15].

5.1.3 Relation Between QRI and MRI

QRI is considered an evolution of MRI, integrating quantum mechanics to overcome MRI's limitations. Traditional MRI generates images by detecting radiofrequency signals emitted by hydrogen nuclei in a magnetic field, but it is limited by resolution and sensitivity constraints. QRI enhances this process by utilizing quantum coherence, spin entanglement, and quantum sensors, which enable nanoscale resolution and the ability to detect molecular and atomic interactions beyond MRI's capability [16]. In resolution QRI achieves atomic and molecular-level imaging, surpassing MRI's millimeter-scale resolution. Sensitivity of QRI uses quantum sensors for higher sensitivity, detecting weak magnetic signals undetectable by MRI. MRI is widely used in clinical settings, whereas QRI is emerging in molecular diagnostics, neuroimaging, and early disease detection. Figure 5.2 shows the differential outlined features of MRI and QRI [17, 18].

Features	MRI	QRI
Technology base	Nuclear magnetic resonance (NMR)	Quantum coherence, entanglement, quantum sensors
Resolution	Millimeter-scale	Atomic and molecular-level
Sensitivity	Detects hydrogen nuclei	Detects weak magnetic signals at nanoscale
Imaging capability	Soft tissues, organs	Molecular diagnostics, neuroimaging, early disease detection
Clinical status	Widely used	Emerging technology in research
Computation	Classical image processing	Quantum algorithms enhance real-time imaging

Figure 5.2 Features of MRI and QRI—comparison.

5.2 Quantum Resonance Imaging

QRI is not a recognized medical imaging technology like MRI or CT scans but has been referenced in speculative discussions, pseudoscientific claims, and research into quantum effects in imaging. MRI, a well-established technology, relies on quantum mechanics by manipulating the spin states of hydrogen nuclei using magnetic fields and radiofrequency pulses. Its fundamental principles, including nuclear spin, resonance, and relaxation times, are rooted in quantum behavior. Researchers are also developing quantum-enhanced imaging techniques using nitrogen-vacancy (NV) centers in diamonds, which can detect weak magnetic fields for high-precision imaging at the atomic scale [19]. While QRI is often mentioned in speculative or pseudoscientific contexts, with claims of diagnosing health conditions using quantum field interactions, there is no scientific evidence supporting its existence. However, future advancements in quantum computing and entanglement could revolutionize medical imaging by improving resolution and sensitivity and enabling real-time subatomic imaging of biological structures [20]. Although QRI as a technology does not currently exist, quantum principles continue to shape innovations in medical imaging, offering potential breakthroughs in diagnostic capabilities and precision imaging.

5.2.1 Discovery of Quantum Resonance Imaging

QRI is a groundbreaking advancement in medical imaging that uses quantum mechanics concepts to improve diagnostic accuracy. While still in its early phases, QRI is based on decades of research in quantum physics, nuclear magnetic resonance (NMR), and quantum computing. Theoretical foundations and early concepts of QRI trace its origins back to fundamental quantum physics discoveries, specifically the principles of superposition, entanglement, and quantum coherence. In the mid-twentieth century, physicists such as Erwin Schrödinger, Albert Einstein, and Richard Feynman laid the framework for understanding atomic and subatomic behavior [21]. The concept of using quantum mechanics for medical imaging gained traction as researchers looked into how quantum states could be used to improve spatial resolution and sensitivity in imaging techniques. This methodology differed from traditional imaging procedures such as MRI and CT scans, which are based on classical physics concepts.

5.2.2 Development of Quantum Resonance Imaging (QRI)

QRI's development is the result of advances in quantum mechanics and quantum computing. The theoretical foundation was created by early twentieth-century quantum physicists who studied wave-particle duality, quantum entanglement, and spin states. The idea of exploiting quantum states for imaging arose as researchers began to investigate how quantum particles interact with biological tissues on microscopic dimensions. One of the primary inspirations for QRI is the success of MRI, which uses NMR to provide detailed images of soft tissues. However, MRI is constrained by SNR and resolution requirements. Quantum technologies aim to address these limits by leveraging quantum coherence and entanglement, enabling higher-resolution imaging and more detailed molecular-level diagnostics [22]. Key advances for QRI include advances in quantum sensing, which enable the ultrasensitive detection of magnetic and electric forces at the nanoscale, setting the groundwork for quantum-enhanced imaging. Quantum optics research has shown that entangled photons can penetrate biological tissues with minimum scattering, allowing for deeper and more detailed imaging following quantum entanglement. Quantum algorithms have greatly enhanced data analysis and image reconstruction, enabling real-time, high-resolution medical imaging by computing processes.

5.2.3 The Science Behind Quantum Resonance Imaging

QRI uses the principles of quantum entanglement, superposition, and quantum coherence to create new imaging capabilities. Unlike MRI, which uses NMR, QRI uses quantum sensors that interact at the atomic and subatomic levels, allowing for higher sensitivity. The introduction of quantum-entangled particles improves signal detection, allowing for the discovery of minute biological changes at the cellular level. Enhanced resolution in quantum coherence reduces noise and interference, resulting in ultra-high-resolution imaging of tissues and organs. Noninvasive precision unlike ionizing radiation used in QRI has less biological impact, making it suitable for repeated usage. Quantum computing and image processing—Quantum algorithms have greatly enhanced data analysis and image reconstruction, enabling real-time, high-resolution medical imaging. QRI operates by first using a quantum sensor to detect resonance signals from the target. An excitation source then stimulates quantum states using controlled energy pulses. These signals are processed in the quantum processing unit, which enhances precision through quantum entanglement and coherence. The refined data is then analyzed by an AI Analysis system, which reconstructs high-resolution images with enhanced clarity. Finally, the processed

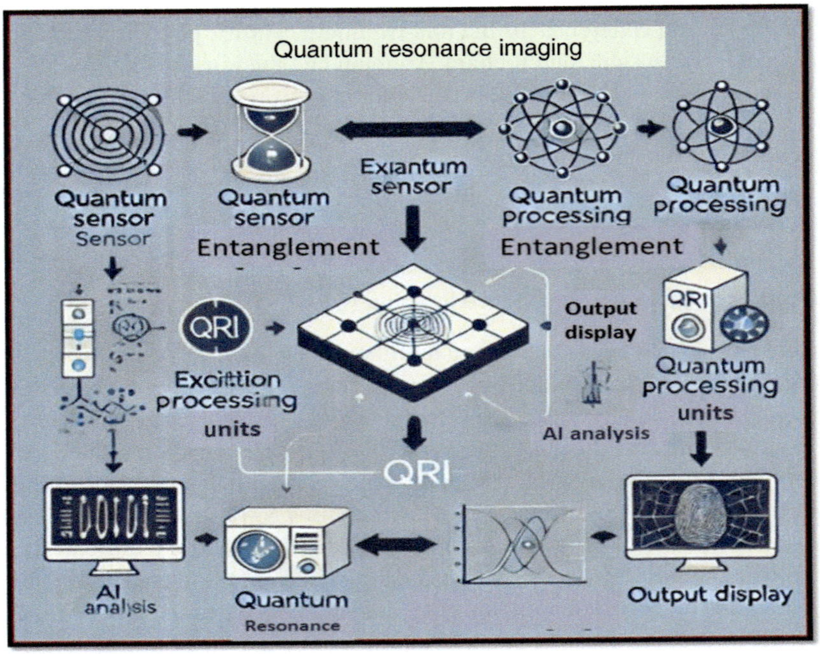

Figure 5.3 Schematic working model of QRI.

imaging results are displayed on the output display, providing valuable insights for medical, scientific, or space exploration applications [23, 24]. Figure 5.3 explains the sequential work carried out in QRI.

The core principle of QRI involves manipulating quantum states of particles, such as electron or nuclear spins, within an external magnetic field. By applying precisely tuned electromagnetic pulses, QRI detects subtle energy transitions linked to specific molecular interactions. Techniques like quantum entanglement amplify weak signals, improving detection limits and enabling imaging of single molecules, protein structures, and even atomic spin distributions, surpassing traditional MRI constraints. QRI includes early disease detection, molecular diagnostics, and neuroimaging with unprecedented precision. As research progresses, QRI is expected to revolutionize medical diagnostics, materials science, and fundamental quantum research. This emerging technology has potential applications in multiple fields like materials science, security, environmental monitoring, astrophysics, and medicine.

5.3 Advance Technologies in Quantum Resonance Imaging

QRI advances materials science by visualizing quantum states in superconductors, semiconductors, and nanomaterials, aiding in zero-resistance conductivity for quantum computing [25, 26]. In nanoelectronics, nanoparticle-based MRI contrast agents improve imaging sensitivity, with T1 agents enhancing resolution and T2 agents aiding molecular imaging [27]. Superparamagnetic iron oxides (SPIONs) are FDA-approved for imaging liver, spleen, and bone marrow. Surface modifications, such as ligand exchange and micelle encapsulation, enhance biocompatibility. Activatable MRI contrast agents respond to stimuli like pH and RF saturation, improving specificity [28]. Nanostructure-based CEST agents further refine MRI applications through tunable RF pulse activation. QRI enhances security by detecting quantum states in communication channels, enabling hack-proof data transfer through quantum key distribution (QKD). It supports military-grade cybersecurity by securing communication via quantum protocols. QKD ensures secure key generation, error correction, and privacy amplification, protecting against active adversaries in cyber warfare. QRI enhances surveillance by detecting concealed weapons and explosives and aids forensic investigations by analyzing microscopic evidence.

It assists in locating missing individuals, even in deep water and mine shafts, and tracks decomposition plumes [29]. Its sensitivity and specificity make it valuable for identifying human remains and forensic analysis [30]. QRI detects trace pollutants in air, water, and soil, aiding environmental monitoring [31]. Brain imaging studies link air pollution to reduced cortical thickness, with PM10 and NO_2 affecting the frontal, temporal, and parietal lobes. Pollutants also impact subcortical structures, highlighting environmental toxins' neurological effects and climate science applications [32]. Figure 5.4 simply shows the other beneficial applications of QRI in different developmental areas along with medical imaging in effective mode.

Agriculture uses QRI to analyze soil quality, plant metabolism, and agricultural diseases by detecting biochemical alterations. It improves precision agriculture by optimizing production projections and minimizing pesticide use. Advances in QRI technology are making plant and food imaging more accessible, overcoming technical difficulties and allowing for larger applications beyond specialized facilities for practical usage [33]. QRI supports

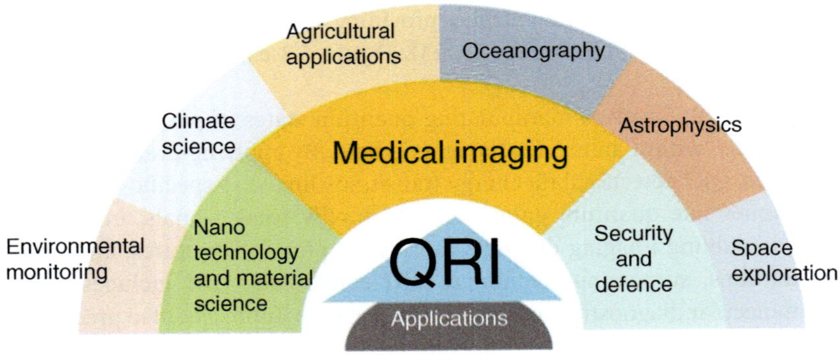

Figure 5.4 Distinct applicable mode of QRI.

oceanography by monitoring water chemistry, microplastic contamination, and deep-sea mineral deposits. MRI and MRS are essential tools in marine biology for evaluating physiological adaptations in aquatic organisms [34]. In astrophysics, QRI detects quantum interactions in space radiation and helps with telescopic imaging of dark matter and black holes. It could also detect biosignatures on exoplanets. However, obstacles include high costs, integration difficulty, and ethical concerns. QRI has vast potential in space exploration. Portable QRI scanners could monitor astronaut health in deep space, detect alien biosignatures, and study microgravity's effects on human cells. It may also analyze exoplanet atmospheres and help detect dark matter interactions at subatomic levels, advancing astrophysics and cosmology [8]. Future potential includes the integration of AI and quantum computing, portability through miniaturization, and applications in medicinal imaging.

5.4 QRI in Medical and Healthcare Applications

Medical imaging, including MRI, CT, PET, and ultrasound, is essential in modern healthcare but struggles with increasing data complexity. Conventional computing faces limitations in speed and scalability when processing large datasets. Quantum computing, leveraging qubits that exist in multiple states offers a revolutionary approach to handling complex imaging tasks efficiently. This emerging technology could enhance image reconstruction, detect subtle biomarkers, and improve diagnostic accuracy [35]. By integrating quantum computing with machine learning, healthcare could achieve earlier diagnoses, reduced costs, and better patient outcomes. Despite challenges, this fusion of quantum computing and medical imaging has the potential to transform healthcare. Figure 5.5 schematically explains the power of QRI in medical sectors.

5.4.1 Disease Diagnosis and Early Detection

Quantum sensing technology enhances disease prevention, diagnosis, and treatment by detecting weak electromagnetic signals from the body. The Copenhagen Center for Biomedical Quantum Sensing, funded by the Novo Nordisk Foundation, aims to advance these technologies. Researchers from leading institutions will develop ultrasensitive

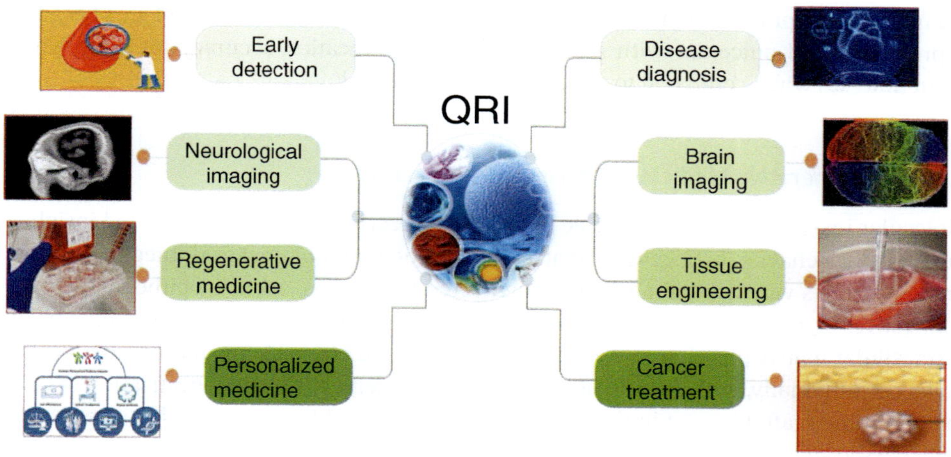

Figure 5.5 Supremacy of quantum resonance imaging in medical imaging.

quantum sensors to detect diseases earlier and with greater precision [35]. These sensors, leveraging quantum superpositions and entanglement, could surpass standard diagnostic methods. Applications include detecting heart and brain anomalies, metabolic disorders, and iron deficiencies. Collaborating with hospitals, researchers aim to translate these advancements into clinical use, improving global healthcare and disease prevention on an unprecedented scale.

Quantum sensors are revolutionizing healthcare by enabling early disease detection, precise biomarker monitoring, and improved diagnostics. Their exceptional sensitivity allows them to detect minute biological changes, surpassing conventional sensors in accuracy and reliability. Applications include quantum dots for bioimaging, which enhance protein tracking for disease detection, and quantum-enhanced magnetocardiography using SQUIDs for high-resolution cardiac imaging [36]. These advancements pave the way for more timely interventions and effective treatments. By offering unprecedented precision in diagnostics, quantum sensors are transforming medical imaging, improving patient outcomes, and redefining the landscape of modern healthcare with cutting-edge technology.

5.4.2 Brain and Neurological Imaging

Quantum learning models, such as AlexNet–quantum, are transforming brain disorder diagnostics by leveraging quantum circuits and dimensionality reduction techniques to analyze neuroimaging data [37]. This hybrid quantum-classical approach enhances classification accuracy by extracting rich feature vectors from high-dimensional MRI scans, outperforming traditional methods. The model is further validated through quantum simulators, demonstrating its feasibility in healthcare. By improving diagnostic precision, this approach provides a viable solution for early detection of brain disorders. Future advancements include implementing quantum transfer learning on real quantum hardware and extending hybrid quantum-classical learning for multi-classification tasks in computer vision, further revolutionizing medical diagnostics. The proposed method would aid in the development of realistic and practicable solutions in the healthcare area because it produces good outcomes with much enhanced classification accuracy for identifying brain illnesses when compared to the conventional network.

5.4.3 Regenerative Medicine and Tissue Engineering

Molecular imaging plays a crucial role in tissue engineering by providing detailed insights into cellular behavior, extracellular matrix composition, and biomaterial integration. The imaging targets vary depending on the engineered tissue type, such as epithelial, neural, connective, or muscle tissues. Key molecular targets include gene expression related to extracellular matrix proteins, protein concentrations, and biomechanical properties of the matrix. Additionally, molecular imaging can track the degradation and localization of biomaterials and scaffolds used in tissue engineering. This enables researchers to assess how well a tissue construct integrates and functions within the body.

Molecular imaging, in addition to structural imaging, aids in the evaluation of biological pathways and the tissue microenvironment, which is crucial for understanding tissue regeneration under various physiological situations, such as acute vs chronic inflammation. This approach enables researchers to track biological responses and detect potential failure sites in created tissues. By combining molecular imaging with conventional anatomical and functional imaging, such as imaging blood vessels, blood flow, and angiogenesis receptors, scientists can gain a full understanding of tissue growth and vascularization [38]. This integrated approach improves the evaluation of engineered tissues, ensuring their effectiveness in clinical applications and paving the way for advanced regenerative medicine solutions.

5.4.4 Personalized Medicine

The primary objective of cell- and tissue-based therapies is to restore damaged or dysfunctional tissues, but many of these approaches have yet to be successfully translated into clinical practice. Advanced molecular imaging strategies can enhance regenerative medicine by improving assessment techniques and facilitating the transition from research to patient applications. Traditionally, regenerative status is evaluated using tissue sections, which require endpoint assays involving animal sacrifice in preclinical models or biopsies in both preclinical and clinical studies. However, these methods have several limitations, including their invasive nature, lack of quantitative analysis, and potential loss of critical information during tissue processing. To overcome these challenges, advanced imaging techniques can provide real-time, noninvasive, and quantitative assessments of tissue regeneration. Molecular imaging can track cellular behavior, extracellular matrix remodeling, and biomaterial integration over time, allowing researchers to monitor regenerative progress without the need for destructive sampling. Additionally, imaging methods can help detect variations between subjects, preserving individualized data that would otherwise be lost in traditional biopsy-based analysis. By offering detailed insights into tissue dynamics and healing processes, molecular imaging enhances the reliability and precision of cell- and tissue-based studies, ultimately accelerating the development of effective regenerative therapies for clinical use.

5.4.5 Cancer Treatment and Monitoring

Imaging cancer metabolism is crucial for understanding tumor growth and malignancy, as cancer cells exhibit altered biochemical pathways such as glycolysis, the pentose phosphate pathway (PPP), and the Tricarboxylic Acid Cycle (TCA) cycle. Monitoring these metabolic changes *in vivo* significantly enhances cancer diagnosis and treatment planning [13]. A key discovery in this field was made by Otto Warburg, who observed that cancer cells consume more glucose and produce higher levels of lactate compared to healthy tissues. The finding of the Warburg consequence indicated that tumor metabolism differs fundamentally from normal tissue metabolism, prompting substantial research into metabolic imaging. Based on Warburg's results, several magnetic resonance (MR) techniques have been developed to visualize and quantify metabolic changes in cancer tissues in real time [39].

These imaging methods help detect metabolic shifts, track tumor progression, and assess responses to therapies. PET using fluorodeoxyglucose (FDG-PET) is one such widely used approach that highlights glucose uptake in tumors, providing critical insights into cancer aggressiveness. Advanced MR-based techniques, including hyperpolarized carbon-13 MR spectroscopy, further enhance metabolic imaging by offering real-time visualization of key metabolic pathways. These innovations enable more precise and personalized cancer treatment strategies, improving patient outcomes [40].

5.5 Advantages of QRI

QRI offers various benefits over standard imaging modalities, making it a promising development in medical diagnosis. One of its most notable advantages is better resolution, which enables molecular and atomic-level imaging that significantly outperforms the millimeter-scale resolution of MRI and CT scans. These ultra-precise imaging capabilities allow for the discovery of minute abnormalities, resulting in earlier illness diagnosis and more tailored treatment strategies. Another significant advantage is that QRI is noninvasive and radiation-free, whereas CT and PET scans expose patients to ionizing radiation. This makes QRI safer for repeated use, especially in fragile groups including children, pregnant women, and patients who need frequent monitoring [41]. Besides, QRI has quick imaging capabilities, which may reduce scan time while retaining great picture quality. Faster imaging improves patient comfort while reducing the possibility of motion artifacts, resulting in more accurate and trustworthy diagnostic outcomes. Moreover, QRI's integration with quantum computing and artificial intelligence enhances data processing and real-time image reconstruction, hence improving diagnostic accuracy. QRI, with its unprecedented sensitivity and precision, has the potential to revolutionize medical imaging by allowing for earlier illness detection, advanced neuroimaging, and more effective monitoring of therapy responses, ultimately leading to better patient outcomes.

5.6 Challenges of QRI

Despite its tremendous potential, QRI faces a number of hurdles that must be overcome before it can be broadly adopted in clinical settings. One of the main impediments is technological constraints. Current quantum technologies, such as quantum sensors, entanglement-based imaging, and quantum coherence techniques, are still in their early phases. Refining these technologies to provide reliable, reproducible, and high-quality imaging suited for medical diagnosis necessitates considerable advances in quantum hardware and software. Additionally, cost and accessibility are significant barriers. The large initial expenditure required for developing, manufacturing, and maintaining QRI systems may limit their adoption, particularly in resource-constrained healthcare settings. Unlike conventional MRI, which is already extensively integrated into hospitals, QRI would necessitate a whole new infrastructure, trained personnel, and continuing support, raising expenses even higher. Another significant problem is regulatory compliance. Before QRI can be utilized in clinical practice, it must be rigorously validated to ensure medical safety. To ensure accuracy, reliability, and patient safety, significant testing, clinical trials, and regulatory approval from agencies such as the FDA and EMA will be required. Addressing these issues will be critical to the successful transition of QRI from experimental research to a functional, widely available diagnostic tool.

5.7 Future Prospects of QRI

QRI holds enormous potential for revolutionizing medical diagnostics by providing unprecedented precision, sensitivity, and safety. QRI has the potential to become a routine tool in healthcare as quantum technologies advance, allowing for earlier disease diagnosis, more accurate diagnostics, and highly personalized treatment strategies. One of the most important future initiatives is to improve technological feasibility, which includes refining quantum sensors, improving quantum coherence stability, and incorporating quantum computing for real-time imaging processing. Researchers are continuously striving to address scalability issues, guaranteeing that QRI can be adapted for widespread clinical application while retaining excellent accuracy and efficiency. Furthermore, lowering costs and increasing accessibility will be critical for its adoption in mainstream healthcare, necessitating cooperation among quantum physicists, engineers, and medical specialists to produce cost-effective solutions. Another key consideration is regulatory approval and clinical validation, which will necessitate extensive testing and clinical trials to establish QRI as a safe and trustworthy imaging technique. Finally, QRI has the potential to revolutionize medical imaging by enabling molecular-level diagnostics, improving early illness identification, and improving patient outcome. As research advances, QRI may change the future of medical diagnostics, ushering in a new era of precision medicine.

5.8 Conclusion

QRI has enormous potential to transform medical diagnostics by incorporating quantum mechanics into imaging techniques. It provides exceptional precision, increased sensitivity, and noninvasive, radiation-free imaging capabilities. QRI could dramatically improve early illness detection, personalized treatment, and neuroimaging by identifying small molecular and cellular changes that typical imaging approaches miss. Quantum computing accelerates data processing and image reconstruction, allowing for real-time, high-resolution imaging. Despite its potential, hurdles such as technological refinement, high costs, and regulatory permissions must be overcome before widespread clinical use occurs. Continued research and collaboration among quantum physicists, medical practitioners, and regulatory organizations will be critical in translating QRI's theoretical advances to practical healthcare applications. As quantum technologies advance, QRI has the potential to transform medical imaging, providing safer, faster, and further precise imaging solutions that could enhance global medical outcomes.

References

1 Israel, Y., Tenne, R., Oron, D., and Silberberg, Y. (2017). Quantum correlation enhanced super-resolution localization microscopy enabled by a fibre bundle camera. *Nature Communications* 8 (1): 1–5. https://doi.org/10.1038/ncomms14786.

2 Rong, J.; Liu, Y. Advances in medical imaging techniques. *BMC Methods* 2024, 1, 10. https://doi.org/10.1186/s44330-024-00010-7.

3 Soto, J.A., Cordova, T., and Sosa, M. (2017). Quantum-mechanical aspects of magnetic resonance imaging. *Revista Mexicana de Física E* 63: 48–55.

4 Geva, T. (2006). Magnetic resonance imaging: historical perspective. *Journal of Cardiovascular Magnetic Resonance* 8 (4): 573–580. https://doi.org/10.1080/10976640600755302. PMID: 16869310.

5 Kabasawa, H. (2022). MR Imaging in the 21st Century: Technical Innovation over the First Two Decades. *Magnetic Resonance in Medical Sciences* 21 (1): 71–82. https://doi.org/10.2463/mrms.rev.2021-0011. PMID: 33867419; PMCID: PMC9199974.

6 Cai, K., Haris, M., Kogan, A.F. et al. (2012). Magnetic Resonance Imaging of Glutamate. *Nature Medicine.* https://doi.org/10.1038/nm.2615.

7 Terreno, E., Castelli, D.D., Viale, A., and Aime, S. (2010). Challenges for Molecular Magnetic Resonance Imaging. *Chemical Reviews* 3019–3042.

8 Jun, Y., Lee, J., and Cheon, J. (2008). Chemical Design of Nanoparticle Probes for High Performance Magnetic Resonance Imaging. *Angewandte Chemie International Edition* 47 (28): 5122–5133. https://doi.org/10.1002/anie.200701674.

9 Tirotta, I., Dichiarante, V., Pigliacelli, C. et al. (2015). ^{19}F Magnetic Resonance Imaging (MRI): From Design of Materials Clinical Applications. *Chemical Reviews* 115 (2): 1106–1129.

10 Pellico, J., Ellis, C.M., and Davis, J.J. (2009). Nanoparticle-Based Paramagnetic Contrast Agents for Magnetic Resonance Imaging. *Contrast Media & Molecular Imaging* https://doi.org/10.1155/2019/1845637.

11 Kozlowska, D., Foran, P., MacMahon, P. et al. (2009). Molecular and magnetic resonance imaging: The value of immunoliposomes. *Advanced Drug Delivery Reviews* 61 (15): 1402–1411. https://doi.org/10.1016/j.addr.2009.09.003. PMID: 19796661.

12 Haris, M., Yadav, S.K., Rizwan, A. et al. (2015).Molecular magnetic resonance imaging in cancer. *Journal of Translational Medicine* 23 (13): 313. https://doi.org/10.1186/s12967-015-0659-x. PMID: 26394751; PMCID:PMC4579835.

13 Louie, A.Y., Huber, M.M., Ahrens, E.T. et al. (2000). In vivo visualization of gene expression using magnetic resonance imaging. *Nature Biotechnology* 18 (3):321–325. https://doi.org/10.1038/73780.

14 Ferrero, V., Fiorina, E., Morrocchi, M. et al. (2018). Online proton therapy monitoring: clinical test of a Silicon-photodetector-based in-beam PET. *Scientific Reports* 8 (1): 1–8. https://doi.org/10.1038/s41598-018-22325-6.

15 Ozcan, H.N., Oğguz, B., Haliloğglu, M. et al. (2015). Imaging patterns of fatty liver in pediatric patients. *Diagnostic and Interventional Radiology* 21 (4): 355–360. https://doi.org/10.5152/dir.2015.14505. PMID: 26027765; PMCID:PMC4498433.

16 Heidemann, R.M., Ozsarlak, O., Parizel, P.M. et al. (2003). A brief review of parallel magnetic resonance imaging. *European Radiology* 13 (10):2323–2337. https://doi.org/10.1007/s00330-003-1992-7. PMID: 12942278.

17 Yu, C.J., von Kugelgen, S., Laorenza, D.W., and Freedman, D.E. (2021). A Molecular Approach to Quantum Sensing. *ACS CentralScience* 7 (5): 712–723. https://doi.org/10.1021/acscentsci.0c00737. PMID: 34079892;PMCID: PMC8161477.

18 Zhang, S., Bian, K., and Jiang, Y. (2023). Perspective: nanoscale electric sensing and imaging based on quantum sensors. *Quantum Frontiers* 2: 19. https://doi.org/10.1007/s44214-023-00045-x.

19 Burgler, B., Sjolander, T.F., Brinza, O. et al. (2023). All-optical nuclear quantum sensing using nitrogen-vacancy centers in diamond. *npj Quantum Information* 9 (1): 1–7. https://doi.org/10.1038/s41534-023-00724-6.

20 Chen, S., Li, W., Zheng, X. et al. (2022). Immunomagnetic microscopy of tumor tissues using quantum sensors in diamond. *National Academy of Sciences of the UnitedStates of America* 119 (5): e2118876119. https://doi.org/10.1073/pnas.2118876119.

21 Willke, P., Yang, K., Bae, Y. et al. (2019). Magnetic resonance imaging of single atomson a surface. *Nature Physics* 15 (10): 1005–1010. https://doi.org/10.1038/s41567-019-0573-x.

22 Chow, J.C.L. (2024). *Quantum Computing in Medicine. Medical Sciences* 12 (4): 67. https://doi.org/10.3390/medsci12040067.

23 Aslam, N., Zhou, H., Urbach, E.K. et al. (2023). Quantum sensors for biomedical applications. *Nature Reviews Physics* 5 (3): 157–169. https://doi.org/10.1038/s42254-023-00558-3.

24 Zrazhevskiy, P., Sena, M., and Gao, X. (2010). Designing multifunctional quantum dots for bioimaging, detection, and drug delivery. *Chemical Society Reviews* 39 (11):4326–4354. https://doi.org/10.1039/b915139g.

25 Feder, T. (2024). New center for quantum sensing focuses on medical applications. *Physics Today* 77 (5): 24–26. https://doi.org/10.1063/pt.esph.sudu.

26 Qiu, Z., Hamo, A., Vool, U. et al. (2022). Nanoscale electric field imaging with an ambient scanning quantum sensor microscope. *npj Quantum Information* 8 (1): 1–7. https://doi.org/10.1038/s41534-022-00622-3.

27 Estelrich, J., Sanchez-Martin, M.J., and Busquets, M.A. (2015). Nanoparticles in magnetic resonance imaging: from simple to dual contrast agents. *International Journal ofNanomedicine* 6 (10): 1727–1741. https://doi.org/10.2147/IJN.S76501.

28 Wahajuddin, A. (2012). Superparamagnetic iron oxide nanoparticles: magnetic nanoplatforms as drug carriers. *International Journal of Nanomedicine* 7: 3445–3471. https://doi.org/10.2147/IJN.S30320.

29 Solaiman, S. (2025). Enhancing Quantum Key Distribution Security Through Hybrid Protocol Integration. *Symmetry* 17 (3): 458. https://doi.org/10.3390/sym17030458.

30 Ubelaker, D.H., Shamlou, A., and Kunkle, A. (2018). Contributions of forensic anthropology to positive scientific identification: a critical Review. *Forensic Sciences Research* 4 (1): 45–50. https://doi.org/10.1080/20961790.2018.1523704.

31 Cedillo-Pozos, A., Ternovoy, S.K., and Roldan-Valadez, E. (2020). *Imaging methods used in the assessment of environmental disease networks: a brief review for clinicians. Insights Into Imaging* 11 (1): 18. https://doi.org/10.1186/s13244-019-0814-7.

32 Ko, J., Sohn, J., Noh, Y. et al. (2024). Effects of Ambient Air Pollution on Brain Cortical Thickness and Subcortical Volume: A Longitudinal Neuroimaging Study. *Neuroepidemiology* 30: 1–11. https://doi.org/10.1159/000539467.

33 Tom, C.N., Edet, J.A., Erokare, T.E. et al. (2024). Precision Agriculture for Maximizing Crop Production. *Iconic Research and Engineering* Journals 8 (2): 384–392.

34 Bock, C., Wermter, F.C., and Mintenbeck, K. (2017). MRI and MRS on preserved samples as a tool in fish ecology. *Magnetic Resonance Imaging* 38:39–46. https://doi.org/10.1016/j.mri.2016.12.017.

35 Ajlouni, N., Ozyavaş, A., Takaoğglu, M. et al. (2023). Medical image diagnosis based on adaptive Hybrid Quantum CNN. *BMC Medical Imaging* 23 (1): 126. https://doi.org/10.1186/s12880-023-01084-5.

36 Le, N. and Kim, K. (2023). Current Advances in the Biomedical Applications of Quantum Dots: Promises and Challenges. *International Journal of Molecular Sciences* 24 (16): 12682. https://doi.org/10.3390/ijms241612682.

37 Alsharabi, N., Shahwar, T., Rehman, A.U., and Alharbi, Y. (2022). Implementing Magnetic Resonance Imaging Brain Disorder Classification via AlexNet–Quantum Learning. *Mathematics* 11 (2):376. https://doi.org/10.3390/math11020376.

38 Orka, N.A., Awal, M.A., Lio, P. et al. (2025). Quantum deep learning in neuroinformatics: a systematic review. *Artificial Intelligence Review* 58: 134. https://doi.org/10.1007/s10462-025-11136-7.

39 Liberti, M.V. and Locasale, J.W. (2016). The *Warburg Effect: How Does it Benefit Cancer Cells? Trends in Biochemical Sciences* 41 (3): 211–218. https://doi.org/10.1016/j.tibs.2015.12.001.

40 Zhu, A., Lee, D., and Shim, H. (2011). Metabolic positron emission tomography imaging in cancer detection and therapy response. *Seminars in Oncology* 38 (1): 55–69. https://doi.org/10.1053/j.seminoncol.2010.11.012.

41 Shams, M., Choudhari, J., Reyes, K. et al. (2023). The Quantum-Medical Nexus: Understanding the Impact of Quantum Technologies on Healthcare. *Cureus* 15 (10): e48077. https://doi.org/10.7759/cureus.48077.

6

Applications of Quantum Biosensing in Oncology

Hülya Silah[1] *and Bengi Uslu*[2]

[1] *Faculty of Science, Department of Chemistry, Bilecik Seyh Edebali University, Bilecik, Türkiye*
[2] *Faculty of Pharmacy, Department of Analytical Chemistry, Ankara University, Ankara, Türkiye*

6.1 Introduction

Cancer remains an utmost public health anxiety owing to its high death rate, prompting intense scientific research. Cancer, described by the uncontrolled dissemination and proliferation of abnormal cells, continues to be the second most usual reason for mortality worldwide, following cardiovascular diseases [1–3]. Cancer can emerge in thereabout two hundred different kinds, such as cancers of the breast, lung, ovary, bladder, colon, bone, prostate, and neuron [4]. However, the increasing prevalence of life-threatening diseases such as cancer has intensified the requirement for early-stage and highly sensitive biomarker monitoring. This expanding demand is accelerating progressions in detection sensitivity through innovative biosensing strategies and signal amplification methods [5]. These cancer biomarkers, also, can be utilized for diagnosis and prognosis of disease and determination of the cancer therapy effectiveness [6].

Within this broad research area, the identification of various biomarkers such as protein biomarkers has a supreme role, especially in advancing early diagnostic strategies and comprehending the mechanisms underlying pathological progress [1]. Accurate, sensitive, and selective recognition of different biomarkers is a crucial act in biomedical diagnostic procedures, proteomic analyses, and systems biology investigations [5].

The early-stage detection of cancer disease by courtesy of the recognition of biomarkers existent in human biological fluid samples such as urine, blood, serum, plasma, sputum, or stool constitutes a basic approach in oncological diagnostic studies. This primary step is vital for simplifying targeted and timely therapeutic procedures, diminishing the risk of metastasis, and finally contributing to the decline of cancer-related mortality rates [7, 8].

Cancer biomarkers that are biological molecules found in blood, other body tissues, or fluids play an essential role. These biomarkers can mark normal or pathogenic processes or pharmacologic responses to therapeutic interventions [9]. Detecting and monitoring cancer biomarkers such as alpha-fetoprotein (AFP), carcinoembryonic antigen (CEA), cancer antigen 125 (CA125), tissue polypeptide antigen (TPA), human epidermal growth factor receptor 2 (HER2-ECD), and prostate-specific antigen (PSA) makes it possible for clinicians

Quantum Biosensing in Medical Diagnostics, First Edition. Edited by J. G. Manjunatha.

to acquire early diagnoses, define prognoses, monitor disease recurrence, and personalize treatment strategies. Moreover, advancements in molecular biology, genetics, medicine, and nanotechnology have facilitated the refinement of highly selective and sensitive biosensor systems for real-time, noninvasive biomarker detection, amplifying their clinical practicality [10, 11]. Consequently, integrating biomarker monitoring into oncological experience is necessary for improving diagnostic precision and therapeutic results.

Biosensor systems are analytical platforms that couple a biological recognition constituent—such as antibodies, enzymes, cells, or nucleic acids—with a physicochemical transducer to create a measurable signal corresponding to the existence or concentration of a specific target analyte species. These apparatuses are designed to detect/determine biological incidents with high sensitivity and specificity, often in real-time, and are capable of transforming a biological interplay into an electrical, thermal, or optical signal depending on the transduction principle operated. With their potential for portability, miniaturization, and rapid response, biosensor systems have become necessary in medical diagnostics, environmental surveillance, food quality control, and biotechnological studies. In the designing of a biosensor, it is indispensable to prioritize attributes such as scalability and robustness, ensuring that implement performance remains stable under varying environmental conditions and is minimally thrilled by user handling. Besides, the fabrication procedures should be cost-effective, straightforward, and able to deliver high reproducibility across plural batches. Achieving a low detection limit (LOD) is equally critical, as it enables the identification of target analytes at trace concentration levels. Additionally, the biosensor must represent high sensitivity, allowing for clear distinction between closely spaced concentration values through discernible signal variations [12].

Recently, there has been growing attention in the use of electrochemical biosensor systems for monitoring cancer biomarkers because of their fast detection abilities, low sample volume requirements, and exceptional sensitivity. These biosensor systems can detect a variety of targets, including circulating tumor cells, protein-based antigens, metabolites, carbohydrate-related markers, and nucleic acid sequences [13, 14].

Developments in nanomaterials, such as carbon-based nanostructures, quantum dots, and metal nanoparticles, have appreciably improved biosensor efficiency by increasing signal strength, decreasing detection limits, and allowing complicated analysis [15]. Moreover, the combination of biosensor systems into point-of-care (POC) systems has paved the way for decentralized and personalized healthcare resolutions.

Quantum dot (QD) materials are semiconductor nanocrystals with nonpareil size-dependent optical and electronic features, such as tunable and narrow emission spectra, high quantum efficiencies, and conspicuous photostability [16, 17]. These properties have placed QDs as potent vehicles in biomedical enforcements, especially in the area of oncological studies. The surface of QDs can be easily modified with various biomolecules such as peptides, antibodies, or nucleic acids to authorize specific binding to tumor or cancer cell microenvironment targets. In oncological explorations, QDs have been put in application for early cancer diagnosis, in vivo tumor imaging, and also real-time monitoring of drug delivery, providing high sensitivity and multiplexing abilities that surpass conventional fluorescent dyes [18]. These supremacies make QDs hopeful candidates for developing molecular diagnostics, image-guided surgery, and personified cancer therapy.

The present chapter has addressed the quickly developing research about electrochemical biosensor-based quantum materials in the analytical and sensing applications area of oncological studies. Quantum-based materials can be utilized as an alternate to various electrode materials owing to their ability to eliminate the restrictions of them, like high cost, low stability, and hard storage conditions. Although quantum-based materials in oncological biosensor systems are still in the early stages of exploration, utilization of them has been investigated considerably in varied determination and treatment techniques for many cancer markers.

6.2 Quantum Dots and Their Unique Physicochemical Properties

QDs are nanoscale semiconductor crystals that surpass conventional fluorophore materials in biological implementations because of their extraordinary optical features. Their wide excitation spectra, sharp emission peaks, and size-tunable fluorescence make them especially well-suited for sensitive biosensing systems and multiplexed detection strategies [19].

QD materials are nanoscale semiconductor crystals that display discrete energy levels on account of quantum confinement, causing inimitable electronic and optical features that are extremely interdependent on their composition and size. Their individual characteristics—including broad absorption spectra, high quantum yield, tunable and narrow emission wavelengths, and superb photostability—have authorized their combination into a wide variety of technological and scientific fields. In biomedical studies, QDs have remarkably developed diagnostic and bioimaging methods, permitting multiplexed identification of plural analytes within complicated biological environments [16, 20]. Because of their extra brightness and resistance to photobleaching compared to conventional fluorophore materials, QDs are widely employed mainly in fluorescence microscopy, real-time cellular tracking, and immunoassays [21, 22]. In biosensing implementations, QDs are frequently functionalized with biological molecules such as DNA, antibodies, or aptamers to selectively identify disease biomarkers, making possible ultrasensitive and specific detection in infectious disease screening, cancer diagnostics, and environmental monitoring [23, 24]. Moreover, QD-based Förster resonance energy transfer (FRET) systems have been designed for dynamic biosensing of molecular interplays and enzyme activities at the nanoscale level [25]. Beyond life sciences, QDs are getting momentum in optoelectronic and energy-related enforcements, including quantum dot light-emitting diodes (QD-LEDs), photodetectors, lasers, and third-generation solar cells, through the agency of their high absorption coefficients, tunable bandgap, and efficient charge carrier dynamics [26, 27]. Latest progressions have also discovered their potency in secure communication and quantum computing systems on account of their coherent spin features and single-photon emission abilities [28]. In light of all this information, it can be said that QDs symbolize one of the most transformative and versatile groups of nanomaterials, with comprehensive multidisciplinary implications ranging from clinical diagnostics to energy harvesting and quantum data science.

Multicolor QDs have emerged as influential vehicles in biomedical imaging, proposing the ability to simultaneously identify many cancer biomarkers, ranging from tens to hundreds, in blood-based diagnostic analysis. These technological evolutions highlight the critical role of precise and early cancer detection in enhancing therapeutic accomplishment and alleviating the global health impact of cancer [29, 30].

QDs expose diverse features that can be differentiated based on determinants such as their chemical composition and structural configuration. The chemical makeup of QDs comprises core-only structures, core–shell architectures, and alloyed variants. Materials such as cadmium, lead, and zinc, as well as selenide and sulfide compounds, are generally exploited in biosensing and fluorescence imaging implementations through the agency of their impressive luminescent and electrochemical characteristics. Nevertheless, despite these superior features, the leaching of toxic metal ions from such QDs gives rise to significant environmental and biological risks. In response to these apprehensions, research efforts have focalized on developing biocompatible, heavy metal-free QDs. Alternatives such as carbon, germanium, graphene, and silica-based QDs have appeared as safer options. Among these, graphene and carbon QDs have garnered substantial recognition in sensing technologies because of their low toxicity, superior electrical conductivity, enhanced solubility across various solvents, unusual chemical stability, abundant functionalization sites at their edges, large surface area, and versatile feasibility [31, 32]. QDs are nanoscale semiconductor crystals, usually constituted of a nucleus made from group II–VI or III–V elements, surrounded by a shell of another semiconductor material. Their diameters normally range from 2 to 10 nm, encompassing thereabouts 200 to 10,000 atoms, placing their size in a parallel range to that of large biomolecules such as proteins [33].

Graphene quantum dots (GQDs), a coming-into-view member of the carbon class, have taken remarkable notice in biomedical application areas through the agency of their extraordinary mechanical, electrical, and chemical traits, along with speedy electron mobility and marvelous electrochemical behavior. These quantum dots, which are considered zero-dimensional derivatives of graphene sheets, combine the extraordinary characteristics of nanomaterials with the intrinsic advantages of graphene-based structures. Recent research shows that integrating these quantum dots into nanocomposite material systems serves up an encouraging strategy for influential biomolecule monitoring through novel biosensor systems [6, 34, 35]. Especially, these nanomaterials have recently been utilized in the designing of electrochemical biosensors, thanks to their distinctive physicochemical features such as extensive surface-to-volume ratio, minimal toxicity, excellent biocompatibility, and simplistic synthesis techniques. Moreover, GQDs display electrochemical attributes akin to graphene material, while their plethora of hydrophilic edges and functional groups with oxygen (–COOH, –OH) considerably augments their capability for immobilizing immunoreagents on the surface of them [36].

QDs, including semiconductor, graphene, and carbon-based groups, each put forward distinct advantages and limitations. Semiconductor QDs stand out for their strong fluorescence efficiency, high photostability, and resistance to degradation; even so, their possible toxicity and poor clearance in biological systems limit their in vivo use. Carbon-based QDs are environmentally friendly, cost-effective, and uncomplicated to synthesize, but they frequently require harsh reagents and display broad size variation. Graphene-based QDs combine good stability and rapid synthesis with eco-friendly properties, yet they face safety concerns from strong oxidants and difficulties in achieving high-yield production [37].

In contrast to traditional semiconductor and metal-based QDs, carbon dots have recently emerged as a discrete category of carbon nanomaterials, drawing prominent research interest. Carbon-based nanomaterials are great for modification of electrode surfaces owing to the electrochemical activity of them. Carbon dots can separate the created charge in the interface; thus, they can supply an electron transfer pathway to the surface of the electrode. The electrochemical (electron transfer) properties of carbon dots depend on the interplay between the carbon core, functional groups, and doped heteroatoms [38].

6.3 Quantum Dots in Oncology: Advanced Applications in Electrochemical Biosensing

Electrochemical biosensor systems supply prominent usefulness when compared to conventional diagnostic and screening platforms, with the inclusion of fast reply times, high sensitivity and specificity, portability, uncomplicated fabrication, minimal training requirements, and conformity for on-site or home-based testing. Their low signal interference, user-friendly nature, and relevance with miniaturized analyzing systems make them supreme applicants for POC implementations [39].

Tumor markers are molecular indicators that can be detected in tissue and blood samples, and raised levels of them are frequently associated with the development of certain cancer species. Attaining early diagnosis of cancer seriously depends on the highly sensitive detection of these disease biomarkers [40]. Carcinoembryonic antigen (CEA) is a heavily glycosylated protein with an approximate molecular mass of 200 kDa and has an important part as a generally utilized tumor biomarker. Its elevated expression has been noticed in diverse malignancies, including gastric, colorectal, breast, pancreatic, and hepatic cancers [41, 42]. MicroRNAs, or miRNAs, are short, single-stranded, non-coding RNA molecules that arrange numerous critical biological processes, including cell differentiation, proliferation, metabolic pathways, and programmed cell death. Dysregulated expression of miRNAs in biological fluids has been closely linked to the initiation and progression of many kinds of cancer, such as colorectal, pancreatic, breast, and lung cancers [43]. AFP is a well-known tumor biomarker, and raised levels in the human body are often associated with the progress of malignancies such as primary hepatocellular carcinoma, yolk sac tumors (both gonadal and extragonadal), testicular neoplasms, and gastric cancer [44]. Currently, PSA is recognized as the most dependable serum biomarker for diagnosing prostate cancer. In healthy males, levels of serum PSA typically range between 0.1 and 4 ng/mL. Elevated PSA concentration levels are frequently indicative of pathological conditions such as prostate disorders or physical trauma to the gland [45].

Among diversified transduction mechanisms, electrochemical biosensing strategies have received noteworthy attention in oncology-related enforcements, where accurate and early biomarker detection is censorious for effective diagnosis and treatment monitoring. Particularly, the integration of nanomaterials such as QDs into electrode architecture has further amplified the electroanalytical performance of these sensor systems by raising electron transfer kinetics, ensuring large surface areas for biomolecule immobilization, and serving up unique signal amplification abilities [46]. To illustrate the advancements in this area, Table 6.1 summarizes a selection of recent electrochemical biosensor studies focusing on

Table 6.1 Some selected electrochemical sensor studies for the determination of biomarkers.

Analyte	Probe	Linear range	Limit of detection	Interferences	Applications	References
AFP	Ab_1-HRP-Ab_2-AuNPs-streptavidin-labeled QDS	0.5–30.0 pg/mL	0.2 pg/mL	CEA, CA125, PSA	Calf serum samples	[5]
AFP	HRP/ZnSe/Azure I/nanoAu/PEDOT/Pt	5×10^{-5}–250 ng/mL	1.1 fg/mL	HBsAg, BSA, glucose, CEA, L-cysteine	Human serum samples	[47]
AFP	PAMAM-CDs/Au	100 fg/mL–100 ng/mL	0.025 pg/mL	CEA, BSA, IgG, Glu, UA, AA	Clinical serum samples	[38]
CA15-3	CysA/Au NSs/GQDs/GCE	0.16–125 U/mL	0.11 U/mL	BSA, PSA, p53, HSA	Cancer cell lysate samples	[8]
CA 15-3	CdTe@MPA QDs		0.027 U/mL	CEA, BSA, CA-125	Human serum samples	[48]
CDH-17	GQDs/MWCNTs-HRP-DAb-CDH-17-CAb-Phe-WE2-dSPCE	0.11–10 ng/mL	0.03 ng/mL	BSA, p53, TNF, Hb, endoglin, FGFR-4	Raw lysate samples of colon and breast cancer cells	[36]
CEA	BSA/ApI/GQD/AuNP/NG/GCE	1.0×10^{-5}–200.0 ng/mL	3.2×10^{-6} ng/mL	AFP, CEA, HSA, CA 15-3	Human serum samples	[6]
CEA	MB-substrate/GQDs-IL-NF/GCE	0.5 fg/mL–0.5 ng/mL	0.34 fg/mL	PSA, BSA, MUC1	Clinical human serum samples	[1]
CEA	Pt||PANI-Au/N,S-GQDs/anti-CEA	0.5–1000 ng/mL	0.01 ng/mL	AFP, Tau, Hb, L-Cys, L-glutamate	Human serum samples	[34]
CEA	PtPd/N-GQDs@Au	5.0 fg/mL–50.0 ng/mL	2.0 fg/mL	HBS, PSA, IgG, BSA	Serum samples	[42]
CEA	CEA/BSA/Ab_1/GQDs-AuNP/P5FIn/erGO/GCE	0.1 pg/mL–10.0 ng/mL	3.78 fg/mL	AFP, CEA, Chol, Glu, HSA, UA	Human serum samples	[40]
HER2-ECD	CdSe@ZnS QDs/SPCE	10–150 ng/mL	2.1 ng/mL	Cystatin C, CA 15-3	Human serum samples	[49]

IL-13Rα2	GQDs/MWCNTs-HRP-DAb-IL-13sRα2-Biotin-Strep-CAb-Phe-WE1-dSPCE	4.92–100 ng/mL	1.4 ng/mL	BSA, p53, TNF, Hb, endoglin, FGFR-4	Raw lysate samples of colon and breast cancer cells	[36]
miRNA-21	AuNPs/GQDs/GO/SPCE	0.001–1000 pM	0.04 fM	N. S.	Human serum samples	[43]
miRNA-155	AuNPs/GQDs/GO/SPCE	0.001–1000 pM	0.33 fM	N. S.	Human serum samples	[43]
miRNA-210	AuNPs/GQDs/GO/SPCE	0.001–1000 pM	0.28 fM	N. S.	Human serum samples	[43]
PCA 3	ITO/Au–GQDs	1 μM–100 fM	211 fM	CYFRA-21, cTnI	N. S.	[50]
PSA	GS–QD–Ab2	0.005–10 ng/mL	3.0 pg/mL	BSA, IgG, lysozyme, glucose	Patient serum samples	[51]
PSA	GCE/AuNPs@MWCNTs-GQDs/EDC-NHS/Ab/BSA/PSA	1–10,000 pg/mL	0.48 pg/mL	CEA, AFP, IgG	Human serum sample	[45]
PSA	GCE-GQDs-CoPc(ππ)-Aptamer	0.034–0.057 ng/mL	0.018 ng/mL	BSA, L-cysteine, glucose.	N. S.	[52]
p16 gene	GQDs	0.5×10^{-9} M–7.5×10^{-9} M	0.10 pM	N. S.	N. S.	[53]
p53 gene	Au/GQDs/P-Cys/GNP-Streptavidin-HRP/BSA/Ab/BSA/p53	0.000197–0.016 pM	0.065 fM	N. S.	Cancer cell lysate samples	[7]

Ab, antibody; AFP, α-fetoprotein; Ap, aptamer; BSA, bovine serum albumin; CA125, carcinoma antigen 125; CA15-3, carbohydrate antigen; CEA, carcinoembryonic antigen; Chol, cholesterol; CoPc, cobalt phthalocyanine; CDH-17, cadherin 17; cTnI, cardiac troponin I; CYFRA 21-1, cytokeratin 19 fragment; DPASV, differential pulse anodic stripping voltammetry; ECL, electrochemiluminescence; EDC, 1-ethyl-3-(3-dimethylaminopropyl); EIS, electrochemical impedance spectroscopy; FGFR-4, fibroblast growth factor receptor 4; GCE, glassy carbon electrode; Glu, glucose; GO, graphene oxide; GQDs, graphene quantum dots; GS, graphene sheets; Hb, hemoglobin; HBS, hepatitis B surface antigen; HER2-ECD, human epidermal growth factor receptor 2; HRP, horseradish peroxidase; HSA, human serum albumin; IgG, immunoglobulin G; IL, ionic liquid; IL-13Rα2, interleukin-13 receptor subunit alpha-2; ITO, indium tin oxide; L-Cys, L-cysteine; MB, methylene blue; MWCNT, multi-walled carbon nanotubes; MUC1, mucin 1; NF, nafion; NHS, N-hydroxysuccinimide; NPs, nanoparticles; NS, not statement; N-GQDs, nitrogen-doped graphene quantum dots; PAMAM, polyamidoamine; PANI, polyaniline; PCA 3, prostate cancer antigen 3; P-Cys, poly-L-cysteine; PEDOT, poly (3,4-ethylenedioxythiophene); P5FIn/erGO, poly(5-formylindole)/reduced graphene oxide nanocomposite; PSA, prostate-specific antigen; SPCE, screen-printed carbon electrode; QDs, quantum dots; Tau, Tau protein; TNF, tumor necrosis factor.

the detection of clinically relevant biomarkers, highlighting their design strategies, analytical performance metrics, and application scopes.

An innovative amperometric enzyme-based immunosensor system with raised sensitivity was designed for the identification of AFP. The immunosensor construction comprised a layer-by-layer modification of a Pt electrode utilizing ZnSe QDs, Azure I dye, Au nanoparticles, and poly(3,4-ethylenedioxythiophene) (PEDOT). First, citrate-stabilized Au nanoparticles were deposited onto an electrochemically polymerized PEDOT film, synthesized in the ionic liquid 1-butyl-3-methylimidazolium tetrafluoroborate ([Bmim][BF4]). Azure I dye, behaving as a redox mediator, was subsequently anchored to the Au NP/PEDOT composite and served as a sensor platform for ZnSe QD attachment. The anti-AFP antibodies were then immobilized on the ZnSe-modified layer, followed by the application of horseradish peroxidase (HRP) to block nonspecific binding sites and enhance signal amplification. The combined electron-transfer facilitation by nanoAu/PEDOT and Azure I, along with the stabilizing effect of the ultrasmall ZnSe QDs on biomolecular activity, leads to a highly sensitive immunosensor with rapid response. The immunosensor achieved an LOD as low as 1.1 fg/mL and demonstrated a remarkably wide dynamic AFP range for detection (5×10^{-5} to 250 ng/mL) [47].

By Kumar et al., a highly sensitive and eco-friendly electrochemical sensor was designed employing GQDs for the early detection of the breast cancer biomarker cluster of differentiation-44 (CD44) antigen (Figure 6.1). The GQDs were prepared via a green and cost-effective electrochemical exfoliation synthesis process making use of waste dry batteries. Transmission electron microscopy approved the generation of spherical GQDs with an average diameter of 4.75 nm. Electrochemical analyses, including CV and electrochemical impedance spectroscopy (EIS), demonstrated the suitability of GQDs for biosensing implementations on account of their favorable electroactive features. The GQDs supplied a high surface area for effective antibody immobilization, authorizing the construction of a sensitive biosensing interface. Differential pulse voltammetry (DPV) was used to evaluate the analytical performance of the biosensor. The developed apparatus exhibited an impressive LOD of 2.11 fg/mL across a dynamic linear range of 0.1 pg/mL to 100 ng/mL in phosphate-buffered saline (PBS). In spiked serum samples, the biosensor achieved a detection limit of 2.71 fg/mL with a linear response between 1.0 pg/mL and 100 ng/mL. These outcomes highlight the potential of electrochemically synthesized GQDs as robust nanomaterials for biomedical diagnostic applications [54].

Yang and coworkers synthesized graphene sheets (GS) functionalized with QDs and exploited them as labeling agents in the development of a sandwich-type electrochemical immunosensor for PSA detection. The capture antibody, anti-PSA, was immobilized directly onto the surface of GS. The resulting immunosensor demonstrated a broad linear detection range of 0.005–10 ng/mL, an extraordinarily LOD of 3 pg/mL, and indicated high specificity, reproducibility, and stability. Thriving PSA determination in human serum samples approved the practical applicability of the sensor platform, suggesting its strong potential for disposal in clinical diagnostics [51].

According to Nxele and Nyokong's study, an innovative electrochemical aptasensor was generated for the sensitive determination of PSA, exploiting electrode surfaces functionalized with varied assemblies of cobalt phthalocyanine (CoPc), a specific aptamer, and GQDs. The analyzing strategy utilized both differential pulse voltammetry (DPV) and

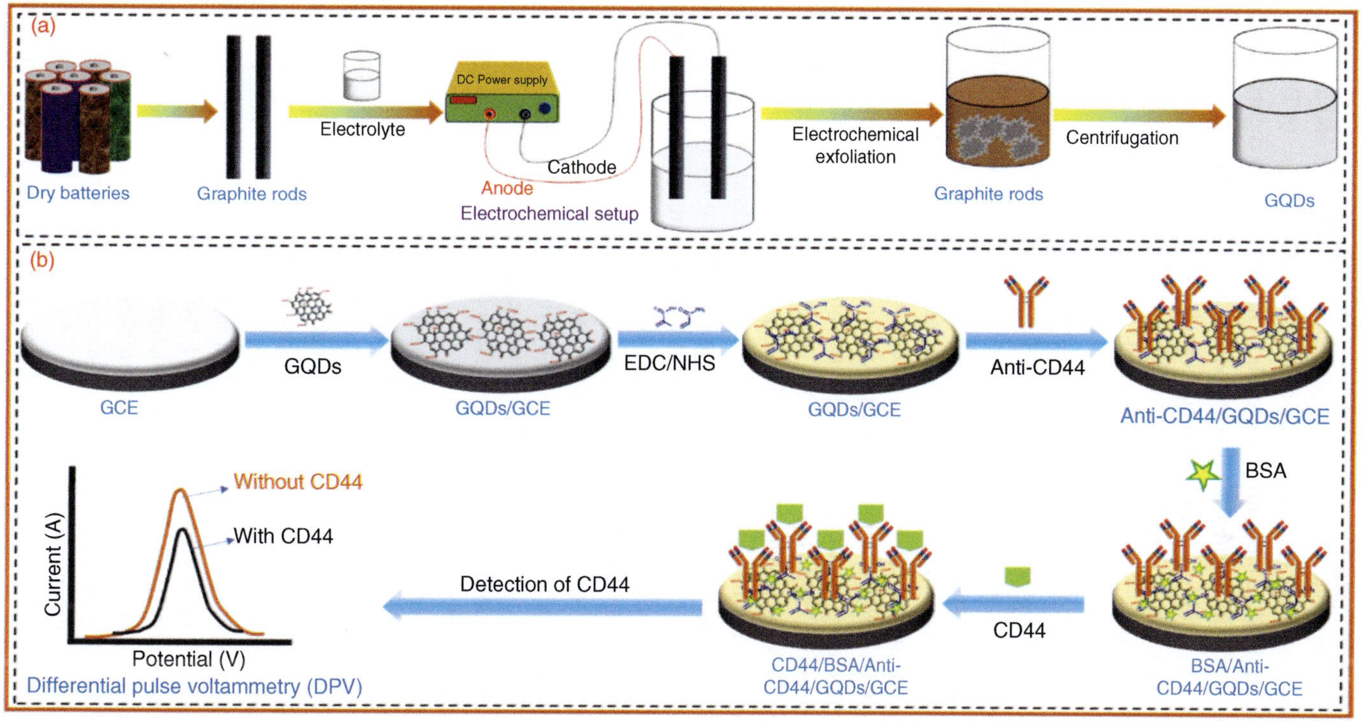

Figure 6.1 Illustration of (a) the synthesis route of GQDs and (b) the operational principle of the constructed electrochemical biosensor [54]. Kumar et al, 2022 / MDPI/CC BY 4.0.

electrochemical impedance spectroscopy (EIS). For both electrochemical techniques, a linear response was achieved across the concentration range of PSA as 1.2–2.0 pM. The indicator electrode is prepared as integrating GQDs, a push–pull CoPc, and a PSA-specific aptamer. This design attained an LOD as low as 0.66 pM, making use of DPV, which is notably lower than clinically concerning PSA levels in male patients screened for prostate cancer. The sensor also represented ideal selectivity toward PSA, maintaining accuracy in the existence of potential interfering molecules such as glucose, BSA, and L-cysteine. Additionally, the aptasensor demonstrated satisfactory reproducibility, stability, and repeatability, highlighting its potential as a dependable platform for early prostate cancer diagnostics [52].

Ganganboina and Doong designed a new label-free impedimetric immunosensor utilizing nanowires composed of N, S-doped GQDs integrated with gold–polyaniline (N,S-GQDs@Au-PANI) for the sensitive quantification of CEA. The N,S-GQDs were prepared via an uncomplicated hydrothermal pyrolysis route, while the Au-PANI nanowires were synthesized through an interfacial polymerization process. These prepared nanowires, measuring 30–50 nm in diameter, were subsequently functionalized with 2–9 nm N,S-GQDs through Au–thiol interplays, forming a bifunctional sensing interface capable of enhancing electrochemical responsiveness and immobilizing anti-CEA antibodies. The resulting N,S-GQDs@Au-PANI structure demonstrated unique electrical conductivity, making possible efficient electron transfer. Upon exposure to CEA, the genesis of antigen–antibody complexes appreciably increased the charge transfer resistance, enabling dependable, label-free impedimetric detection. The immunosensor system represented a broad dynamic range from 0.5 to 1000 ng/mL and an impressive LOD of 0.01 ng/mL. In addition to superb selectivity against other biomarkers and amino acids, the immunosensor showed strong operational stability. Notably, prosperous implementation of human serum analysis yielded recovery rates ranging from $96.0 \pm 2.6\%$ to $103 \pm 3.8\%$, underscoring the platform's practical utility. These outcomes validate the N,S-GQDs@Au-PANI nanowires as a highly effective electrochemical transducer for early-stage CEA detection, offering an encouraging foundation for the future development of high-performance impedimetric sensors [34].

Prostate cancer is among the most common malignancies in men, frequently diagnosed at later stages, contributing to high mortality rates. To address the requirement for early detection, Raj and coworkers' study presented a novel electrochemical biosensor operating on an indium tin oxide (ITO) electrode modified with a gold GQD (Au–GQD) nanocomposite (Figure 6.2). A prostate cancer antigen 3 (PCA3)-specific DNA capture probe was covalently attached to the Au–GQDs, licensing the selective hybridization of target biomarker sequences. The designed biosensor demonstrated an impressive detection limit of 211 fM, with a wide linear PCA 3 range between 1 μM and 100 fM and a swift detection time of just 5 minutes. Furthermore, the sensor retained functionality for approximately 19 ± 1 days and revealed high specificity against non-target interference species. This platform offers a promising route for early prostate cancer diagnostics, potentially improving patient outcomes through timely intervention [50].

Huang and coworkers constructed an electrochemical aptasensor for the ultrasensitive detection of CEA, operating a Pb^{2+}-dependent DNAzyme-driven signal amplification strategy combined with a composite film implicating GQDs, ionic liquid, and Nafion

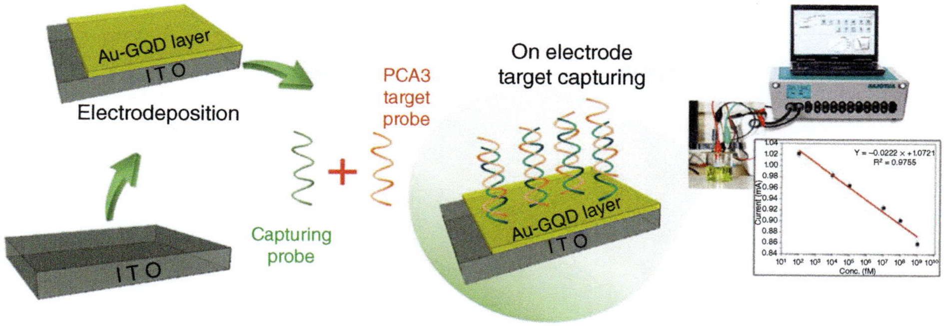

Figure 6.2 Illustration of step-by-step manufacturing of developed sensor for PCA3 detection strategy [50]. Raj et al, 2024 /MDPI/CC BY 4.0.

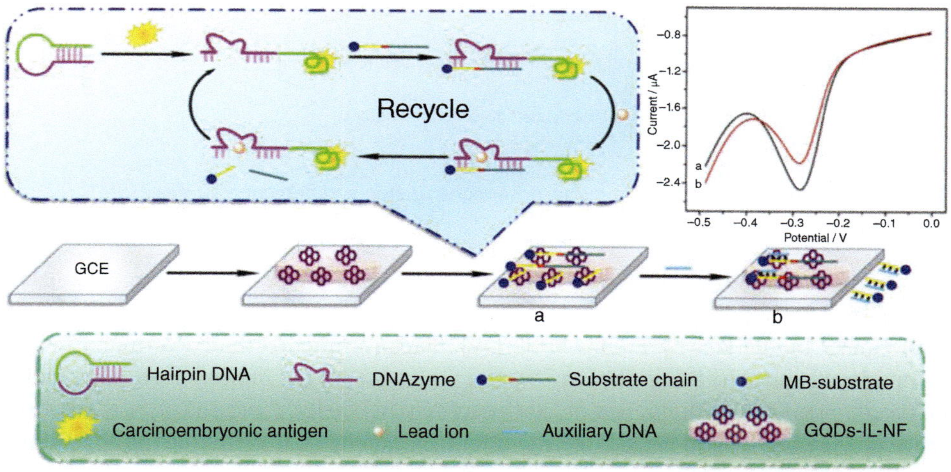

Figure 6.3 Schematic demonstration of the GQDs-IL-NF nanomatrix and DNAzyme-assisted recycling of the target-aptamer complex for sensitive electrochemical determination of CEA [1].

(GQDs-IL-NF) (Figure 6.3). The sensor employed a specially designed hairpin-structured DNA probe integrating both a DNAzyme domain and a CEA-specific aptamer sequence. Upon target binding, the structure opened, activating the DNAzyme-mediated catalytic reaction that created plenty of single-stranded DNA products. These DNA strands interacted with the GQDs-IL-NF-modified GCE via π–π stacking, permitting efficient attachment of methylene blue-labeled substrate DNA (MB-substrate), which manufactured the initial electrochemical signal. Under optimized experimental parameters, the variation in current directly corresponded with the CEA concentration, providing a linear dynamic range between 0.5 fg/mL and 0.5 ng/mL and an impressively low detection threshold of 0.34 fg/mL. The aptasensor also demonstrated reliable performance in human serum sample analysis, highlighting its powerful potential for clinical diagnostic implementations [1].

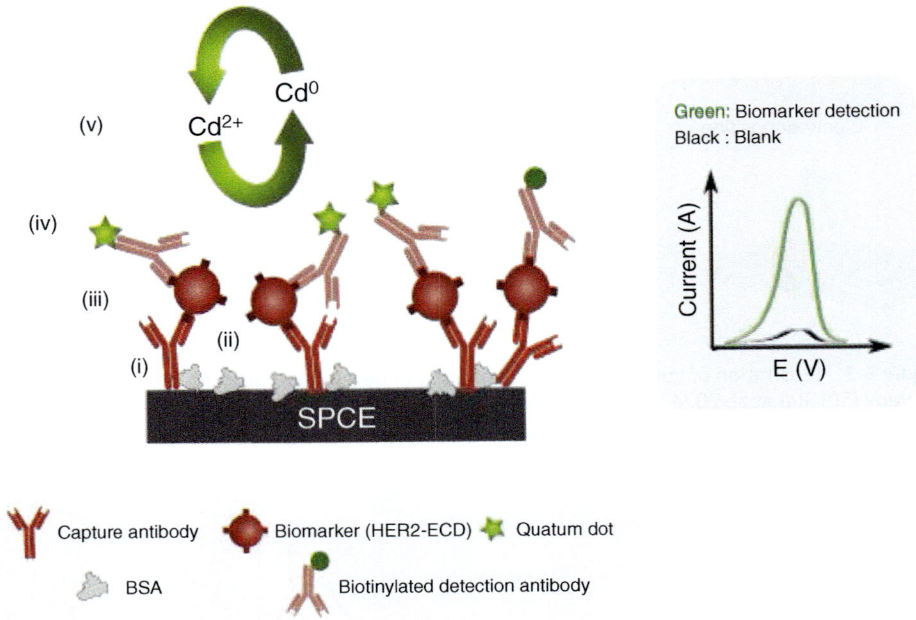

Figure 6.4 Representation of the immunosensor construction and detection strategy for HER2-ECD [49].

An occupant-friendly, single-use electrochemical immunosensor system has been designed for the rapid identification of the extracellular domain HER2-ECD, a major breast cancer biomarker, by Freitas et al. [49]. Screen-printed carbon electrode (SPCE) was employed as the sensing platform, where a sandwich-type immunoassay was built. The mechanism of detection relied on electroactive core/shell CdSe@ZnS QDs as labeling agents (Figure 6.4), with signal measurement executed via differential pulse anodic stripping voltammetry (DPASV). In the proposed method, the entire assay was completed in 2 hours, requiring less than 30 minutes of hands-on time. The sensor indicated a linear response to HER2-ECD concentrations ranging between 10.0 and 150.0 ng/mL, with a limit of detection (LOD) of 2.1 ng/mL, which is almost seven times lower than the clinical cut-off threshold (15 ng/mL). The analytical output was validated utilizing spiked human serum samples, confirming both the selectivity and precision of the developed immunosensor. Cross-reactivity investigations with CA15-3 and diverse serum proteins further demonstrated the sensor's specificity for HER2-ECD detection [49].

6.4 Conclusion

Because of the restrictions of standard diagnostic approaches for various cancer species, new technologies have obtained notice for their early-stage sensitivity and multiplexing potential. Especially quantum dots, with their improved optical properties and modifiable surfaces, have come into view as encouraging non-invasive apparatus for accurate and stable detection of cancer-related molecular targets [55]. QD-based sensing systems offer a cost-effective, fast, and user-friendly alternative to conventional, labor-intensive diagnostic methods. Nevertheless, limitations such as their poor water solubility and inherent toxicity must be investigated to exactly realize their potency in clinical diagnostics and targeted drug delivery [56, 57]. Incorporating quantum dots into biosensor platforms outstandingly upgrades signal amplification, minimizes background interference and erroneous signals, and boosts electron transfer yield, finally leading to more precise and dependable detection [14]. The latest advancements in electrochemical biosensor systems integrated with QDs and graphene-based materials have outstandingly evolved the early, noninvasive monitoring of cancer by making possible sensitive recognition of morphological and molecular variances. On the other hand, to provision clinical translation, there remains a necessity for optimized multiplexed sensing platforms capable of trustworthily detecting assorted biomarkers—including exosomes—in liquid biopsy samples for personalized cancer diagnostics [39].

References

1 Huang, J.-Y., Zhao, L., Lei, W. et al. (2018). A high-sensitivity electrochemical aptasensor of carcinoembryonic antigen based on graphene quantum dots-ionic liquid-nafion nanomatrix and DNAzyme-assisted signal amplification strategy. *Biosensors and Bioelectronics* 99: 28–33. https://doi.org/10.1016/j.bios.2017.07.036.

2 Kabir, M.Z., Tayyab, H., Erkmen, C. et al. (2024). Comprehensive views toward the biomolecular recognition of an anticancer drug, leflunomide with human serum albumin. *Journal of Biomolecular Structure and Dynamics* 42 (14): 7257–7271. https://doi.org/10.1080/07391102.2023.2239931.

3 Iannazzo, D., Espro, C., Celesti, C. et al. (2021). Smart biosensors for cancer diagnosis based on graphene quantum dots. *Cancers* 13 (13): 3194. https://doi.org/10.3390/cancers13133194.

4 Sadighbayan, D., Sadighbayan, K., Tohid-Kia, M.R. et al. (2019). Development of electrochemical biosensors for tumor marker determination towards cancer diagnosis: recent progress. *TrAC Trends in Analytical Chemistry* 118: 73–88. https://doi.org/10.1016/j.trac.2019.05.014.

5 Zhang, H., Liu, L., Fu, X., and Zhu, Z. (2013). Microfluidic beads-based immunosensor for sensitive detection of cancer biomarker proteins using multienzyme-nanoparticle amplification and quantum dotslabels. *Biosensors and Bioelectronics* 42: 23–30. https://doi.org/10.1016/j.bios.2012.10.076.

6 Shekari, Z., Zare, H.R., and Falahati, A. (2019). Electrochemical sandwich aptasensor for the carcinoembryonic antigen using graphene quantum dots, gold nanoparticles and nitrogen doped graphene modified electrode and exploiting the peroxidase-mimicking activity of a G-quadruplex DNAzyme. *Microchimica Acta* 186 (8): 530. https://doi.org/10.1007/s00604-019-3572-9.

7 Hasanzadeh, M., Baghban, H.N., Shadjou, N., and Mokhtarzadeh, A. (2018). Ultrasensitive electrochemical immunosensing of tumor suppressor protein p53 in unprocessed human plasma and cell lysates using a novel nanocomposite based on

poly-cysteine/graphene quantum dots/gold nanoparticle. *International Journal of Biological Macromolecules* 107: 1348–1363. https://doi.org/10.1016/j.ijbiomac.2017.11.006.

8 Hasanzadeh, M., Tagi, S., Solhi, E. et al. (2018). An innovative immunosensor for ultrasensitive detection of breast cancer specific carbohydrate (CA 15-3) in unprocessed human plasma and MCF-7 breast cancer cell lysates using gold nanospear electrochemically assembled onto thiolated graphene quantum dots. *International Journal of Biological Macromolecules* 114: 1008–1017. https://doi.org/10.1016/j.ijbiomac.2018.03.183.

9 Henry, N.L. and Hayes, D.F. (2012). Cancer biomarkers. *Molecular Oncology* 6 (2): 140–146. https://doi.org/10.1016/j.molonc.2012.01.010.

10 Sarsenbayeva, A. (2025). Molybdenum-based electrochemical sensors for breast cancer biomarker detection: advances and challenges. *Comprehensive Reviews in Food Science and Food Safety* 24 (9): 15432–15457. https://doi.org/10.1080/10408347.2025.2487581.

11 Tovar-Lopez, F.J. (2023). Recent progress in micro- and nanotechnology-enabled sensors for biomedical and environmental challenges. *Sensors* 23 (12): https://doi.org/10.3390/s23125406.

12 Quesada-González, D. and Merkoçi, A. (2025). Quantum dots for biosensing: classification and applications. *Biosensors and Bioelectronics* 273: 117180. https://doi.org/10.1016/j.bios.2025.117180.

13 Brycht, M., Kaczmarska, K., Uslu, B. et al. (2016). Sensitive determination of anticancer drug imatinib in spiked human urine samples by differential pulse voltammetry on anodically pretreated boron-doped diamond electrode. *Diamond and Related Materials* 68: 13–22. https://doi.org/10.1016/j.diamond.2016.05.007.

14 Liu, M., Song, Y., Liu, M. et al. (2025). The application of functional nanomaterials-based electrochemical biosensors in detecting cancer biomarkers: a review. *Molecules* 30 (13): 2708. https://doi.org/10.3390/molecules30132708.

15 Dincer, C., Bruch, R., Costa-Rama, E. et al. (2019). Disposable sensors in diagnostics, food, and environmental monitoring. *Advanced Materials* 31 (30): https://doi.org/10.1002/adma.201806739.

16 Medintz, I.L., Uyeda, H.T., Goldman, E.R., and Mattoussi, H. (2005). Quantum dot bioconjugates for imaging, labelling and sensing. *Nature Materials* 4 (6): 435–446. https://doi.org/10.1038/nmat1390.

17 Michalet, X., Pinaud, F.F., Bentolila, L.A. et al. (2005). Quantum dots for live cells, in vivo imaging, and diagnostics. *Science* 307 (5709): 538–544. https://doi.org/10.1126/science.1104274.

18 Jamieson, T., Bakhshi, R., Petrova, D. et al. (2007). Biological applications of quantum dots. *Biomaterials* 28 (31): 4717–4732. https://doi.org/10.1016/j.biomaterials.2007.06.018.

19 Pandey, S.N., Afzal, M., Goyal, A. et al. (2025). Quantum dot biosensors for Glioblastoma: merging nanotechnology with precision oncology. *Inorganic Chemistry Communications* 179: 114671. https://doi.org/10.1016/j.inoche.2025.114671.

20 Resch-Genger, U., Grabolle, M., Cavaliere-Jaricot, S. et al. (2008). Quantum dots versus organic dyes as fluorescent labels. *Nature Methods* 5 (9): 763–775. https://doi.org/10.1038/nmeth.1248.

21 Bruchez, M., Moronne, M., Gin, P. et al. (1998). Semiconductor nanocrystals as fluorescent biological labels. *Science* 281 (5385): 2013–2016. https://doi.org/10.1126/science.281.5385.2013.

22 Wagner, A.M., Knipe, J.M., Orive, G., and Peppas, N.A. (2019). Quantum dots in biomedical applications. *Acta Biomaterialia* 94: 44–63. https://doi.org/10.1016/j.actbio.2019.05.022.

23 Algar, W.R. and Krull, U.J. (2008). Quantum dots as donors in fluorescence resonance energy transfer for the bioanalysis of nucleic acids, proteins, and other biological molecules. *Analytical and Bioanalytical Chemistry* 391 (5): 1609–1618. https://doi.org/10.1007/s00216-007-1703-3.

24 Freeman, R., Liu, X., and Willner, I. (2012). Chemiluminescent and chemiluminescent resonance energy transfer (CRET) detection of DNA using catalytic nucleic acid-functionalized quantum dots. *Chemical Society Reviews* 41 (10): 4067–4085. https://doi.org/10.1021/ja202639m.

25 Clapp, A.R., Medintz, I.L., and Mattoussi, H. (2006). Förster resonance energy transfer investigations using quantum-dot fluorophores. *ChemPhysChem* 7 (1): 47–57. https://doi.org/10.1002/cphc.200500217.

26 Colvin, V.L., Schlamp, M.C., and Alivisatos, A.P. (1994). Light-emitting diodes made from cadmium selenide nanocrystals and a semiconducting polymer. *Nature* 370 (6488): https://doi.org/10.1038/370354a0.

27 Mimona, M.A., Rimon, M.I.H., Zohura, F.T. et al. (2025). Quantum dot nanomaterials: empowering advances in optoelectronic devices. *Chemical Engineering Journal Advances* 21: 100704. https://doi.org/10.1016/j.ceja.2025.100704.

28 Michler, P., Kiraz, A., Becher, C. et al. (2000). A quantum dot single-photon turnstile device. *Science* 290 (5500): 2282–2285. https://doi.org/10.1126/science.290.5500.2282.

29 Noreen, S., Ishaq, I., Saleem, M.H. et al. (2025). Electrochemical biosensing in oncology: a review advancements and prospects for cancer diagnosis. *Cancer Biology & Therapy* 26 (1): https://doi.org/10.1080/15384047.2025.2475581.

30 Sanko, V. and Kuralay, F. (2023). Label-free electrochemical biosensor platforms for cancer diagnosis: recent achievements and challenges. *Biosensors* 13 (3): 333. https://doi.org/10.3390/bios13030333.

31 Farzin, M.A. and Abdoos, H. (2021). A critical review on quantum dots: from synthesis toward applications in electrochemical biosensors for determination of disease-related biomolecules. *Talanta* 224: https://doi.org/10.1016/j.talanta.2020.121828.

32 Jessy Mercy, D., Girigoswami, K., and Girigoswami, A. (2024). A mini review on biosensor advancements-emphasis on quantum dots. *Results in Chemistry* 7: https://doi.org/10.1016/j.rechem.2023.101271.

33 Peng, C.-W. and Li, Y. (2010). Application of quantum dots-based biotechnology in cancer diagnosis: current status and future perspectives. *Journal of Nanomaterials* 2010 (1): https://doi.org/10.1155/2010/676839.

34 Ganganboina, A.B. and Doong, R.-A. (2019). Graphene quantum dots decorated gold-polyaniline nanowire for impedimetric detection of carcinoembryonic antigen. *Scientific Reports* 9 (1): 7214. https://doi.org/10.1038/s41598-019-43740-3.

35 Ganganboina, A.B., Dutta Chowdhury, A., and Doong, R. (2018). N-Doped graphene quantum dots-decorated V_2O_5 nanosheet for fluorescence turn off–on detection of cysteine. *ACS Applied Materials & Interfaces* 10 (1): 614–624. https://doi.org/10.1021/acsami.7b15120.

36 Serafín, V., Valverde, A., Garranzo-Asensio, M. et al. (2019). Simultaneous amperometric immunosensing of the metastasis-related biomarkers IL-13Rα2 and CDH-17 by using grafted screen-printed electrodes and a composite prepared from quantum dots and carbon nanotubes for signal amplification. *Microchimica Acta* 186 (7): 411. https://doi.org/10.1007/s00604-019-3531-5.

37 Mousavi, S.M., Hashemi, S.A., Yari Kalashgrani, M. et al. (2022). The pivotal role of quantum dots-based biomarkers integrated with ultra-sensitive probes for multiplex detection of human viral infections. *Pharmaceuticals* 15 (7): https://doi.org/10.3390/ph15070880.

38 Gao, Q., Han, J., and Ma, Z. (2013). Polyamidoamine dendrimers-capped carbon dots/Au nanocrystal nanocomposites and its application for electrochemical immunosensor. *Biosensors and Bioelectronics* 49: https://doi.org/10.1016/j.bios.2013.05.048.

39 Tabish, T.A., Hayat, H., Abbas, A., and Narayan, R.J. (2021). Graphene quantum dot–based electrochemical biosensing for early cancer detection. *Current Opinion in Electrochemistry* 30: 100786. https://doi.org/10.1016/j.coelec.2021.100786.

40 Nie, G., Wang, Y., Tang, Y. et al. (2018). A graphene quantum dots based electrochemiluminescence immunosensor for carcinoembryonic antigen detection using poly(5-formylindole)/reduced graphene oxide nanocomposite. *Biosensors and Bioelectronics* 101: https://doi.org/10.1016/j.bios.2017.10.021.

41 Thomas, P., Gangopadhyay, A., Steele, G. et al. (1995). The effect of transfection of the CEA gene on the metastatic behavior of the human colorectal cancer cell line MIP-101. *Cancer Letters* 92 (1): https://doi.org/10.1016/0304-3835(95)03764-N.

42 Yang, Y., Liu, Q., Liu, Y. et al. (2017). A novel label-free electrochemical immunosensor based on functionalized nitrogen-doped graphene quantum dots for carcinoembryonic antigen detection. *Biosensors and Bioelectronics* 90: https://doi.org/10.1016/j.bios.2016.11.029.

43 Pothipor, C., Jakmunee, J., Bamrungsap, S., and Ounnunkad, K. (2021). An electrochemical biosensor for simultaneous detection of breast cancer clinically related microRNAs based on a gold nanoparticles/graphene quantum dots/graphene oxide film. *The Analyst* 146 (12): 4000–4009. https://doi.org/10.1039/D1AN00436K.

44 Zhong, X., Zhang, M., Guo, L. et al. (2021). A dual-signal self-checking photoelectrochemical immunosensor based on the sole composite of MIL-101(Cr) and CdSe quantum dots for the detection of α-fetoprotein. *Biosensors and Bioelectronics* 189: 113389. https://doi.org/10.1016/j.bios.2021.113389.

45 Ghanavati, M., Tadayon, F., and Bagheri, H. (2020). A novel label-free impedimetric immunosensor for sensitive detection of prostate specific antigen using Au nanoparticles/MWCNTs- graphene quantum dots nanocomposite. *Microchemical Journal* 159: 105301. https://doi.org/10.1016/j.microc.2020.105301.

46 Uslu, B. (ed.) (2021). *Electroanalytical Applications of Quantum Dot-Based Biosensors.* Elsevier.

47 Liu, K., Zhang, J., Liu, Q., and Huang, H. (2013). Electrochemical immunosensor for alpha-fetoprotein determination based on ZnSe quantum dots/Azure I/gold nanoparticles/poly (3,4-ethylenedioxythiophene) modified Pt electrode. *Electrochimica Acta* 114: 448–454. https://doi.org/10.1016/j.electacta.2013.10.018.

48 Sousa, M.P., Piloto, A.M.L., Pereira, A.C. et al. (2022). New *quantum*-dot-based fluorescent immunosensor for cancer biomarker detection. *Chem* 10 (12): 518. https://doi.org/10.3390/chemosensors10120518.

49 Freitas, M., Neves, M.M.P.S., Nouws, H.P.A., and Delerue-Matos, C. (2020). Quantum dots as nanolabels for breast cancer biomarker HER2-ECD analysis in human serum. *Talanta* 208: 120430. https://doi.org/10.1016/j.talanta.2019.120430.

50 Raj, D., Kumar, A., Kumar, D. et al. (2024). Gold–graphene quantum dot hybrid nanoparticle for smart diagnostics of prostate cancer. *Biosensors* 14 (11): 534. https://doi.org/10.3390/bios14110534.

51 Yang, M., Javadi, A., and Gong, S. (2011). Sensitive electrochemical immunosensor for the detection of cancer biomarker using quantum dot functionalized graphene sheets as labels. *Sensors and Actuators B: Chemical* 155 (1): 357–360. https://doi.org/10.1016/j.snb.2010.11.055.

52 Nxele, S.R. and Nyokong, T. (2021). The electrochemical detection of prostate specific antigen on glassy carbon electrode modified with combinations of graphene quantum dots, cobalt phthalocyanine and an aptamer. *Journal of Inorganic Biochemistry* 221: https://doi.org/10.1016/j.jinorgbio.2021.111462.

53 Joshi, P. and Waghmode, S. (2016). Graphene quantum dot-based on-chip electrochemical DNA hybridization sensor for pancreatic cancer. *Reports in Electrochemistry* 6: 31–40. https://doi.org/10.2147/rie.s83253.

54 Kumar, N., Yadav, S., Sadique, M.A., and Khan, R. (2022). Electrochemically exfoliated graphene quantum dots based biosensor for CD44 breast cancer biomarker. *Biosensors* 12 (11): 966. https://doi.org/10.3390/bios12110966.

55 Singh, R.D., Shandilya, R., Bhargava, A. et al. (2018). Quantum dot based nano-biosensors for detection of circulating cell free miRNAs in lung carcinogenesis: from biology to clinical translation. *Frontiers in Genetics* 9: https://doi.org/10.3389/fgene.2018.00616.

56 Elakkiya, V., Menon, M.P., Nataraj, D. et al. (2017). Optical detection of CA 15.3 breast cancer antigen using CdS quantum dot. *IET Nanobiotechnology* 11 (3): 268–276. https://doi.org/10.1049/iet-nbt.2016.0012.

57 Malik, P., Gulia, S., and Kakkar, R. (2013). Quantum dots for diagnosis of cancers. *Advanced Materials Letters* 4 (11): https://doi.org/10.5185/amlett.2013.3437.

7

Quantum Biosensors for Cardiovascular Disease Detection

Seydanur Yücer[1], Begüm Sarac[1], and Fatih Ciftci[1,2]

[1] *Department of Biomedical Engineering, Fatih Sultan Mehmet Vakıf University, Istanbul, Turkey*
[2] *Biomedical Electronic Design Application and Research Center (BETAM), Fatih Sultan Mehmet Vakıf University, Istanbul, Turkey*

7.1 Introduction

Cardiovascular diseases (CVD), including ischemic heart disease (IHD) and stroke, are major contributors to global mortality and disability. According to the 2019 Global Burden of Disease (GBD) Study, both the prevalence and mortality rates of CVD have risen significantly. Between 1990 and 2019, the number of CVD cases nearly doubled, and the number of CVD-related deaths reached 18.6 million. Additionally, the years lived with disability (DALYs) have substantially increased, with DALYs due to IHD rising to 182 million and those due to stroke reaching 143 million. The burden of these diseases continues to grow, especially in countries with lower income levels. To address this growing challenge, it is essential to adopt cost-effective policies and interventions [1]. Cardiovascular diseases (CVD) are major global health concerns, contributing to high rates of mortality and morbidity and significantly impacting patients' quality of life. The timely detection of these conditions is crucial for effective treatment. Despite advancements, the early diagnosis of CVD continues to present significant challenges. These difficulties arise because many patients do not exhibit clear symptoms, and by the time they seek medical attention, the conditions are often already in advanced stages. In recent years, there have been notable improvements in the management of CVDs. Minimally invasive surgical procedures and noninvasive diagnostic tools have become key treatment options, particularly for patients who face high surgical risks. Treatments like transcatheter aortic valve implantation (TAVI) and transcatheter mitral valve implantation (TMVI) offer promising outcomes with reduced risks. Additionally, mechanical circulatory support devices help improve the quality of life and extend survival for those with severe heart failure. However, to fully capitalize on these advancements, early detection and the application of appropriate clinical strategies are essential for directing patients toward the most effective treatments. As a result, the primary challenge in managing CVDs lies in integrating established clinical practices with modern diagnostic techniques and treatment approaches [2]. The principles of quantum mechanics [3]. They play a revolutionary role in the field of health, especially in early diagnosis of diseases and real-time health monitoring. They can detect diseases

Quantum Biosensing in Medical Diagnostics, First Edition. Edited by J. G. Manjunatha.

such as cancer, Alzheimer's, and Parkinson's through low concentrations of biomolecules, thus accelerating the diagnosis process and providing more accurate results. They also contribute to a better understanding and management of neurological and CVDs by monitoring brain activity and heart rhythm more precisely. Quantum biosensors, which offer higher accuracy compared to traditional sensors, minimize false positive and negative results and prevent unnecessary treatments and tests [4]. Thanks to their small, portable, and low-cost nature, they make home health monitoring and remote diagnosis possible, making healthcare more accessible [5]. They are expected to offer groundbreaking innovations in biomedical research and personalized medicine thanks to technologies such as NV (nitrogen-vacancy) centers, quantum entanglement, and atomic magnetometers [6].

7.2 Cardiovascular Diseases: Current Diagnostic Challenges

7.2.1 Common Types of Cardiovascular Diseases

Cardiovascular diseases, which affect the heart and blood vessels, can be life-threatening. Key conditions such as coronary artery disease (CAD), hypertension, heart failure, arrhythmia, peripheral artery disease, heart valve disorders, and aortic aneurysm pose significant risks to heart health. Early detection and proper treatment of these diseases are crucial for enhancing quality of life.

7.2.1.1 Coronary Artery Disease (CAD)

Coronary artery disease (CAD) is caused by the buildup of atherosclerotic plaques in the coronary arteries and typically develops without symptoms. Also referred to as coronary heart disease (CHD), it encompasses conditions like stable angina, acute coronary syndrome (ACS), and silent myocardial ischemia. ACS includes conditions such as unstable angina and heart attacks, which are symptomatic. CAD is the leading cause of death globally, with a higher prevalence in low- and middle-income nations. In 2015, CAD was responsible for 8.9 million deaths and 164 million disability-adjusted life years (DALYs) worldwide. Individuals who survive a heart attack are at a greater risk for recurrent events and have a mortality rate five to six times higher than those without CAD. The condition occurs when blockages in the coronary arteries reduce the blood and oxygen supply to the heart muscle.

7.2.1.2 High Blood Pressure (Hypertension)

Hypertension is defined by consistently elevated systolic blood pressure (SBP) of 130 mmHg or higher, or diastolic blood pressure (DBP) of 80 mmHg or higher. It is estimated that around 116 million adults in the United States and over 1 billion adults globally are affected by this condition. Hypertension is linked to a higher risk of CVD, such as CHD, heart failure, stroke, and increased mortality [7].

7.2.1.3 Heart Failure

Heart failure is a common cardiovascular disorder marked by the heart's reduced capacity to circulate blood effectively, leading to inadequate oxygen and nutrient supply throughout the body. This condition is linked to significant rates of morbidity and mortality. There are two main types of heart failure: heart failure with reduced ejection fraction (HFrEF) and heart failure with preserved ejection fraction (HFpEF). The condition is frequently caused by underlying issues such as IHD and high blood pressure. The progression of heart failure involves multiple factors, including disrupted energy metabolism, mitochondrial dysfunction, and decreased efficiency in cardiac energy utilization. In particular, HFrEF is often characterized by impaired energy production in the heart, which can result in myocardial injury and compromised ventricular function. Emerging treatments that focus on improving cardiac energy metabolism hold promise for advancing heart failure care [8].

7.2.1.4 Arrhythmia (Irregular Heartbeat)

Arrhythmias refer to irregular heart rhythms that occur when the heart beats outside of its normal sinus rhythm. The electrical impulses that regulate the heartbeat typically originate in the sinoatrial (SA) node, pass through the atrioventricular (AV) node, and travel along the His bundle before reaching the Purkinje fibers. Disruptions at any point in this conduction system can lead to arrhythmias. These conditions are commonly categorized by heart rate: bradyarrhythmias involve a heart rate slower than 60 beats per minute, while tachyarrhythmias describe rates exceeding 100 beats per minute. As a common cardiovascular disorder, arrhythmias pose a considerable clinical concern.

7.2.1.5 Peripheral Artery Disease (PAD)

Peripheral artery disease (PAD) is a common cardiovascular condition, primarily affecting the lower limbs, and is often caused by atherosclerosis. It can either be asymptomatic or present with symptoms like intermittent claudication and chronic limb ischemia. The severity of PAD and appropriate treatment options are determined using classification systems such as Rutherford, TASC, and GLASS, which focus on vascular anatomy. PAD is linked to other cardiovascular and cerebrovascular disorders and is known to elevate the risk of mortality. A low ankle-brachial index (ABI) of ≤ 0.9 is associated with a higher risk of death. However, improvements in medical care and treatment strategies have led to a reduction in the death rate associated with PAD [9].

7.2.1.6 Heart Valve Diseases

Valvular heart disease arises when the heart valves become narrowed (stenosis) or fail to close properly (regurgitation), disrupting normal blood flow. Common conditions include aortic valve stenosis, mitral regurgitation, aortic regurgitation, and tricuspid regurgitation. If left untreated, these conditions can lead to severe complications like heart failure, atrial fibrillation, and pulmonary edema. Timely diagnosis and proper treatment are essential for managing these diseases effectively.

- Aortic valve stenosis disease (AVSD) is the most commonly observed valve disease in developed countries. The most prevalent form of this condition is calcific aortic valve stenosis, which results from the thickening and calcification of the aortic valve. Symptoms of AVSD may include angina, shortness of breath (dyspnea), syncope, and heart failure. These symptoms usually appear when the valve area measures less than 1 cm^2. Rheumatic heart disease (RHD) is the leading cause of AVSD worldwide. However, in regions like Europe and North America, the condition is more frequently linked to calcific diseases and congenital bicuspid aortic valves.

- Rheumatic heart disease (RHD) remains the most common cause of valve disease globally. RHD typically follows an episode of streptococcal pharyngitis and is associated with the molecular mimicry of the bacterial M-protein with cardiac antigens. The mitral and aortic valves are most commonly affected in RHD. The disease can lead to severe complications, including atrial fibrillation, heart failure, pulmonary edema, and cardiorenal syndrome. RHD manifests differently in various age groups, with regurgitation (insufficiency) being more common in younger patients and stenosis (narrowing) more frequent in older adults.
- Mitral regurgitation (MR) is a widespread cardiac condition, especially among the elderly. It results primarily from myxomatous degeneration and mitral valve prolapse. In developing countries, RHD is a common cause of MR. Acute MR presents suddenly with shortness of breath and pulmonary edema, often following a myocardial infarction (heart attack). Chronic MR, in contrast, may remain asymptomatic for a long time but can cause significant issues if not addressed early.
- Aortic regurgitation (AR) ranks as the fourth most common valve disorder worldwide, with prevalence varying by age. AR is not strongly correlated with atherosclerosis. Acute AR may lead to cardiogenic shock, while chronic AR often has a long asymptomatic phase, later progressing to symptoms like dyspnea. During physical examination, AR may be detected through a soft or absent second heart sound, a blowing diastolic murmur, and a wide pulse pressure.
- Tricuspid regurgitation (TR), although less common, is a serious condition that results in right heart failure symptoms, including neck vein distention, fluid retention, and liver and abdominal swelling. The prevalence of TR varies by country and has been increasing in recent years, partly due to the growing incidence of heart failure and the use of intracardiac devices.

These diseases are common forms of CVD that can cause severe complications if not treated. Early diagnosis and appropriate management are essential to effectively address these conditions and prevent further health issues [10].

7.2.1.7 Aortic Aneurysm

An aortic aneurysm is a chronic condition characterized by the abnormal widening of the aorta, usually without noticeable symptoms. Despite being asymptomatic, it poses a significant health threat due to the potential for rupture, which can be life-threatening. The condition involves the localized expansion of the aortic artery caused by a weakening of the aortic wall. Factors such as inflammatory responses and protease activation contribute to the development of the aneurysm. At present, there are no effective medications to treat this condition, with surgical and endovascular interventions being the primary treatment options. Early detection is difficult, and without timely intervention, the risk of rupture leads to a high mortality rate [11].

7.2.2 Traditional Diagnostic Methods and Their Limitations

Cardiovascular disease (CVD) is the primary cause of death globally, responsible for over 31% of all fatalities. Diseases such as CAD, myocardial infarction (MI), heart failure, and arrhythmias present considerable difficulties in terms of diagnosis, treatment, and management. Traditional diagnostic techniques, such as electrocardiography (ECG), echocardiography, digital subtraction angiography (DSA), and cardiac magnetic resonance imaging (CMR), play a crucial role in evaluating the structure and function of the heart in clinical settings [12].

Electrocardiography (ECG) is one of the most frequently used methods for diagnosing heart disease. It is a straightforward and affordable technique often employed to identify myocardial ischemia, acute myocardial infarction (AMI), and arrhythmias. However, its diagnostic scope is limited to detecting cardiomyopathy and cannot identify structural heart damage or issues with heart function. Echocardiography, in contrast, provides a clear view of the heart's structure, such as the thickness of the heart walls, the size of the chambers, and the function of the valves. It also helps assess the movement of the heart walls and offers a more accurate evaluation of heart function. However, its resolution is relatively low. Digital subtraction angiography (DSA) is considered the gold standard for diagnosing vascular stenosis and can determine the degree of blockage in the coronary arteries. It also allows for interventional procedures to open blocked vessels when necessary. However, due to its invasive nature, it carries certain risks. Cardiac magnetic resonance imaging (CMR) is a scanning technique that uses magnetic fields to evaluate the heart's location and size, as well as the structure of the coronary arteries and pericardium. While it provides detailed information, it is more time-consuming and expensive compared to other imaging techniques. As a result, there is growing interest in advanced imaging methods that can offer more comprehensive insights into the underlying pathophysiology and assist in guiding targeted treatments [13].

7.2.3 Need for More Sensitive and Rapid Detection

Cardiovascular diseases (CVDs) are the leading cause of death worldwide and pose a significant public health challenge. Traditional diagnostic methods often fall short in capturing the intricate and evolving nature of cardiovascular pathophysiology, highlighting the need for continuous real-time monitoring and efficient diagnostic solutions. This review explores how wearable sensors, point-of-care (POC) systems, signal processing, and artificial intelligence (AI) can enhance diagnostic accuracy and facilitate proactive management of cardiovascular health. Wearable sensor technologies are revolutionizing health monitoring by incorporating miniaturized electronics and advanced algorithms. Devices like ECG monitors, pulse oximeters, and continuous glucose monitors (CGM) measure critical parameters such as heart rate variability, blood oxygen levels, and bioimpedance analysis using sophisticated signal processing techniques. These innovations have the potential to transform how cardiovascular health is managed. Single-electrode ECG setups and electrode arrays enhance the precision of data acquisition, enabling accurate cardiac monitoring, identifying localized cardiac abnormalities, and providing comprehensive spatial data. Additionally, flexible strain sensors, such as those made from carbon nanotube and graphene nanoparticle composites using 3D printing, have shown impressive sensitivity and stability, allowing for reliable physiological monitoring. Single-electrode ECG configurations and electrode arrays offer enhanced data acquisition, allowing for more precise cardiac monitoring and detection of localized abnormalities, while providing comprehensive spatial data. Flexible strain sensors, such as those made from carbon nanotube and graphene nanoparticle composites created through 3D printing, exhibit high sensitivity and stability, making them ideal for accurate physiological monitoring. Lab-on-chip devices and microfluidic technologies are accelerating decentralized diagnostic approaches, enabling rapid biomarker detection near the patient. When combined with CRISPR-based diagnostic platforms, these systems improve the sensitivity and specificity of POC tests. Moreover, the integration of artificial intelligence allows for real-time analysis of complex data, making clinical care information more interpretable and actionable. Rapid testing technologies leverage advanced signal processing and feature extraction methods to convert raw sensor data into meaningful clinical insights. Techniques such as Fourier transform and wavelet analysis enhance signal accuracy and reduce noise. Additionally, deep learning models like Convolutional Neural Networks (CNN) and Recurrent Neural Networks (RNN) have proven to significantly improve diagnostic accuracy and reliability, particularly in CVD detection [14].

7.3 Quantum Biosensors: Principles and Mechanisms

7.3.1 Fundamental Quantum Concepts

At the dawn of the twentieth century, quantum mechanics emerged as a theoretical framework to describe the behavior of atoms and subatomic particles. This field introduced principles and observations that contradicted classical physics, offering a radically different perspective on the nature of reality.

- *Wave-particle duality and the uncertainty principle:* A key principle of quantum mechanics is wave-particle duality, which states that fundamental particles like photons and electrons can simultaneously exhibit characteristics of both waves and particles. This phenomenon is mathematically represented by the wave function and has significantly altered our understanding of matter and energy. Another fundamental concept is Heisenberg's Uncertainty Principle, which asserts that the exact position and momentum of a particle cannot be determined simultaneously. This inherent limitation reveals the probabilistic nature of quantum systems and distinguishes them from classical mechanics.
- *Quantum entanglement:* Entanglement is a quantum effect where two or more particles become interconnected, meaning the state of one particle instantaneously influences the other, regardless of the distance between them. This non-local correlation defies classical physics and plays a crucial role in quantum computing and secure communication technologies.
- *Quantum superposition:* Superposition refers to the ability of a quantum system to exist in multiple states at once. For instance, an electron can occupy two energy levels simultaneously rather than just one. This principle underpins the extraordinary computational power of quantum computers, as quantum bits (qubits) can process multiple possibilities at the same time.
- *Quantum tunneling:* Quantum tunneling is a phenomenon where particles can pass through barriers that would be impossible to overcome according to classical physics. The likelihood of tunneling depends on the characteristics and thickness of the barrier.

These core principles of quantum mechanics play a crucial role in shaping the limits of modern physics. Concepts like entanglement, superposition, and tunneling are driving groundbreaking advancements in areas such as quantum computing, high-precision measurement techniques, and next-generation information processing technologies [15].

7.3.2 Sensing Elements and Working Mechanisms of Quantum Biosensors

The sensing elements and operating mechanisms of quantum biosensors allow for the highly sensitive detection of biomolecules using advanced technologies like NV centers, quantum dots, and optical sensing. Methods such as spin and magnetic sensing enhance the sensors' sensitivity, playing a crucial role in monitoring biological processes.

7.3.2.1 Nitrogen-Vacancy (NV) Centers in Diamonds

The nitrogen-vacancy (NV) center is a point defect found in diamond crystals, consisting of a nitrogen atom substituting a carbon atom adjacent to a lattice vacancy. This defect exhibits pyramidal symmetry and can align along four distinct crystallographic directions. Depending on its electronic charge, the NV center exists in two primary states. In its neutral form (NV^0), it has no net charge and possesses a total electronic spin of 1/2. When it captures an extra electron, it becomes negatively charged (NV^-), with a total spin of 1. Additionally, the NV center can be found in a state where its spin is effectively zero. The NV^- state is especially significant for quantum sensing applications due to its favorable electronic and spin properties. When illuminated with a green laser, typically at a wavelength of 532 nm, the NV^- center is optically excited to a higher energy state. It subsequently relaxes back to its ground state, emitting red fluorescence with wavelengths ranging from approximately 637 to 800 nm. During this optical cycle, the spin state of the NV^- center remains stable, allowing for precise spin control. The excited state has a higher probability of transitioning to a metastable singlet state, denoted as ^1E, which has a lifetime of around 300 ns. This metastable state relaxes non-radiatively, resulting in weak infrared emission at roughly 1042 nm. The process reduces the overall fluorescence intensity, with reductions of up to 30% observed in single NV centers and about 1–2% in large NV ensembles. This fluorescence variation forms the basis for initializing and reading out the spin state of the NV^- center, making it a powerful tool in quantum sensing technologies [16].

7.3.2.2 Quantum Dots

Multiplex detection utilizing optical techniques has emerged as a significant approach in biosensor technologies, offering enhanced sensitivity, specificity, and improved signal-to-noise ratios when compared to traditional methods. Photoluminescent nanocrystalline materials, including semiconductor quantum dots (QDs) and carbon dots (CDs), play a crucial role in these systems by enabling the simultaneous identification of multiple analytes through their photoluminescent (PL) capabilities. Their unique optical features—such as exceptional brightness, high photoluminescence quantum efficiency, resistance to photobleaching, and broad absorption spectra—facilitate multi-wavelength emission and the concurrent detection of various target molecules. Moreover, QDs exhibit narrow and symmetric emission peaks, which enhance spectral resolution and minimize signal overlap, while their extended PL lifetimes allow for high temporal resolution and the observation of dynamic biological events. The surface chemistry of QDs can be precisely modified to attach biomolecules like antibodies, oligonucleotides, and proteins, as well as polymers such as molecularly imprinted polymers (MIPs), providing them with high specificity in target recognition. In certain cases, a single quantum dot can be used to monitor multiple analytes, where advanced data processing and mathematical algorithms are essential to accurately interpret overlapping signals and ensure reliable analysis [17].

Recent developments in quantum sensing have propelled transducer-based sensing technologies to the forefront, offering exceptional sensitivity, precision, and spatial resolution. Due to their atomic-level tunability, molecular qubits and their collective systems are emerging as promising candidates for detecting chemical analytes. In this study, we showcase the quantum sensing of lithium ions at room temperature using organic radicals integrated into a microporous metal-organic framework (MOF). These organic radicals exhibit electron spin coherence at room temperature and function as qubits with microwave addressability. The MOF's high surface area facilitates the accessibility of guest analytes to the organic qubits, allowing for precise identification of lithium ions and their concentration to be quantitatively measured. These measurements are performed using relaxation and high-resolution spectroscopic techniques based on electron paramagnetic resonance (EPR) spectroscopy. The detection method demonstrated here is also applicable to other metal ions with non-zero nuclear spin, highlighting its broad potential for various detection applications [18].

7.3.2.3 Spin and Magnetic Sensing

Researchers have developed an innovative approach for detecting magnetic noise using a quantum sensor based on hexagonal boron nitride (hBN). Unlike conventional sensors, which are limited by their 3D crystal structures that constrain sensitivity, hBN can be exfoliated into a 2D material, offering enhanced performance. The study initially created defects in the hBN material and measured the spin relaxation time. Upon introducing paramagnetic Gd3+ ions, a decrease in the spin relaxation time was observed, signaling the presence of magnetic noise. Furthermore, the team successfully demonstrated the suspension of hBN nanopowder in a solution for additional spin measurements. These findings highlight the potential of hBN as a highly sensitive quantum sensor, paving the way for its application in a range of sensing technologies. This research brings us closer to realizing a fully 2D, ultra-sensitive quantum sensor. This technique offers significant potential, especially for the use of biosensors as sensitive sensing elements, but further applications in biological analyses need to be developed [19].

A novel method for detecting the components of the residual magnetic field within the shielding of spin-polarized potassium (K) atoms is introduced. This technique, based on EPR, enables the measurement of all three components of the residual magnetic field. The Larmor precession frequency is the dominant factor influencing the field experienced by the spin-polarized K atoms, with lower frequencies being beneficial for achieving the spin-exchange relaxation-free (SERF) regime. The measurements are carried out by utilizing population reduction, optical pumping, and detecting the peak of the K EPR spectrum optically. To independently determine the three components of the residual field, a cross-modulation method is used, which modulates the transverse field frequency and the strength of the longitudinal field. The average measured values of the field components, along with their associated uncertainties, are 9.60 ± 0.06, 2.25 ± 0.03, and 2.83 ± 0.17 nT, with their respective axes corresponding to the negative directions [20].

7.3.2.4 Optical Detection

This chapter emphasizes the exceptional photoluminescence characteristics of carbon quantum dots (CQDs), which hold significant promise for sensing applications in optical sensor technology. By undergoing physical and chemical modifications, CQDs exhibit enhanced physicochemical properties that lead to improved sensing performance. The mechanisms of interaction between CQDs and chemical analytes, such as dipole-dipole interactions and charge transfer, are examined. These interactions modify the fluorescence behavior of CQDs, resulting in stronger fluorescence signals and better optical sensor efficiency. Additionally, integrating CQDs into diverse optical sensing techniques further contributes to the advancement of optical sensor technology [21].

7.3.3 Comparison with Conventional Biosensors

Quantum biosensors (QBs) represent an emerging and rapidly evolving area of research. As advancements in technology continue, QBs are expected to have more practical applications in the fields of biotechnology and healthcare. Unlike traditional biosensors, QBs utilize quantum mechanical phenomena, such as quantum interference and entanglement, to detect biomolecules even at extremely low concentrations, including at the single-molecule level. This innovative approach holds the potential to overcome the limitations commonly associated with conventional biosensing techniques. Conventional sensors, for instance, are prone to drift—an accumulation of errors caused by factors like noise and manufacturing imperfections. These sensors are also limited by the standard quantum limit (SQL), which restricts the accuracy of measurements due to inherent noise and measurement errors. On the other hand, QBs are capable of measuring various physical properties, such as electric, magnetic, and thermal fields, with exceptionally high sensitivity. Their compact size allows them to capture data that traditional sensors cannot, offering significant advantages in terms of precision. Moreover, the nanoscale design of QBs enables them to be miniaturized further, making them ideal for use in restricted environments, such as lab-on-a-chip (LOC) platforms or portable diagnostic devices. The fast response times of QBs also make them well-suited for real-time monitoring of biological processes, a critical feature for applications like medical diagnostics. With ongoing advancements in quantum technology and physics, it is expected that next-generation biological sensors with significantly improved performance will soon become available. This text describes the advantages of quantum biosensors over conventional biosensors and their potential in new biotechnological applications [22].

For example, Peng et al. demonstrated operation below the SQL, providing enhanced sensitivity and a lower detection limit (DL) when compared to traditional biosensors. This sensor incorporates gold-coated polaritons (SPPs) alongside a multimode fiber (MMF) and single mode fiber (SMF) configuration. The 50 nm gold layer was deposited using a plasma sputtering technique. Variations in Bovine Serum Albumin (BSA) concentration were monitored by analyzing changes in the number of transmitted photons. The biosensor achieved a detection limit of 0.971 μg/mL and a sensitivity of 0.0342/(mg/mL). These performance metrics surpass those of conventional Surface Plasmon Resonance

(SPR) sensors. Additionally, the sensor offers key advantages, including a rapid response time of five seconds, label-free detection, and minimal photodamage due to its operation at low photon intensities. Experimental outcomes align with quantum theoretical predictions, highlighting the potential of quantum biosensors for highly sensitive applications such as single-molecule detection, biomedical engineering, and early disease diagnostics. Nonetheless, challenges remain; the fabrication process for quantum biosensors tends to be more complex and costly compared to conventional alternatives. Furthermore, ensuring the stability and long-term reliability of quantum materials is a critical hurdle that must be addressed before these sensors can be deployed widely in practical applications [23] (Table 7.1).

Table 7.1 The main differences between quantum biosensors and conventional biosensors.

Feature	Quantum biosensors	Traditional biosensors	References
Detection sensitivity	Extremely high; can detect at the single-molecule level	Medium to high; typically detects bulk samples	[22]
Working principle	Utilizes quantum phenomena (spin, superposition, entanglement)	Based on biochemical, electrochemical, or optical principles	[22]
Measurement speed	Generally faster and can provide real-time data	Measurement time varies depending on the method used	[5]
Technology used	Quantum dots, superconducting circuits, NV centers in diamonds	Enzymes, antibodies, fluorescent dyes, electrodes	[16]
Measurement accuracy	High accuracy with minimal error	Accuracy depends on the type of biological and physical sensors used	[24]
Sensitivity to external factors	More sensitive to environmental factors (temperature, magnetic fields, etc.)	Generally more stable and less affected by external factors	[25]
Applications	Cancer detection, virus identification, DNA analysis, neurological disease diagnosis	Blood glucose monitoring, pathogen detection, food safety analysis	[26]
Cost	High; requires advanced laboratory environments	Relatively lower; mass production is possible	[24]
Commercial availability	Still in early research and development stages	Widely used in healthcare, food, and environmental sectors	[27]
Operating environment	Typically requires a controlled laboratory environment	Portable versions available, suitable for field use	[28]
Measurement repeatability	High; consistent results under the same conditions	Medium to high; varies depending on biological material	[29]
Energy consumption	May require more energy	Low; often battery-powered portable systems are available	[30]

7.4 Design and Functionality of Quantum Biosensors for Cardiovascular Biomarkers

7.4.1 Key Cardiovascular Biomarkers for Detection

7.4.1.1 Troponin (TnC), (TnI), (TnT)

Troponins are regulatory proteins that play a key role in both skeletal and cardiac muscle contraction. Cardiac troponins (cTns) form a complex made up of three subunits: troponin C, troponin I (cTnI), and troponin T (cTnT). While troponin C is present in both cardiac and skeletal muscle, cTnI and cTnT are exclusive to cardiac muscle. Within myocytes, troponins are distributed across two main pools: the free cytosolic pool (free TnI) and the sarcomeric pool (TnT). Following myocardial injury, troponins from both pools can be detected in the bloodstream, exhibiting different release patterns. cTnT is typically released more gradually and can be detected for up to two weeks post-injury, while cTnI is released more rapidly within 1–2 hours. In recent years, cTns have become important as prognostic biomarkers for patients with chronic heart failure (HF) [31].

7.4.1.2 B-type Natriuretic Peptide (BNP)

B-type natriuretic peptide (BNP) is recognized as a valuable and promising biomarker for heart failure due to its significant role in diagnosis, characterized by its quick release kinetics and well-established cut-off threshold. Brain natriuretic peptide (BNP) is primarily synthesized in the left ventricular myocardium and is regulated through post-translational modifications. First isolated from pig brains in 1988, BNP is a small protein released by the heart when excess fluid causes the ventricles to stretch. The human BNP is a 3.5 kDa peptide formed by the cleavage of pro-BNP into two parts: NT-proBNP (N-terminal fragment) and BNP (C-terminal fragment). Pro-BNP consists of 77–108 amino acids, and further processing of pro-BNP generates the biologically active BNP peptide, which consists of 32 amino acids. BNP acts as a regulatory molecule within the renin-angiotensin-aldosterone system (RAAS), promoting diuresis, natriuresis, and vasodilation by increasing the glomerular filtration rate (GFR). Additionally, it enhances sodium excretion from the kidneys, which leads to increased urine output, reduced blood volume, and decreased blood pressure [32].

7.4.1.3 High-Sensitivity C-Reactive Protein (hs-CRP)

C-reactive protein (CRP) is a member of the pentraxin family of proteins, consisting of five homopentamer subunits, each containing 206 amino acids. The molecular weight of the entire CRP molecule is approximately 120 kDa, and its structure can be determined through X-ray crystallography. The purpose of these studies is to explore the diagnostic relevance of CRP, the effectiveness of various analytical techniques, and recent advancements in CRP measurement methods. CRP plays a crucial role in the immune system's response to pathological conditions like bacterial infections, systemic inflammation, and organ dysfunction. Elevated CRP levels are indicative of these health issues and aid in their diagnosis. Current diagnostic methods for CRP include both traditional laboratory approaches and simpler bedside devices, such as lateral flow and immunoturbidimetric tests. Despite their utility, these methods have certain limitations, which is why biosensors and bioassay devices are being developed as more accurate, efficient alternatives for CRP detection [33].

7.4.1.4 Myoglobin

Myoglobin (Mb) is an oxygen-binding protein found in muscles and serves as a crucial biomarker for cardiovascular health. It plays a significant role in the quick detection of heart-related conditions. A cost-effective, reliable, and durable analytical device for paper-based potentiometric detection has been developed for POC testing. The device utilizes a molecular printing technique to create biomimetic antibodies on carboxylated multi-walled carbon nanotubes (MWCNT-COOH). These sensors are designed specifically for the detection of Mb, demonstrating high levels of accuracy and selectivity. The sensors are capable of selectively detecting Mb at low detection limits, outperforming other biomolecules. This innovative approach provides an efficient means for the mass production of disposable, affordable paper-based potentiometric sensors for use in clinical diagnostics [34].

7.4.1.5 Creatine Kinase-MB (CK-MB)

Creatine kinase (CK)-MB, first identified as a cardiac biomarker in 1979, is commonly used to assess reinfarction. Its levels rise after symptom onset and normalize within 32–72 hours, but its low cardiac specificity limits its diagnostic value. Similarly, myoglobin, an oxygen-binding protein present in multiple muscle types, serves as a marker of myocardial cell injury. It increases in the bloodstream within 2–3 hours and peaks at 4–6 hours post-damage, yet its lack of specificity for heart muscle reduces its reliability. Despite these limitations, measuring CK-MB and myoglobin alongside troponin helps evaluate the extent of cardiac damage, offering valuable insights when combined with imaging and pathology studies [35] (Table 7.2).

7.4.2 Quantum Materials and Detection Techniques

Quantum materials are crucial in enhancing biosensor performance. By integrating these materials with advanced detection methods, they enable increased sensitivity and lower detection thresholds.

7.4.2.1 Quantum Materials

- *Quantum dots:* Low-dimensional materials are attracting considerable attention in the development of functional devices, with their quantum effects enabling advancements in nanoelectronics. The performance of these materials is governed by the quantum confinement (QC) of charge carriers and the de Broglie wavelength theory. Quantum dots (QDs), which are 0D semiconductor nanoparticles, offer the ability to finely tune their electrical and optical properties due to their size-dependent energy gaps. QDs are often referred to as "artificial atoms" because the electrons within atomic clusters are fully confined, resulting in a discrete energy spectrum. This characteristic provides significant advantages over bulk materials. QDs have shown considerable promise in applications such as gas sensors, where they play crucial roles in gas adsorption and detection. Additionally, the growing demand for real-time environmental pollution monitoring and health tracking is driving the need for more advanced sensing technologies [41].
- *Carbon quantum dots (CQDs):* They have gained an important place among the next generation of quantum materials due to their quantum confinement effects (QCE) and

Table 7.2 The clinical use of cardiovascular biomarkers, which diseases they help to diagnose, the time of elevation, and their clinical interpretation, and it explains the clinical importance of each biomarker and how it is used.

Biomarker	Source tissue	Primary clinical use	Time to initial rise	Clinical utility	Normal reference range	Clinical interpretation	References
Troponin (I, T)	Cardiac muscle	Diagnosis of Acute Myocardial Infarction (AMI)	3–6 hours	Essential for diagnosing myocardial injury; critical for early detection of AMI	<0.01–0.04 ng/mL (varies by test)	Highly specific for myocardial injury; remains elevated for several days.	[36]
BNP	Cardiac ventricles	Diagnosis and prognosis in heart failure	Within minutes to hours	Used to diagnose heart failure and assess severity; helps in predicting prognosis	<100 pg/mL	Elevated in heart failure; reflects ventricular wall stress. Useful for diagnosis and prognosis.	[37]
hs-CRP	Liver	Cardiovascular risk assessment	4–6 hours	Used to assess inflammation and cardiovascular risk, particularly for predicting future cardiovascular events	<1 mg/L (low risk)	Marker of systemic inflammation; higher levels linked with increased cardiovascular risk	[38]
Myoglobin	Skeletal and cardiac muscle	Early marker of myocardial injury	1–3 hours	Useful for early detection of myocardial injury but lacks specificity	<85 ng/mL	Early but non-specific marker; also rises with skeletal muscle damage.	[39]
CK-MB	Cardiac muscle	Diagnosis and monitoring of AMI	3–12 hours	Used for detecting myocardial damage and reinfarction due to its shorter half-life	Male: <5 ng/mL Female: <3 ng/mL	Less specific than troponin; used for detecting reinfarction due to its shorter half-life.	[40]

unique optoelectronic properties. Carbon quantum dots (CQDs) are a class of fluorescent carbon nanomaterials with a zero-dimensional (0D) structure, typically measuring less than 10 nm in size. Initially identified by Xu et al. in 2004 while purifying single-walled carbon nanotubes (SWCNTs), their development was further advanced by Sun et al. in 2006 through the process of laser ablation. CQDs exhibit remarkable properties, including high photoluminescence, elevated quantum yield (QY), excellent thermal and optical stability, ease of chemical modification, and good biocompatibility. Additionally, they can be produced at a low cost from bio-based sources. Due to these characteristics, CQDs have found significant applications in diverse fields, including electrochemical and biosensing, bioimaging, drug and gene delivery, energy conversion and storage, photocatalysis, electrocatalysis, and corrosion prevention [42].

- *Graphene quantum dots (GQDs):* GQDs, carbon nanodots (CNDs), and carbonized polymer dots (CPDs) exhibit a wide range of properties. GQDs, specifically, feature a distinct graphene lattice and possess advantageous characteristics such as chemical stability, fluorescent activity, photostability, low cytotoxicity, high solubility, and biocompatibility. These materials generate luminescence emissions influenced by factors like the QCE and edge effects. The surface groups, conjugated π structures, and water solubility of GQDs provide potential for applications in fields such as bioimaging, sensors, photovoltaics, and energy storage. Additionally, their high solubility in organic solvents and enhanced solubility in aqueous solutions present significant opportunities for biotechnological and medical uses. However, the quantum yield of GQDs, which is the ratio of absorbed light to emitted light via fluorescence, is often limited by challenges such as low luminescence quantum yield, inconsistent fluorescence emissions, and poorly understood luminescence mechanisms. Improvements in the surface chemistry of GQDs and adjustments to their photoluminescent properties through oxidation and organic modifications can help enhance their quantum yield. Recent studies have reported quantum yields in GQDs and functionalized GQDs ranging from 2% to 46%. For instance, single-layer GQDs have achieved a quantum yield of 46%, while multilayer GQDs exhibit a lower yield of 2.29%. GQDs doped with magnesium and sulfur (Mg-S/GQDs) demonstrate an impressive fluorescence quantum yield of 53%. The variation in quantum yield is primarily influenced by the morphology of the GQDs and the impact of oxygen-containing surface groups. Fluorescence spectroscopy tests, with quinine sulfate serving as the reference standard, have provided valuable insights into the quantum yield of GQDs. These findings highlight the potential applications of GQDs in fields such as bioimaging and sensor technology [43].
- *Quantum ring fluids and superfluids:* Quantum fluids, such as superfluid helium-4 (He II), are composed of two intermixing fluid components: the superfluid (non-viscous) and the normal fluid (viscous, consisting of thermal quasi-particles). A key feature of superfluids is the presence of topological defects called quantized vortices. In three-dimensional space, these vortices appear as slender tubes, each with a fixed circulation. The movement of these vortices is responsible for phenomena like quantum turbulence, Bose-Einstein condensation, and dissipation in type II superconductors. As vortices shift, mutual friction between the normal fluid and the superfluid is induced, highlighting the importance of understanding the dynamics of vortices. Various models have been developed to describe the motion of vortices, often incorporating the

Navier-Stokes equation. These models help explain observations in He II turbulence by characterizing the interaction between the normal and superfluid components. Recently, the S2W model has been advanced, allowing for direct calculation of the mutual friction coefficient. This model has yielded results that align with experimental data, offering deeper insights into vortex behavior. These developments could significantly enhance our understanding of vortex dynamics and contribute to further research in quantum two-fluid systems [44].

7.4.2.2 Detection Techniques

7.4.2.2.1 Fluorescence Detection Nitrite (NO_2) is commonly found in both food and the environment, and excessive consumption of NO_2 can pose serious health risks. As a result, it is crucial to detect NO_2 quickly and accurately. Traditional detection methods often face challenges such as the need for expensive instruments and complex procedures. Current standard techniques, including the Griess assay and 2,3-diaminonaphthalene assay, suffer from slow detection rates and poor solubility in water. However, CQDs, which have recently been developed, offer several advantages, including easy synthesis, low cost, high quantum yield, excellent photostability, tunable emission properties, good water solubility, and low toxicity. These characteristics make CQDs ideal for use in fluorescence-based NO_2 detection. This review provides an overview of the synthesis methods for CQDs, highlights recent advancements in fluorescence detection techniques, and discusses the challenges and future directions in this field [45].

Carbon quantum dots (CQDs) were synthesized using a microwave-assisted approach with orange peel as the carbon source and applied for the sensitive and selective detection of *Escherichia coli* O157:H7 (*E. coli*). These CQDs exhibited strong green fluorescence when excited at 420 nm. A fluorescent probe, CQDs-MNP, was created by combining aptamer-labeled CQDs and complementary DNA-labeled magnetic nanoparticles (cDNA-MNPs). The fluorescence intensity of the CQDs-MNPs decreased upon the addition of *E. coli*, establishing a linear correlation between fluorescence intensity and *E. coli* concentration. This method enabled the development of a fluorescence-based detection system with a detection range of 500–10^6 CFU/mL and a detection limit of 487 CFU/mL. Furthermore, the method was applied to analyze milk samples contaminated with *E. coli*, and the results were consistent with those obtained using the plate-counting method. This fluorescent probe shows significant potential for ensuring food safety and quality [46].

Liu et al. presented the development of a dual-channel "naked eye" colorimetric and ratio fluorescence probe for detecting curcumin and hypochlorite (ClO), utilizing titanium carbide quantum dots (Ti_3C_2 MQDs). Ti_3C_2 MXene quantum dots emit fluorescence in the 350–600 nm range, which aligns with the UV absorption of curcumin, promoting fluorescence resonance energy transfer (FRET). Upon adding ClO, the phenolic and methoxy groups in curcumin undergo oxidation to form quinones, restoring the fluorescence of Ti_3C_2 MQDs. Additionally, the probe allows for the visual detection of both curcumin and ClO− with the naked eye. The linear detection range for curcumin spans from 0.05 to 10 μM, with a detection limit of 20 nM. For ClO−, the detection ranges are 25–150 μM and 150–275 μM, with a detection limit of 5 μM. This work introduces the first dual-channel "naked eye" colorimetric and ratio fluorescence detection method for curcumin and ClO using Ti_3C_2 MQDs [47].

Yi et al. created a new nanoprobe by modifying molybdenum disulfide quantum dots (MoS QDs) with 3-aminophenyl boronic acid and further functionalizing them using hydropropyl-β-cyclodextrin (β-CD). The nanoprobe was analyzed using various techniques, including transmission electron microscopy (TEM), X-ray photoelectron spectroscopy (XPS), fluorescence, and UV spectroscopy. For the detection of parathion-methyl (MP), the compound was hydrolyzed into p-nitrophenol (p-NP) under alkaline conditions. Due to the host-guest recognition ability of β-CD, p-NP is able to enter the β-CD cavity, resulting in fluorescence quenching of the nanoprobe. This principle enabled the development of an enzyme-free platform for MP detection. The β-CD-MoS QDs nanoprobe demonstrates a broad detection range of 0.01–18.0 ppm and a low detection limit of 3.3 ppb for MP. Moreover, the nanoprobe exhibits excellent selectivity for MP, making it suitable for MP detection in real-world samples [48].

John et al. synthesized carbon quantum dots (CQDs) from the rhizomes of *Acorus calamus*, achieving a quantum yield (QY) of 15%. These CQDs were utilized in four distinct applications: fluorescence detection of morin, catalytic reduction of rhodamine B (Rh B) and sunset yellow (SY) dyes, creation of fluorescent ink for writing and drawing, and investigation of their anticancer and apoptosis effects on SKMEL-28 human skin cancer cell lines. The CQDs' optical, structural, and morphological properties were characterized using various techniques. The average particle size was found to be 6 nm, with blue fluorescence emission when exposed to UV light at 360 nm. The developed nanosensor, based on static and inner filter effects (IFE), exhibited high selectivity for morin, with a linear detection range of 0–2 μM and a detection limit of 96 nM. A $NaBH_4$-CQD mixture was used to catalyze the reduction of Rh B and SY dyes within seconds, following a pseudo-first-order kinetic process with a rate constant of 0.0116/minutes. Cytotoxicity tests revealed the anticancer potential of CQDs against SKMEL-28 cells, with a low LD50 value of 102 μg/mL. Additionally, apoptosis studies using the double staining method showed both live and dead cells in the treated samples. This study demonstrates the wide range of applications and potential uses of CQDs [49].

Chen et al. developed a ratiometric fluorescence sensing system for pesticide detection using a combination of blue-emitting nitrogen-doped CQDs (N-CQDs) and red-emitting copper nanoclusters (CuNCs). This system allows for the selective detection of two common pesticides, thiram and paraquat, by adjusting the solvent environment. The detection limits were found to be 7.49 nM for thiram and 3.03 nM for paraquat. Furthermore, a strong linear relationship was observed for thiram within the range of 10–500 nM and for paraquat within 5–100 nM. Notably, the system also enables fast and visual detection of these pesticides through smartphone-based colorimetric analysis, offering a practical solution for real-time monitoring of pesticide residues in food [50].

7.4.2.2.2 Electrical Sensing Feng et al. developed an electrochemical sensor for nitrite detection using a green synthesis approach. Xylan-based CQDs were employed as an *in situ* reducing agent to prepare gold nanoparticles (Au@CQDs). The sensor was fabricated by integrating the Au@CQDs with MXene, a material known for its excellent electrical conductivity, to create Au@CQDs-MXene nanocomposites. These nanocomposites were then loaded onto a glassy carbon electrode to form the electrochemical sensor. The resulting sensor demonstrated high sensitivity, stability, a wide linear detection range, and excellent selectivity. These advantages were attributed to the high catalytic properties of the gold nanoparticles and CQDs, the large surface area of MXene, and the superior electrical conductivity of both MXene and gold nanoparticles. Under optimal conditions, the sensor exhibited a linear detection range from 1 to 3200 μM, with a detection limit of 0.078 μM (S/N = 3), surpassing the sensitivity of many sensors reported in literature using differential pulse voltammetry (DPV). Additionally, the sensor successfully detected nitrite in tap water and pickled vegetables, showcasing its potential for practical applications in nitrite detection [51].

Tran et al. developed an electrochemical biosensor for the detection of breast cancer cells (MCF-7) in human serum, utilizing nitrogen-doped GQDs (N-GQDs) and phytohemagglutinin-L (PHA-L). NGQDs were synthesized using passion fruit juice as a green carbon source via a microwave-assisted hydrothermal method and were conjugated with PHA-L through an amide bond. This dual functionalization enhanced the electrical conductivity of the sensor and allowed PHA-L to act as a nanocargo, facilitating the efficient binding of MCF-7 cells. The sensor exhibited a broad linear detection range of 5–106 cells/mL in phosphate-buffered saline (PBS) and 20–106 cells/mL in human serum, with detection limits as low as 1 and 2 cells/mL, respectively. Additionally, the sensor demonstrated excellent selectivity against six different interferents and maintained long-term stability even after 80 days of storage. This makes the sensor a highly effective tool for the electrochemical detection of low concentrations of cancer cells, offering potential for early diagnosis and treatment of breast cancer and other malignant tumors [52].

7.4.2.2.3 Magnetic Detection Techniques Webb et al. explored the potential of non-invasive and non-contact methods to measure the magnetic fields generated by electrical signals from action potentials, which holds significant promise in the biomedical field. Traditional magnetic field detection techniques often require cryogenically cooled superconducting detectors, which come with several limitations, such as high costs, low flexibility, poor portability, and restricted spatial and temporal resolution. Here, an alternative approach is proposed that uses nitrogen vacancy centers in diamond to detect magnetic fields produced by action potentials. The first magnetic field measurements were conducted using diamond sensors, with mouse muscle activation via blue light in an optogenetic setup, achieving a sensitivity of 50 $pT/\sqrt{Hz}$. These measurements were performed in a standard, unshielded laboratory environment, and the signals were successfully recovered using digital signal processing techniques. Although the sensitivity is currently not on par with electrophysiology, this work demonstrates the feasibility of detecting action potentials through magnetic fields using diamond quantum sensors, marking a significant step toward microscopic imaging of electrical activity in biological tissues [53].

7.4.2.2.4 Temperature and Gas Detection This research emphasizes the urgent need for advancements in gas sensing technology, particularly highlighting the potential of low-dimensional semiconductor structures to transform traditional gas sensors. Quantum dots, due to the confinement of charge carriers in all three spatial dimensions, exhibit unique electrical and optical properties. The field of quantum dot chemical sensors is rapidly evolving, becoming an increasingly important area within modern sensing technologies. Quantum dot-based structures have demonstrated promising performance in chemically resistant sensors. However, their integration into monitoring systems is still in its early stages. Studies suggest that it is crucial to explore the impact of synthesis methods and additives on the sensing capabilities of these structures. Additionally, the materials' response and selectivity must be assessed, taking into account how the quantum dots' band gap changes with size. This chapter provides an overview of the progress in using semiconductor quantum dots for chemical gas sensors, focusing on advancements in metal oxide, chalcogenide, and CQD production. It also examines how different preparation methods affect the structural characteristics, chemical nature, size reduction, and electronic properties of quantum dots and their influence on sensing performance [41].

Wongrat et al. introduced a highly selective ammonia (NH_3) sensor based on ZnO nanostructures integrated with GQDs. The ZnO:GQD heterojunctions were designed to detect ammonia vapor. Various characterization methods, including FE-SEM, EDS, FTIR, TEM, and XPS, were employed to analyze the surface morphology and chemical composition. The results demonstrated that ZnO:GQD sensors displayed significant sensitivity for ammonia detection at room temperature, largely due to the enhanced oxygen density from carboxyl and hydroxyl groups, which increased the density of H molecules. These sensors showed excellent selectivity and sensitivity to ammonia under aerobic conditions, with an optimum sensor response of 6047 at a 1000 ppm ammonia concentration [54].

Xue et al. explored the potential of quantum dots (QDs) for mid-infrared (IR) photodetectors with high-temperature (HOT) capabilities. By utilizing quantum confinement in all three dimensions, electron thermal generation is minimized, reducing the energy gap caused by thermal carriers. The solution processability of colloidal nanocrystals, along with their surface dipoles, allows for precise doping control. As a result, HOT mid-IR photodetectors achieve background-limited performance. Additionally, the detector demonstrates excellent performance with high external quantum efficiency and sensitivity, yielding promising results in applications like spectrometry, chemical sensing, and thermal imaging. This work highlights the growing interest in cost-effective, high-performance mid-IR photodetection technologies [55].

7.4.2.2.5 Biosensor Iannazzo et al. highlighted the potential of nanomaterial-based biosensors incorporating GQDs for the early and precise detection of cancer biomarkers. GQDs are ideal for sensor development due to their unique chemical and physical properties, large surface areas, and ease of functionalization with biomolecules. This review explores advanced methods for synthesizing and functionalizing GQDs, enabling them to selectively identify cancer-specific biomarkers, including antigens, enzymes, hormones, cancer-related byproducts, pH variations, and biomolecules found on cancer cell surfaces. Furthermore, these biosensors, including optical, electrochemical, and chemiluminescent types, can be used not only to diagnose various cancer types but also to assess the effectiveness of anticancer treatments. The excellent linear detection range and low detection limits of these biosensors underscore their significant clinical potential in cancer diagnosis and management [56].

Pandit et al. functionalized superparamagnetic iron oxide nanoparticles (SPIONs) with Aptamer I to detect *E. coli* cells. The *E. coli*-SPION complexes were isolated using magnetic separation from sludge. Confirmation of detection was carried out using confocal laser scanning microscopy (CLSM) with CdTe-MPA quantum dots (QDs) conjugated with Aptamer II. Additionally, the ATmega 328P-based prototype biosensor demonstrated both quantitative and qualitative capabilities for *E. coli* detection. This biosensor was able to detect bacterial concentrations as low as 1×10^2 CFU. Featuring a UV LED, an LCD display, and an ATmega328P microcontroller, the biosensor provides high-resolution and sensitive *in situ* detection of *E. coli* in water and food samples [57].

7.4.3 Wearable and Implantable Quantum Biosensors

The advancement of personalized diagnostics and CVD treatment is increasingly dependent on portable biosensing technologies. These devices address the limitations of conventional, hospital-centric healthcare by enabling the monitoring of biomarkers present in bodily fluids, such as sweat. Despite their promise, challenges like insufficient sensitivity, limited selectivity, and lack of specificity remain significant obstacles to their broader adoption. Wei et al. have developed an innovative portable biosensor capable of the simultaneous detection of sodium ions (Na^+), ascorbic acid, and neuropeptide Y in sweat, which are key indicators for evaluating cardiovascular health. The sensor integrates CQD-modified working electrodes, leveraging their expansive reactive surface area to reduce interfacial impedance and enhance detection sensitivity. To ensure high specificity, the sensor employs a combination of selective ion recognition, enzyme-mediated catalysis, and immune-response-based mechanisms tailored to each target biomarker. This system successfully demonstrated the real-time, concurrent monitoring of these three biomarkers in sweat samples collected from healthy volunteers engaged in everyday activities, such as physical exercise and dietary intake. These findings highlight the potential of our portable quantum biosensor as a powerful tool for noninvasive, personalized healthcare, facilitating continuous metabolic monitoring and improving CVD management [26].

Chen et al. introduced a gas sensing model utilizing CuS quantum dots (QDs) and Bi_2S_3 nanosheets (NSs). Simulation results reveal that CuS QDs and Bi_2S_3 NSs possess significant adsorption capabilities and function as charge transport pathways. A sensor based on an artificial neural network was designed to achieve highly sensitive detection of NO_2 gas. Experimental findings demonstrate that CuS QDs exhibit a high adsorption capacity, contributing significantly to NO_2 detection. Additionally, Bi_2S_3 NSs are employed for efficient charge collection and transport. The resulting neuron-inspired sensor showcases excellent performance, featuring a fast response time and low detection limits. The wearable device enables real-time detection of NO_2 [58].

Mukasa et al. presented QuantumDock, an innovative computational framework designed for non-invasive health monitoring through wearable sweat sensors and MIPs. By utilizing density functional theory to optimize selectivity, QuantumDock aids in selecting the best monomer and cross-linker combinations. Graphene-based wearable devices have been successfully used to monitor biomarkers, such as phenylalanine, marking the first successful demonstration of this approach for personalized health applications [59].

7.5 Recent Advances and Applications

7.5.1 Breakthrough Studies in Quantum Biosensors for CVD Diagnosis

Cardiac troponin-I (cTnI) serves as a key biomarker for diagnosing AMI. Yola et al. presented an advanced electrochemical biosensor based on boron nitride quantum dots (BNQDs) for detecting cTnI in plasma samples. To assess the biosensor's characteristics, scanning electron microscopy (SEM), TEM, X-ray diffraction (XRD), cyclic voltammetry (CV), and electrochemical impedance spectroscopy (EIS) were utilized. Experimental findings revealed that the biosensor exhibited high sensitivity, successfully detecting cTnI within a 0.01–5.00 ng/mL range, with an impressive limit of detection (LOD) of 0.0005 ng/mL. Additionally, selectivity tests confirmed that the sensor effectively distinguished cTnI from other biomarkers, including cardiac myoglobin (MYG), BSA, and cardiac troponin T (cTnT). Further evaluation demonstrated the biosensor's stability, reproducibility, and reusability, ensuring its reliability in clinical applications. Ultimately, the developed biosensor was successfully applied to plasma samples, achieving high recovery rates. These findings highlight the potential of quantum biosensors in CVD diagnostics, offering a novel approach for early detection [60].

Kim et al. designed an innovative electrochemical immunosensor using triangular CQDs (TCQDs) stabilized with gold nanoparticles (AuNPs) and functionalized with boron nitride nanosheets (HO-BNNS). The biosensor was systematically examined through advanced characterization methods, including XRD, XPS, TEM, and CV. The fabricated quantum biosensor demonstrated a broad detection range of 0.0001–100 ng/mL, with an exceptionally low LOD of 0.0013 ng/mL. To enhance electrochemical sensitivity, ferrocene carboxylic acid (Fc-COOH) was employed as a signal amplifier. Performance evaluations revealed high specificity, outstanding reproducibility, and strong reliability, along with efficient biomarker detection in human serum samples. These findings highlight the significant potential of quantum biosensors in the early diagnosis of CVDs [61].

Karaman et al. presented the development of an electrochemical immunosensor designed for the early detection of AMI. The precise measurement of cTnI levels in human blood plasma plays a crucial role in ensuring accurate diagnosis. The proposed sensor integrates Ce-doped SnO_2/SnS_2 (Ce-SnO_2/SnS_2) nanostructures with nitrogen- and boron-doped GQDs (N,B-GQDs) to enhance sensitivity. A sandwich-type biosensing platform was established using a capture antibody (anti-cTnI-Ab_1) and a secondary antibody (anti-cTnI-Ab_2) for selective detection. For sensor characterization, CV, DPV, and EIS were employed. Experimental results demonstrated high sensitivity for cTnI detection, with an exceptionally low LOD of 2.00 fg/mL. The outstanding sensitivity and selectivity of this quantum biosensor highlight its potential as a powerful diagnostic tool for AMI, offering rapid and reliable detection capabilities [62].

Acute myocardial infarction (AMI) is a critical cardiovascular condition resulting from inadequate blood flow to the heart muscle. Hasabnis et al. presented the development of an electrochemical voltammetric monitoring system for AMI diagnosis, utilizing polyaminophenol-based thin films. To enhance sensitivity, a molecularly imprinted polymer (MIP) was integrated with gold nanoparticles (AuNPs) and GQDs, creating a nanocomposite designed to detect cTnI. The sensor's performance was evaluated using techniques such as CV, square wave voltammetry (SWV), EIS, ATR-FTIR, and fluorescence microscopy. The findings indicate that the biosensor provides excellent sensitivity, selectivity, and reproducibility for cTnI detection. Additionally, it demonstrated long-term stability for up to 30 days and a rapid detection time of 18 minutes, making it a promising tool for early diagnosis and POC applications. This research highlights the potential of quantum biosensors in advancing CVD diagnostics [63].

Wang et al. designed an ultrasensitive electrochemiluminescence (ECL) immunosensor for the detection of cTnI. The integration of MoC@C nanospheres with $Ru(bpy)_3^{2+}$ significantly enhanced the ECL signal, ensuring superior conductivity and chemical stability. Additionally, BNQDs were employed as co-reactants, further improving the sensor's efficiency. The biosensor demonstrated remarkable specificity, stability, and reproducibility, achieving a low detection limit of 0.03 pg/mL and a broad linear range of 0.1 pg/mL–50 ng/mL. Furthermore, its successful application in human serum analysis highlights the strong potential of this quantum biosensor in the early diagnosis of CVDs [64].

Mansuriya et al. modified an electrochemical nano-immunosensor with GQDs and developed gold nanoparticles (AuNPs) for the early diagnosis of AMI. This enzyme-free sensor, designed using a screen-printed gold electrode (SPGE), offers high sensitivity for cTnI biomarker detection. The sensor was characterized by SWV, CV, EIS, and amperometry techniques, providing a low LOD of 0.1 pg/mL and a fast analysis time of 10.5–13 minutes. Analyses by atomic force microscopy (AFM), TEM, and SEM showed that the nanocomposite structure was successfully formed. This study demonstrates the innovative potential of quantum biosensors in CVD diagnosis in terms of sensitivity, cost-effectiveness, speed, and specificity (Figure 7.1a) [65].

Tuteja et al. designed a label-free EIS immunosensor for the detection of cardiac myoglobin (cMyo). Graphene quantum dots (GQDs) were utilized to construct a sensor platform on SPEs and functionalized with anti-myoglobin antibodies. The sensor's effectiveness was assessed by monitoring charge transfer resistance (Rct) variations, exhibiting a linear response between 0.01 and 100 ng/mL. Specificity evaluations confirmed accurate cMyo detection even in the presence of other biomarkers. With a low LOD of 0.01 ng/mL, the sensor's sensitivity matches ELISA-based methods. This advanced quantum biosensor offers a fast, sensitive, and selective approach for CVD diagnostics [67].

Vasudevan et al. presented the development of an electrochemical biosensor for the detection of Troponin I (cTnI), a key biomarker for diagnosing AMI. The sensor was constructed using lignin-derived graphene (LSG), which was further modified with N-GQDs and silver nanoparticles (Ag NPs). The hydrothermally synthesized nanocomposite exhibited excellent conductivity and biocompatibility. Characterization through EIS demonstrated a low detection limit of 1 fM (30 fg/mL) along with high selectivity, stability, and reproducibility. This novel quantum biosensor holds significant potential for accurate and efficient CVD diagnostics (Figure 7.1b) [66].

Liu et al. developed an ECL biosensor for the detection of cTnI, a key biomarker for diagnosing AMI. The sensor was designed by integrating N, Zn-GQDs onto a graphene oxide (GO)-coated glassy carbon electrode. Through amide bond formation, cTnI antibodies were immobilized on the surface of N, Zn-GQDs to enhance specificity. The biosensor's performance was assessed using ECL and DPV, demonstrating effective

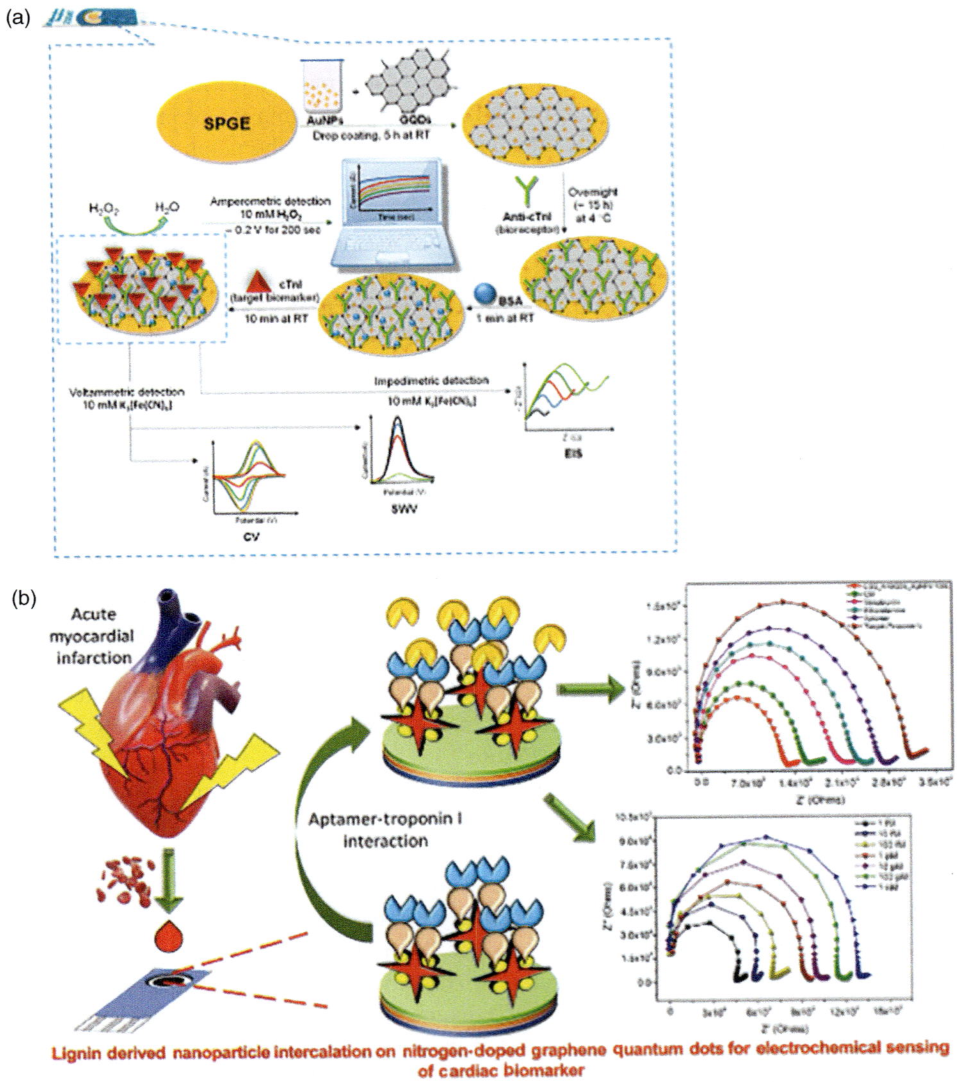

Figure 7.1 (a) Schematic overview illustrates the fabrication process of a label-free electrochemical immunosensor for cTnI detection. This sensor is developed using a screen-printed gold electrode (SPGE) modified with gold nanoparticles (AuNPs) and graphene quantum dots (GQDs). Mansuriya and Altintas [65] / MDPI / CC BY 4.0. (b) The schematic summary illustrates the development of an N-GQD and Ag NP-doped LSG sensor for electrochemical Troponin I detection, achieving a detection limit of 1 fM. Vasudevan et al. [66] / with permission of Elsevier.

detection of cTnI in human serum. Results aligned with reference values from the ELISA method, with a relative standard deviation (RSD) of 9.09–11.1%. This cost-effective and highly sensitive biosensor holds promise for detecting other biomarkers in clinical diagnostics [68].

Gurjar et al. designed an advanced quadruple hybrid nanophotonic assay system for the early detection of CVD by integrating smart biosensors with artificial intelligence (AI). The system incorporates CNDs and graphene nanodots (GNDs) to detect key biomarkers, including circulating cell-free mitochondrial DNA (ccf-mtDNA), exosomes, and NT-proBNP. Characterization through EIS and optical analysis confirmed high sensitivity, with a detection limit as low as 0.369 fg. Additionally, a long short-term memory (LSTM) deep learning model processed biosensor data, achieving an impressive 99.86% diagnostic accuracy. This AI-powered quantum biosensor presents a rapid, cost-effective, and highly precise solution for CVD diagnosis.

Plasmonic- and SPR-based biosensors have become increasingly prominent due to advancements in nano-optics. Their incorporation into POC diagnostic systems has facilitated label-free detection, ensuring enhanced sensitivity and lower detection limits. Wearable plasmonic sensors, in particular, hold significant promise for applications in personalized medicine and global healthcare. By integrating with microsystems and LOC platforms, these sensors are evolving into compact and portable diagnostic tools. Additionally, quantum plasmonic sensors are gaining attention for their potential in chemical and biological detection, especially when combined with microfluidic technologies. Continuous progress in nanophotonics, plasmonics, and microsystem engineering is essential for improving the efficiency and accuracy of biosensors. These technological advancements pave the way for next-generation biosensors by addressing challenges related to quantum fluctuations and noise, enabling ultrasensitive detection, and expanding their role in health monitoring and disease diagnostics [4].

Liu et al. designed a triple cascade quantum-strip (TCQS) biosensor for the rapid, sensitive, and simultaneous detection of heart failure (HF) biomarkers BNP, NT-proBNP, and ST2. Utilizing quantum immunoprobes, the system enabled direct biomarker enrichment from serum samples with high selectivity. To enhance usability, the biosensor was integrated with a smartphone-based portable reader, ensuring POC testing (POCT) compatibility. The system demonstrated high sensitivity (LOD: 0.097, 0.072, and 0.948 ng/mL) and 92.86% accuracy, establishing itself as a miniaturized, portable, and efficient biosensing platform for HF diagnosis, risk assessment, and disease monitoring (Figure 7.2a) [69].

7.5.2 Performance Evaluation in CVD Diagnosis

Cardiovascular diseases (CVDs) are a significant global health issue, causing considerable socioeconomic challenges. The demand for diagnostic technologies with high sensitivity and specificity is increasing due to the importance of early and accurate detection of these diseases.

In this regard, cardiac biomarkers (such as BNP, NT-proBNP, troponin, and ST2) are crucial for the early detection of CVDs. Advanced technologies now enable the detection of these biomarkers with improved sensitivity and a broader dynamic range. Portable and rapid POCT systems are enhancing the possibilities for personalized healthcare and timely intervention. Innovations such as the integration of microfluidic platforms, signal amplification techniques, and next-generation sensor designs are driving progress in this area. However, clinical implementation of these technologies faces challenges, including the complexity of biological samples, reproducibility concerns, and regulatory hurdles.

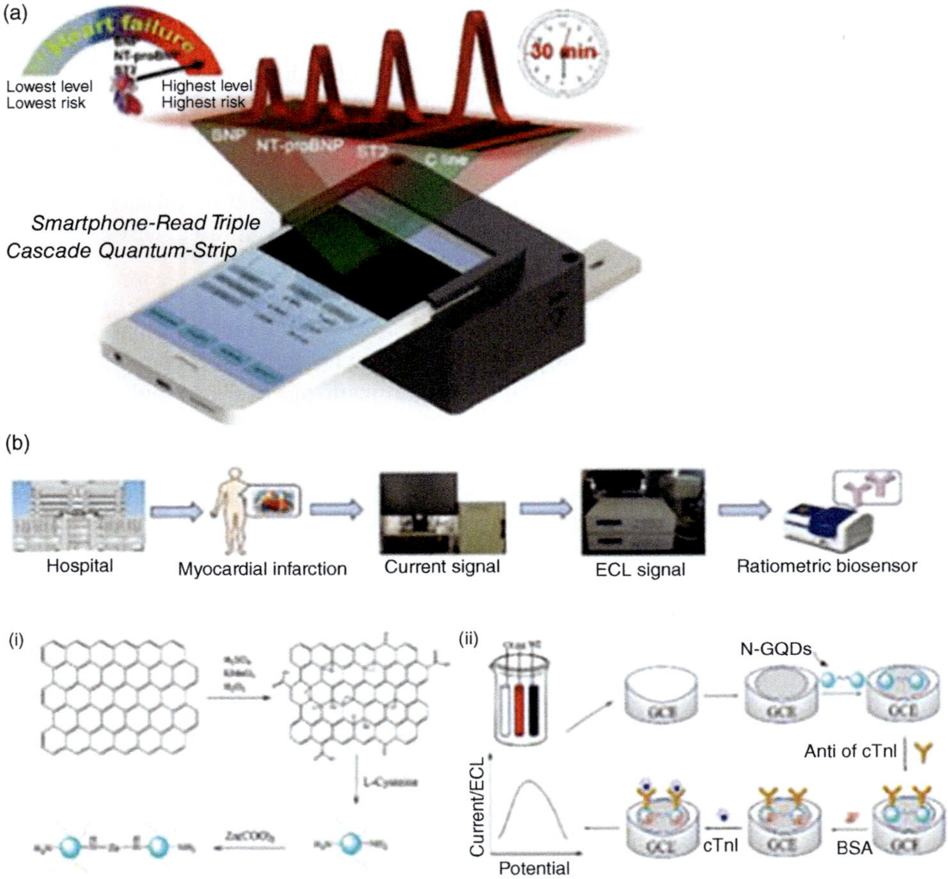

Figure 7.2 (a) Schematic summary shows the development of a triple cascade quantum strip (TCQS) sensing strategy to rapidly and selectively monitor the heart failure (HF) biomarkers BNP, NT-proBNP, and ST2. Liu et al. [69] / with permission of American Chemical Society. (b) Schematic summary: a biosensor using N, Zn-doped graphene quantum dots (N, Zn-GQDs) was constructed to detect cardiac troponin I (cTnI). The sensor showed excellent sensitivity with a detection limit of 4.59 pg/L using electrochemiluminescence and differential pulse voltammetry techniques. Liu et al. [68] / with permission of Elsevier.

In summary, new diagnostic systems with high sensitivity and specificity show great promise for the early detection and effective management of CVDs [70].

Sensitivity is the percentage of individuals who receive a correct positive result among those who actually have the condition. It reflects how effectively a test identifies subjects who truly have the condition of interest. Specificity is the percentage of individuals who are correctly identified with a negative result among those who do not have the condition. It indicates how accurately a test can exclude subjects who do not have the condition of interest [71].

Quantum biosensors (QBs) encounter considerable challenges in detecting target biomolecules within complex biological matrices, particularly when high sensitivity and specificity are required. These two factors are essential performance metrics for ensuring accurate results from QBs. To address issues like non-specific binding and background interference, research should focus on enhancing surface functionalization methods, advancing signal processing techniques, and utilizing machine learning approaches. These advancements aim to optimize crucial parameters, particularly sensitivity and specificity [22].

Nawaz et al. emphasized key performance metrics like sensitivity and accuracy in diagnosing CVDs. The gradient descent optimization (GDO) model demonstrated the highest sensitivity (99.43%) and accuracy (98.54%) for effectively detecting CVDs. These results underscore the crucial role that sensitivity plays in the accurate diagnosis of cardiovascular conditions [72].

Lakshamanakumar et al. highlighted the use of GQDs to enhance the sensitivity of quantum biosensors, particularly in the detection of cardiac biomarkers like CRP. The developed Au/GQDs/anti-CRP biosensor demonstrates the ability to detect CRP at ultralow concentrations, achieving a sensitivity of 58 μA/ng/cm^2 and a detection limit of 0.104 ng/mL. This advancement significantly improves the sensitivity of quantum biosensors for precise and early detection of cardiac biomarkers [73].

This review explores the application of nanomaterial-based biosensors in the diagnosis of CVD, emphasizing the enhancements in sensitivity and specificity for detecting the NT-proBNP biomarker. The incorporation of advanced nanomaterials like quantum dots and graphene derivatives has significantly boosted the sensitivity and specificity, enhancing the accuracy of heart failure (HF) detection. These biosensors enable quicker and more accurate detection, with improved sensitivity and reduced false positives due to high specificity. Despite this, challenges such as clinical integration and associated costs hinder the broader use of these biosensors in healthcare environments. Nevertheless, their potential to improve sensitivity and specificity at the clinical level makes them promising tools for transforming diagnostic approaches in the future [74].

Oranzie et al. explored the application of quantum biosensors for diagnosing CVD. Specifically, an electrochemical aptasensor utilizing MSA-NiSe2 quantum dots (QDs) is examined as a highly effective diagnostic tool for CVD, demonstrating both high sensitivity and high specificity in detecting BNP. The biosensor achieves a remarkable detection limit of 5.45 pg/mL, exhibiting excellent sensitivity and specificity in both physiological buffer and human plasma samples. The unique physical and electronic properties of QDs allow these biosensors to be readily adapted for use in other diagnostic applications, offering promising potential as an efficient diagnostic tool for CVD based on sensitivity and specificity [75].

Liu et al. presented the development of an advanced quantum biosensor designed for the precise detection of CVD biomarkers, emphasizing both high sensitivity and specificity. The biosensor incorporates N, Zn-GQDs to enable the efficient identification of cTnI, a key indicator of cardiac events. Demonstrating excellent sensitivity, the sensor achieved a low detection threshold of 4.59 pg/L. Moreover, it exhibited strong specificity when tested on human serum samples, yielding results consistent with conventional ELISA techniques. These findings highlight the potential of this next-generation quantum biosensor as a reliable tool for the early and accurate diagnosis of CVDs, offering improved performance through enhanced sensitivity and specificity (Figure 7.2b) [68].

Wue et al. presented a quantum biosensor developed for the detection of CVDs, specifically targeting cTnI, a key biomarker for AMI. The biosensor is based on the integration of GQDs and gold nanoclusters, enhancing signal performance through ECL resonance energy transfer (ECL-RET). The biosensor demonstrated exceptional sensitivity, achieving an ultralow detection limit of 354.2 fg/mL. Additionally, it exhibited a broad linear detection range from 500 fg/mL to 20 ng/mL, enabling accurate differentiation of cTnI in complex biological samples while maintaining specificity. These findings highlight the potential of these quantum biosensors, offering high sensitivity for the early and precise diagnosis of CVDs [76].

Quantum biosensors demonstrate exceptional performance in both sensitivity and specificity. They hold great promise for the accurate and swift diagnosis of CVD by providing high sensitivity and low detection thresholds. Additionally, their high specificity makes them a reliable tool for CVD detection, reducing the risk of errors in complex biological samples.

7.5.3 Real-World Applications and Case Studies

Stringer et al. introduced an optical biosensor designed for detecting cTnI. The biosensor employs the FRET technique, utilizing a protein A molecule that induces conformational changes upon binding with the troponin I antigen. Quantum dots serve as the FRET donors, while organic dyes are used as acceptors. The biosensor demonstrates high specificity and sensitivity toward Troponin I, achieving detection limits of 32 nM in PBS and 55 nM in human plasma. With a rapid response time of under one minute, the biosensor offers precise measurements, positioning it as a promising tool for the early detection of myocardial infarction [77].

Song et al. focused on the preparation of WS2 quantum dots (WS2 QDs) using a simple one-step hydrothermal method, resulting in a high quantum yield of 11.23%. The WS2 QDs exhibit photobleaching resistance, salt tolerance, and pH stability. The researchers found that these QDs can be used as highly efficient fluorescent sensors for detecting hemoglobin (Hb) and myoglobin (Myo), with detection limits as low as 260 and 7.6 ng/mL, respectively. In addition, the WS2 QDs were successfully employed to determine Hb/Myo levels in human blood/serum samples, showing promise for clinical applications in diagnosing conditions related to Hb and Myo levels. This study is significant as it is the first example of using transition metal dichalcogenide (TMD) QDs without any modification as a fluorescent sensor for simultaneous detection of Hb and Myo [78].

Xu et al. created a novel nanohybrid probe, termed AuNCs-QDs, for the precise detection of the cardiac biomarker myoglobin (Mb). This probe is formed by integrating gold nanoclusters (AuNCs) with CdSe/ZnS quantum dots (QDs), enabling efficient and sensitive Mb detection. Under UV light, the probe emits a distinct pink fluorescence, with two emission peaks—one from the QDs and the other from the AuNCs. When Mb is present, the fluorescence intensity decreases, accompanied by a noticeable shift in fluorescence color. The probe demonstrates a low detection limit of 4.99 μg/mL and maintains a strong linear relationship up to 0.52 mg/mL. It also shows high specificity for Mb. Additionally, the probe has been successfully used to detect Mb in urine samples, and a smartphone-assisted paper-based strip has been developed for on-site monitoring. This innovative approach has potential applications in various medical diagnostics, providing a valuable tool for biomarker monitoring [79].

Thamminana et al. synthesized water-soluble L-glutamic acid (Glu)-functionalized cesium lead bromide perovskite quantum dots ($CsPbBr_3$ PQDs), specifically $CsPbBr_3$@Glu PQDs, and employed them for Myo detection using a fluorescence "turn-off" method. The synthesized $CsPbBr_3$@Glu PQDs exhibited a remarkable photoluminescence quantum yield of 25%, with an emission peak at 520 nm upon excitation at 380 nm. The fluorescence "turn-off" strategy was developed to detect Myo, with the $CsPbBr_3$@Glu PQDs serving as a straightforward optical probe. This probe demonstrated a broad linear detection range for Myo (0.1–25 μM) and achieved a detection limit of 42.42 nM. The $CsPbBr_3$@Glu PQD-based optical probe showed excellent performance in detecting Myo in serum and plasma samples [80].

Lie et al. presented the development of a portable fluorescent biosensor designed for the rapid and ultrasensitive detection of protein biomarkers. The sensor integrates quantum dots (QDs) with a lateral flow test strip, leveraging the high signal brightness and photostability of QDs alongside the advantages of lateral flow assays to achieve high sensitivity, selectivity, and speed in detection. The biosensor was specifically applied to detect nitrated ceruloplasmin (NCP), a key biomarker associated with CVDs and smoking-related stress response. Through fluorescence intensity measurements, it successfully detected NCP at concentrations as low as 1 ng/mL, while also demonstrating effective performance in human plasma samples across a wide dynamic range, with a detection limit of 8 ng/mL. These findings highlight the strong potential of the quantum dot-based lateral flow test strip for POC diagnostics and on-site biomarker analysis, paving the way for improved early disease detection and monitoring [81].

7.6 Future Perspectives and Challenges

Quantum biosensors offer significant advantages for the precise and selective detection of biomolecules. Semiconductor quantum dots (QDs) are especially valuable due to their characteristics, such as high brightness, excellent photostability, a broad absorption spectrum, a narrow emission spectrum, and tunable size, which contribute to the high efficiency of these biosensors. Additionally, features like a large Stokes shift, versatile surface modification, and unique photoelectrochemical properties enhance the sensitivity of QD-based biosensors. These sensors are recognized for their versatility in detecting biomolecules such as DNA, microRNA, proteins, enzymes, and living cells. By enabling fast, simple, and effective biological analyses, QD-based biosensors hold significant potential in both basic biomedical research and clinical diagnostics. Looking ahead, the development of QD-based biosensors will continue to be shaped by advancements in surface modifications and nanotechnology, aiming to further enhance their efficiency, selectivity, and sensitivity [82].

Quantum biosensors (QBs) encounter several challenges that hinder their widespread adoption and practical application in healthcare. The high development and manufacturing costs, the need for specialized equipment, and complex production processes make QBs less accessible for many medical applications. Integrating QBs into existing medical diagnostic workflows is also a difficult task. Achieving optimal performance, particularly in terms of sensitivity and specificity, can be challenging in real-world settings due to quantum mismatch, which is influenced by external factors such as environmental conditions. Fluctuations in temperature and electromagnetic interference further complicate these issues. Additionally, privacy concerns, data security, safety, and biocompatibility remain significant challenges. Quantum sensor technology is advancing brain imaging techniques. For example, magnetoencephalography is being enhanced with quantum sensors, which allows for the detection of brain activity on a millisecond time scale, even when the person is in motion. These innovations are leading to significant progress in understanding neurological disorders and improving disease detection and monitoring [22]. Quantum biosensors used for detecting CVDs encounter several challenges, including high costs, the need for specialized equipment, and sensitivity to environmental conditions. Moreover, integrating these sensors into existing medical workflows and achieving clinical validation can be complex. Issues like data security and biocompatibility also hinder their broader adoption. Despite these obstacles, if successfully addressed, quantum biosensors hold significant promise for the early detection and treatment of CVDs. The integration of artificial intelligence and machine learning algorithms will significantly enhance disease prediction and facilitate data-driven decision-making in medical diagnostics. To fully unlock the potential of quantum biosensors (QBs) in healthcare, ongoing research, interdisciplinary cooperation, and a thoughtful approach to ethical concerns are essential. Quantum computing has the capacity to revolutionize every industry globally, and with its rapid advancement, many once-theoretical computational processes are now feasible, opening up vast possibilities for innovation across sectors such as aerospace and medicine. Challenges such as data security, doctor-patient relationships, accountability, and cost must be carefully addressed. Blockchain technology, which ensures secure data transmission and surpasses the security standards of existing algorithms like Diffie-Hellman and RSA, offers a promising solution to these data security concerns [83].

7.7 Conclusion

Cardiovascular diseases (CVDs) remain a leading cause of death worldwide, emphasizing the importance of early detection for effective treatment and management. Traditional diagnostic techniques often struggle to identify CVDs in their initial stages, which can delay timely intervention. Quantum biosensors, utilizing the principles of quantum mechanics, present a promising solution to these diagnostic limitations. By leveraging quantum phenomena like superposition, coherence, and entanglement, these sensors offer exceptional sensitivity, enabling the detection of cardiovascular biomarkers even at very low concentrations. This capability could revolutionize the diagnosis and treatment of CVDs, improving patient outcomes. The design of quantum biosensors for cardiovascular biomarker detection focuses on their ability to detect even the smallest changes in biomolecular composition. Recent advancements have enhanced the performance of these sensors by integrating them with nanomaterials and microfluidic technologies, which improve sensitivity and speed. Furthermore, these biosensors can be integrated into wearable devices, providing continuous, real-time monitoring of cardiovascular health.

This integration supports personalized and proactive treatment plans, offering valuable data for patients and healthcare providers in managing heart health effectively. Recent progress in quantum biosensor technology has yielded promising results in both laboratory and clinical settings. These innovations demonstrate the ability to detect cardiovascular biomarkers with greater precision and efficiency than conventional methods. Moreover, the integration of quantum biosensors with artificial intelligence (AI) systems further enhances their diagnostic capabilities, enabling predictive analysis of disease progression and facilitating timely interventions. While the future of quantum biosensors in cardiovascular diagnostics holds great potential, challenges remain, including the reduction of production costs, improving sensor stability in various environmental conditions, and ensuring seamless integration with existing healthcare systems. Additionally, expanding the sensors' ability to detect a wider range of biomarkers and diseases is crucial for their broader clinical application. With ongoing research and development, quantum biosensors have the potential to transform CVD detection, enabling earlier diagnosis and improving overall health outcomes.

References

1 Roth, G.A., Mensah, G.A., Johnson, C.O. et al. (2020). Global burden of cardiovascular diseases and risk factors, 1990–2019: update from the GBD 2019 study. *Journal of the American College of Cardiology* 76: 2982–3021.

2 Alagarsamy, K.N., Mathan, S., Yan, W. et al. (2021). Carbon nanomaterials for cardiovascular theranostics: promises and challenges. *Bioactive Materials* 6: 2261–2280.

3 Chugh, V., Basu, A., Kaushik, A., and Basu, A.K. (2023). Progression in quantum sensing/bio-sensing technologies for healthcare. *ECS Sensors Plus* 2 (1): 15001. https://doi.org/10.1149/2754-2726/acc190.

4 Simone, G. (2024). Trends of biosensing: plasmonics through miniaturization and quantum sensing. *Critical Reviews in Analytical Chemistry* 54: 2183–2208.

5 Chamkouri, H., Si, J., Chen, P. et al. (2024). A first step to develop quantum medicine: radiometers, detectors, and biosensors. *Sensing and Bio-Sensing Research* 44: 100658. https://www.sciencedirect.com/science/article/pii/S2214180424000400.

6 Zhang, T., Pramanik, G., Zhang, K. et al. (2021). Toward quantitative bio-sensing with nitrogen-vacancy center in diamond. *ACS Sensors* 6: 2077–2107.

7 Milošević, M. and Otašević, P. (2022). Treatment-resistant hypertension. *Arhiv za Farmaciju* 72: 1–19.

8 Lopaschuk, G.D., Karwi, Q.G., Tian, R. et al. (2021). Cardiac energy metabolism in heart failure. *Circulation Research* 128: 1487–1513.

9 Nordanstig, J., Behrendt, C.A., Bradbury, A.W. et al. (2023). Peripheral arterial disease (PAD) – a challenging manifestation of atherosclerosis. *Preventive Medicine* 171: 107489.

10 Aluru, J.S., Barsouk, A., Saginala, K. et al. (2022). Valvular heart disease epidemiology. *Medical Sciences (Basel, Switzerland)* 10: 32.

11 Gao, J., Cao, H., Hu, G. et al. (2023). The mechanism and therapy of aortic aneurysms. *Signal Transduction and Targeted Therapy* 8: 55.

12 Saraiva, R.M., Mediano, M.F.F., Mendes, F.S.N.S. et al. (2021). Chagas heart disease: an overview of diagnosis, manifestations, treatment, and care. *World Journal of Cardiology* 13: 654–675.

13 Pang, W., Yuan, C., Zhong, T. et al. (2024). Diagnostic and therapeutic optical imaging in cardiovascular diseases. *iScience* 27 (11). https://doi.org/10.1016/j.isci.2024.111216.

14 RUS, A., Khan, W.U., Khan, M.S., and Cheung, P. (2025). Emerging rapid detection methods for the monitoring of cardiovascular diseases: current trends and future perspectives. *Materials Today Bio* 32: 101663. https://www.sciencedirect.com/science/article/pii/S2590006425002212.

15 Kyriazos, T. and Poga, M. (2024). Quantum concepts in psychology: exploring the interplay of physics and the human psyche. *Bio Systems* 235: 105070.

16 Petrini, G., Moreva, E., Bernardi, E. et al. (2020). Is a quantum biosensing revolution approaching? Perspectives in NV-assisted current and thermal biosensing in living cells. *Advanced Quantum Technologies* 3: 2000066.

17 Castro, R.C., Saraiva, M.L.M.F.S., Santos, J.L.M., and Ribeiro, D.S.M. (2021). Multiplexed detection using quantum dots as photoluminescent sensing elements or optical labels. *Coordination Chemistry Reviews* 448: 214181.

18 Sun, L., Yang, L., Dou, J.H. et al. (2022). Room-temperature quantitative quantum sensing of lithium ions with a radical-embedded metal-organic framework. *Journal of the American Chemical Society* 144: 19008–19016.

19 Robertson, I.O., Scholten, S.C., Singh, P. et al. (2023). Detection of paramagnetic spins with an ultrathin van der Waals quantum sensor. *ACS Nano* 17: 13408–13417.

20 Zou, S., Zhang, H., Chen, X., and Fang, J. (2022). In-situ triaxial residual magnetic field measurement based on optically-detected electron paramagnetic resonance of spin-polarized potassium. *Measurement* 187: 110338.

21 Nazri, N.A.A., Azeman, N.H., Luo, Y., and Bakar, A.A.A. (2021). Carbon quantum dots for optical sensor applications: a review. *Optics and Laser Technology* 139: 106928.

22 Das, S., Mazumdar, H., Khondakar, K.R. et al. (2024). Review—quantum biosensors: principles and applications in medical diagnostics. *ECS Sensors Plus* 3 (2): 25001. https://doi.org/10.1149/2754-2726/ad47e2.

23 Peng, Y., Zhao, Y., Hu, X., and Yang, Y. (2020). Optical fiber quantum biosensor based on surface plasmon polaritons for the label-free measurement of protein. *Sensors and Actuators B: Chemical* 316: 128097.

24 Naresh, V. and Lee, N. (2021). A review on biosensors and recent development of nanostructured materials-enabled biosensors. *Sensors (Switzerland)* 21: 1109.

25 Niu, C., Yao, Z., and Jiang, S. (2023). Synthesis and application of quantum dots in detection of environmental contaminants in food: a comprehensive review. *The Science of the Total Environment* 882: 163565.

26 Wei, J., Zhang, X., Mugo, S.M., and Zhang, Q. (2022). A portable sweat sensor based on carbon quantum dots for multiplex detection of cardiovascular health biomarkers. *Analytical Chemistry* 94: 12772–12780.

27 García de Arquer, F.P., Talapin, D.V., Klimov, V.I. et al. (2021). Semiconductor quantum dots: technological progress and future challenges. *Science (New York, N.Y.)* 373: eaaz8541.

28 Sharma, A., Badea, M., Tiwari, S., and Marty, J.L. (2021). Wearable biosensors: an alternative and practical approach in healthcare and disease monitoring. *Molecules* 26: 748.

29 Spring, S.A., Goggins, S., and Frost, C.G. (2021). Ratiometric electrochemistry: improving the robustness, reproducibility and reliability of biosensors. *Molecules* 26: 2130.

30 Park, Y.S., Roh, J., Diroll, B.T. et al. (2021). Colloidal quantum dot lasers. *Nature Reviews Materials* 6: 382–401.

31 Savic-Radojevic, A., Pljesa-Ercegovac, M., Matic, M. et al. (2017). Novel biomarkers of heart failure. *Advances in Clinical Chemistry* 79: 93–152.

32 Dahiya, T., Yadav, S., Yadav, N. et al. (2021). Monitoring of BNP cardiac biomarker with major emphasis on biosensing methods: a review. *Sensors International* 2: 100103.

33 Pohanka, M. (2022). Diagnoses based on C-reactive protein point-of-care tests. *Biosensors* 12: 344.

34 Almehizia, A.A., Naglah, A.M., Alrasheed, L.S. et al. (2023). Point-of-care paper-based analytical device for potentiometric detection of myoglobin as a cardiovascular disease biomarker. *RSC Advances* 13: 15199–15207.

35 Gerdan, Z., Saylan, Y., and Denizli, A. (2024). Biosensing platforms for cardiac biomarker detection. *ACS Omega* 9: 9946–9960.

36 Tilea, I., Varga, A., and Serban, R.C. (2021). Past, present, and future of blood biomarkers for the diagnosis of acute myocardial infarction—promises and challenges. *Diagnostics* 11: 881.

37 Yoo, B.S. (2014). Clinical significance of B-type natriuretic peptide in heart failure. *Journal of Lifestyle Medicine* 4: 34–38.

38 Krishnamurthy, H.K., Reddy, S., Jayaraman, V. et al. (2024). Association between high-sensitivity C-reactive protein (hs-CRP) levels with lipids and micronutrients. *Cureus* 16 (8): e67268.

39 Asl, S.K. and Rahimzadegan, M. (2022). The recent progress in the early diagnosis of acute myocardial infarction based on myoglobin biomarker: nano-aptasensors approaches. *Journal of Pharmaceutical and Biomedical Analysis* 211: 114624.

40 Kim, J. and Hashim, I.A. (2016). The clinical utility of CK-MB measurement in patients suspected of acute coronary syndrome. *Clinica Chimica Acta* 456: 89–92.

41 Galstyan, V. (2021). "Quantum dots: perspectives in next-generation chemical gas sensors" – a review. *Analytica Chimica Acta* 1152: 238192.

42 Bin, W.J., Lin, M.L., Cong, X. et al. (2018). Raman spectroscopy of graphene-based materials and its applications in related devices. *Chemical Society Reviews* 47: 1822–1873.

43 Kumar, Y.R., Deshmukh, K., Sadasivuni, K.K., and Pasha, S.K.K. (2020). Graphene quantum dot based materials for sensing, bio-imaging and energy storage applications: a review. *RSC Advances* 10: 23861–23898.

44 Tang, Y., Guo, W., Kobayashi, H. et al. (2023). Imaging quantized vortex rings in superfluid helium to evaluate quantum dissipation. *Nature Communications* 14: 2941.

45 Guo, Y., Wang, R., Wei, C. et al. (2023). Carbon quantum dots for fluorescent detection of nitrite: a review. *Food Chemistry* 415: 135749.

46 Hu, X., Li, Y., Xu, Y. et al. (2021). Green one-step synthesis of carbon quantum dots from orange peel for fluorescent detection of *Escherichia coli* in milk. *Food Chemistry* 339: 127775.

47 Liu, M., Bai, Y., He, Y. et al. (2021). Facile microwave-assisted synthesis of Ti_3C_2 MXene quantum dots for ratiometric fluorescence detection of hypochlorite. *Microchimica Acta* 188: 15.

48 Yi, Y., Zeng, W., and Zhu, G. (2021). β-Cyclodextrin functionalized molybdenum disulfide quantum dots as nanoprobe for sensitive fluorescent detection of parathion-methyl. *Talanta* 222: 121703.

49 John, B.K., Mathew, J., and K S, K RE, Mathew B. (2024). Biomass derived carbon quantum dots as a versatile platform for fluorescent sensing, catalytic reduction, fluorescent ink and anticancer agents. *Materials Today Sustainability* 26: 100715.

50 Chen, L., Lu, J., Luo, M. et al. (2022). A ratiometric fluorescent sensing system for the selective and ultrasensitive detection of pesticide residues via the synergetic effects of copper nanoclusters and carbon quantum dots. *Food Chemistry* 379: 132139.

51 Feng, X., Han, G., Cai, J., and Wang, X. (2022). Au@carbon quantum dots-MXene nanocomposite as an electrochemical sensor for sensitive detection of nitrite. *Journal of Colloid and Interface Science* 607: 1313–1322.

52 Tran, H.L., Dang, V.D., Dega, N.K. et al. (2022). Ultrasensitive detection of breast cancer cells with a lectin-based electrochemical sensor using N-doped graphene quantum dots as the sensing probe. *Sensors and Actuators B: Chemical* 368: 132233.

53 Webb, J.L., Troise, L., Hansen, N.W. et al. (2021). Detection of biological signals from a live mammalian muscle using an early stage diamond quantum sensor. *Scientific Reports* 11: 2412.

54 Wongrat, E., Nuengnit, T., Panyathip, R. et al. (2021). Highly selective room temperature ammonia sensors based on ZnO nanostructures decorated with graphene quantum dots (GQDs). *Sensors and Actuators B: Chemical* 326: 128983.

55 Xue, X., Chen, M., Luo, Y. et al. (2023). High-operating-temperature mid-infrared photodetectors via quantum dot gradient homojunction. *Light: Science & Applications* 12: 2.

56 Yazirlioğlu, L. (2021). Sustainable design considerations for emotional durability and product longevity through product care activities by repair enthusiasts. Thesis.

57 Pandit, C., Alajangi, H.K., Singh, J. et al. (2022). Development of magnetic nanoparticle assisted aptamer-quantum dot based biosensor for the detection of *Escherichia coli* in water samples. *Science of The Total Environment* 831: 154857.

58 Chen, X., Wang, T., Shi, J. et al. (2021). A novel artificial neuron-like gas sensor constructed from CuS quantum dots/Bi_2S_3 nanosheets. *Nano Micro Letters* 14: 8.

59 Mukasa, D., Wang, M., Min, J. et al. (2023). A computationally assisted approach for designing wearable biosensors toward non-invasive personalized molecular analysis. *Advanced Materials* 35: e2212161.

60 Yola, M.L. and Atar, N. (2019). Development of cardiac troponin-I biosensor based on boron nitride quantum dots including molecularly imprinted polymer. *Biosensors & Bioelectronics* 126: 418–424.

61 Kim, S.E., Yoon, J.C., Muthurasu, A., and Kim, H.Y. (2024). Functionalized triangular carbon quantum dot stabilized gold nanoparticles decorated boron nitride nanosheets interfaced for electrochemical detection of cardiac troponin T. *Langmuir* 40 (47): 25051–25060. https://doi.org/10.1021/acs.langmuir.4c03281.

62 Karaman, O., Özcan, N., Karaman, C. et al. (2022). Electrochemical cardiac troponin I immunosensor based on nitrogen and boron-doped graphene quantum dots electrode platform and Ce-doped SnO_2/SnS_2 signal amplification. *Materials Today Chemistry* 23: 100666.

63 Hasabnis, G.K. and Altintas, Z. (2024). Cardiac troponin I-responsive nanocomposite materials for voltammetric monitoring of acute myocardial infarction. *ACS Omega* 9 (28): 30737–30750. https://doi.org/10.1021/acsomega.4c03252.

64 Wang, L., Xing, B., Ren, X. et al. (2020). Mo_2C combined with carbon material nanosphere as an electrochemiluminescence super-enhancer and antibody label for ultrasensitive detection of cardiac troponin I. *Biosensors & Bioelectronics* 150: 111910.

65 Mansuriya, B.D. and Altintas, Z. (2021). Enzyme-free electrochemical nano-immunosensor based on graphene quantum dots and gold nanoparticles for cardiac biomarker determination. *Nanomaterials* 11: 578.

66 Vasudevan, M., Remesh, S., Perumal, V. et al. (2023). Lignin derived nanoparticle intercalation on nitrogen-doped graphene quantum dots for electrochemical sensing of cardiac biomarker. *Microchemical Journal* 195: 109405.

67 Tuteja, S.K., Chen, R., Kukkar, M. et al. (2016). A label-free electrochemical immunosensor for the detection of cardiac marker using graphene quantum dots (GQDs). *Biosensors & Bioelectronics* 86: 548–556.

68 Liu, M., Jiang, R., Zheng, M. et al. (2022). A sensitive ratiometric biosensor for determination cardiac troponin I of myocardial infarction markers based on N, Zn-GQDs. *Talanta* 249: 123577.

69 Liu, X., Li, F., Qiu, Y. et al. (2024). Triple cascade quantum-strip for heart failure point-of-care testing. *ACS Sensors* 9: 29–41.

70 Nazir, S. and Iqbal, R.A. (2023). Biosensor for rapid and accurate detection of cardiovascular biomarkers: progress and prospects in biosensors. *Biosensors and Bioelectronics: X* 14: 100388.

71 Monaghan, T.F., Rahman, S.N., Agudelo, C.W. et al. (2021). Foundational statistical principles in medical research: sensitivity, specificity, positive predictive value, and negative predictive value. *Medicina (Lithuania)* 57: 503.

72 Nawaz, M.S., Shoaib, B., and Ashraf, M.A. (2021). Intelligent cardiovascular disease prediction empowered with gradient descent optimization. *Heliyon* 7: e06948.

73 Lakshmanakumar, M., Nesakumar, N., Sethuraman, S. et al. (2022). Development of an electrodeposited graphene quantum dot electrode for the electrochemical detection of C-reactive protein (CRP) biomarker. *ChemistrySelect* 7: e202104511.

74 Lee, Y.Y., Sriram, B., Wang, S.F. et al. (2024). Advanced nanomaterial-based biosensors for N-terminal pro-brain natriuretic peptide biomarker detection: progress and future challenges in cardiovascular disease diagnostics. *Nanomaterials* 14: 153.

75 Oranzie, M., Douman, S.F., Uhuo, O.V. et al. (2022). Chronocoulometric signalling of BNP using a novel quantum dot aptasensor. *The Analyst* 147: 4829–4837.

76 Wu, Z., Liu, S., Li, Y. et al. (2021). Electrochemiluminescence resonance energy transfer system fabricated by quantum state complexes for cardiac troponin i detection. *Sensors and Actuators B: Chemical* 336: 129733.

77 Stringer, R.C., Hoehn, D., and Grant, S.A. (2008). Quantum dot-based biosensor for detection of human cardiac troponin I using a liquid-core waveguide. *IEEE Sensors Journal* 8: 295–300.

78 Song, T., Wang, Q., Yu, H. et al. (2022). One-step hydrothermal synthesis of WS_2 quantum dots as fluorescent sensor for sensitive and selective recognition of hemoglobin and cardiac biomarker myoglobin. *Analytical and Bioanalytical Chemistry* 414: 1623–1630.

79 Xu, Z., Xu, H., Duan, H. et al. (2024). Smartphone-aided fluorescence detection of cardiac biomarker myoglobin by a ratiometric fluorescent AuNCs-QDs nanohybrids probe with high sensitivity. *Journal of Fluorescence* 34: 179–190.

80 Thamminana, B., Patel, M.R., Deshpande, M.P. et al. (2024). Fluorescence turn-off detection of myoglobin as a cardiac biomarker using water-stable L-glutamic acid functionalized cesium lead bromide perovskite quantum dots. *Microchimica Acta* 191 (11): 674. https://doi.org/10.1007/s00604-024-06752-z.

81 Li, Z., Wang, Y., Wang, J. et al. (2010). Rapid and sensitive detection of protein biomarker using a portable fluorescence biosensor based on quantum dots and a lateral flow test strip. *Analytical Chemistry* 82: 7008–7014.

82 Ma, F., Li, C., and chen, Zhang C yang. (2018). Development of quantum dot-based biosensors: principles and applications. *Journal of Materials Chemistry B* 6: 6173–6190.

83 Mazumdar, H., Chakraborty, C., Venkatakrishnan, S.B. et al. (2024). Quantum-inspired heuristic algorithm for secure healthcare prediction using Blockchain technology. *IEEE Journal of Biomedical and Health Informatics* 28: 3371–3378.

8

Neurodiagnostics: Quantum Approaches to Brain Health

Brikshadipa Mandal, Subrata Barick, and Sandeep Chandrashekharappa

Department of Medicinal Chemistry, National Institute of Pharmaceutical Education and Research-Raebareli (NIPER-R), Lucknow, Uttar Pradesh-226002, India

8.1 Introduction

Brain is the most complex part of the human body, as it is controlling the complex mechanisms of the human body [1]. To maintain overall health well-being, it is important to maintain the good health of the human brain and protect it from neurodegenerative disorders like Parkinson's disease (PD), Alzheimer's disease (AD), Huntington disease (HD), and amyotrophic lateral sclerosis (ALS); stroke, trauma, brain tumor, and cancer; infectious diseases like meningitis, encephalitis, and brain abscesses [2, 3]. Brain health can be maintained by reducing stress [4], doing physical activity [5], maintaining good eating habits [6], and maintaining good sleeping habits [7]. The issues related to neurological disorders are increasing day by day, with risk factors being our poor lifestyle, aging, infections, and different injury-related other external factors [8]. To diagnosis of the disease spread, many methods such as CT scans, positron emission tomography (PET) scans, and magnetic resonance imaging (MRI) scans are incorporated for many decades [9]. The main obstruction for compounds to get into the brain from the blood circulatory system or cardiovascular system is the blood–brain barrier (BBB), for which the majority of the drugs, contrast media failed to reach the targeted location [10]. So, it is needed to introduce such material that is small enough to penetrate BBB. Though these conventional diagnostic technologies are very useful, there is still an unmet need to develop newer technology for neurodiagnostics and real-time imaging.

Biosensing is a technique that is used to detect the presence of analytes, even in trace amounts, using a biomolecular component by interacting or binding with the analyte, e.g. detection of curcumin with vanillin in food samples [11], determination of the antihistaminic drug cetirizine in the presence of paracetamol [12], development of an electrochemical method for the determination of tyrosine in the presence of ascorbic acid and uric acid [13], quantification of flavonoids [14], and detection of microbes by voltammetry [15].

The term "quantum" refers to the smallest quantity of any entity engaged in an interaction. Quantum dots (QDs), being a photoluminescence nanomaterial, hold a

Quantum Biosensing in Medical Diagnostics, First Edition. Edited by J. G. Manjunatha.

promising capability in the neurodiagnostic field. QDs were discovered and successfully synthesized by Alexei Ekimov, Moungi Bawendi, and Louis Brus [16] and awarded the Noble Prize in 2023.

The quantum dots (QDs) are inorganic substances, consisting of a radius ranging between 10 and 100 Å, a zero-dimensional nanomaterial. Quantum dots are semiconductors, and their optical properties help in disease diagnosis, biological imaging, biosensing, dose monitoring, and drug delivery [17–20]. It works on the principle of the quantum confinement effect, which leads to discrete electronic states that influence the optical and electronic properties of these nanomaterials. Quantum confinement occurs when the size of a semiconductor material is reduced to dimensions less than the exciton Bohr radius [21]. Quantum confinement is observed in particles having a dimension of confinement smaller than their de Broglie wavelength. Quantum confinement occurs when the charged particles, the electron-hole pair (also known as excitons), are being confined in three dimensions by a potential barrier the spatial restriction causes [22, 23]. Quantum dots are commonly utilized to mark live biological material *in vitro* and *in vivo* in animals for research and immunoassay investigations. According to the theoretical model of the multiband effective mass approximation, size-dependent nanocrystals are 2D nanoplates, 1D nanorods and wires, and 0D quantum dots [24].

8.2 Classes of QDs

There are two classes of QDs discovered so far, described as follows. The core-shell type QDs consist solely of a semiconductor core, typically ranging from 2 to 10 nm in size, without any protective shell. These nanocrystals exhibit unique optical and electronic properties due to quantum confinement effects. Common materials include cadmium selenide (CdSe), cadmium telluride (CdTe), and lead sulfide (PbS). By adjusting their size, the emission wavelengths can be precisely tuned, enhancing their utility in applications such as biological imaging, solar cells, and quantum computing. However, the lack of a protective shell makes them more vulnerable to surface defects and environmental degradation, impacting stability and luminescence efficiency [25]. Core-shell quantum dots (CSQDs) exhibit diverse functionalities based on material composition and band alignment. For instance, CdSe/ZnS (Type I) demonstrates strong carrier confinement and high fluorescence efficiency, making it suitable for bioimaging and LEDs. CdTe/CdSe (Type II) and ZnTe/ZnSe (Type II) facilitate spatial charge separation, enhancing their utility in solar cells and photodetectors. PbS/CdS (Inverse Type II) improves near-infrared (NIR) absorption, benefiting solar cells and infrared imaging. CdSe/CdS (Quasi-Type II) enhances carrier dynamics by allowing partial electron delocalization, useful for LEDs and luminescent solar concentrators [26–29].

The alloyed type QDs are semiconductor nanocrystals synthesized by the amalgamation of two or more materials, resulting in optical and electronic properties that are distinct from those of their constituent semiconductors and bulk materials. These QDs can possess either homogeneous or gradient internal structures, which facilitate the precise tuning of their properties through compositional modifications while maintaining a constant crystallite size. This capability allows for the exploration of novel physical phenomena and enhances the performance of QDs in various applications, including optoelectronics, photovoltaics, and biological imaging [30]. A novel approach to tuning the optical and electronic properties of quantum dots (QDs) focuses on alloying rather than altering particle size. In this study, cadmium selenide telluride (CdSeTe) alloyed QDs were synthesized with homogeneous and gradient internal structures, revealing that composition and internal structure are critical for controlling quantum confinement effects. The research demonstrated a nonlinear relationship between alloy composition and optical absorption/emission, achieving red-shifted light emission up to 850 nm and quantum yield of 60%. These findings suggest that alloyed QDs present significant opportunities for band gap engineering, particularly in NIR fluorescence applications for biomarker detection and *in vivo* imaging [30, 31].

8.3 Photophysical Properties of QDs

Quantum dots are tiny crystals with a changeable number of electrons occupying discrete quantum states. Their electrical characteristics are midway between bulk and discrete basic particles. QDs are nanocrystals of 2–10 nm in diameter, also called zero dimensional materials, having the characteristic of tunable photoluminescence. The bigger the particle, the more the red shift it shows, which means the larger dots will emit larger wavelengths (Figure 8.1). The color of irradiation is dependent upon the dimensions of the quantum dot and the band gap. QDs absorb white light and reflect different chemiluminescence depending on the particle's size, ranging between 400 and 800 nm of wavelength. This spherically shaped nanomaterial possesses a higher density. Band gap energy is inversely proportional to the size of particles. QDs emit wavelengths that are between ultraviolet and infrared electromagnetic light. It possesses a heavy thermal stability, though it is dependent upon the valence shell capacity. These QDs are surface engineered to manipulate semiconduction properties and chemiluminescence properties as well. QDs are composed of II–VI groups or III–V elements. Types of QDs are as follows: 1. II–VI quantum dots, 2. III–V quantum dots, 3. IV–VI quantum dots, 4. silicon quantum dots, 5. graphene quantum dots, 6. carbon quantum dots, and 7. fullerenes [17]. The IUPAC definition of quantum yield (Φ) is the number of a certain event occurring per photon absorbed by the system.

$$\text{Quantum yield } (\emptyset) = \frac{N_{\text{emitted photon}}}{N_{\text{absorbed photon}}}$$

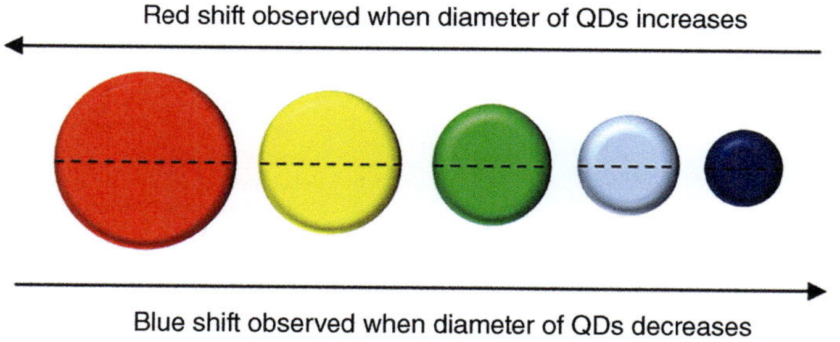

Figure 8.1 Photophysical properties of quantum dots.

8.4 Mechanism of Electroluminescence

The electroluminescence (EL) in QDs refers to the light emission that occurs when energy (as photon) is applied, resulting from the radiative recombination of electrons and holes within the QDs. This process enables efficient light generation without significant heat production, making QDs ideal for applications in displays and lighting.

Luminescence is observed by electron and hole recombination, related to energy relaxation within the material. A photon is emitted when an electron transitions from the excited π^* state to the ground π state, producing discrete light emission. In molecules, the emission of light is determined by the energy gap between HOMO (highest occupied molecular orbital) and LUMO (lowest unoccupied molecular orbital) states. Possessing a few atoms, QDs exert an intermediate energy structure with discrete energy levels rather than bands. Valence band electrons are excited by photons, causing the excitation of that electron to the conduction band, creating a virtually positively charged hollow hole [32]. When the excited electron relaxes, the emission of the relaxation energy reveals different luminescence within the UV to IR range. The more the confinement occurs, the more the excitons tend to jump to the free electron-hole state, resulting in electroluminescence.

8.5 Synthesis of Quantum Dots

These ultrasmall-sized quantum dots can be synthesized by two techniques: the top-down technique and the bottom-up technique (Figure 8.2). Top-down techniques include breaking the bulk materials into nanostructures. Previously reported and well-used methods for fabricating graphene quantum dots (GQDs) are microwave-assisted technique, ultrasonication, solvothermal and hydrothermal methods, electrochemical exfoliations methods, etc. A novel top-down method using self-assembled gold nanoparticles, employing the

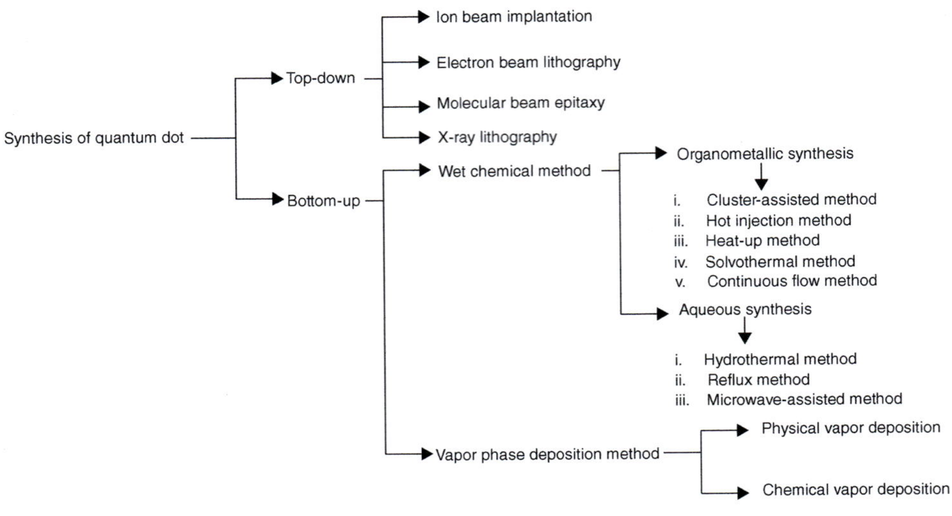

Figure 8.2 Schematic diagram of synthesis of QDs. Adapted from [32].

lithographical method has been developed, and this method results in good control over the size and impurity of GQDs [33]. The molecular beam epitaxy (MBE) synthesis of quantum dots (QDs) involves precise layer-by-layer deposition under ultrahigh vacuum (UHV) conditions. A GaAs or InP substrate is prepared, followed by molecular beam deposition of group III (e.g. Ga, In) and group V (e.g. As, P) elements. The capping layer stabilizes the QDs, and annealing enhances uniformity [34]. A novel method for molybdenum disulfide quantum dots (MoS_2 QDs) was developed by using a hydrothermal approach for fluorescence imaging in SW480 tumor cells [35]. The most biocompatible carbon quantum dots (CQDs) are synthesized by the bottom-up method by the processes: templet method, microwave-assisted method, combustion, thermal oxidation, and various chemical methods [36]. For silicon quantum dots (SiQDs), in the top-down technique, bulk silicon electrochemical etching and milling of silicon oxides have been employed. And in the bottom-up technique, chemical method, i.e. solution-based reduction, Zintl salt formation, laser pyrolysis, and plasma synthesis techniques, has been reported [37]. By the hydrothermal method, cadmium tellurium quantum dots (CdTe QDs) were made as a bioprobe to detect the biomarker 2,6-Pyridinedicarboxylic acid (DPA) [38]. In preparation of lead sulfide quantum dots (PbS QDs), a bottom-up approach has been employed [39]. Reviewed by Mattaussi and coworkers [40], the synthesis of QDs by the top-down method includes ion implantation, electron beam lithography, X-ray lithography, and MBE [41]. In the bottom-up method, colloidal synthesis, hydrothermal and solvothermal technique, etc., monodisperse colloidal quantum dots are synthesized using a high-quality nanocrystal with equal size distributions. The hot-injection method is the most commonly used, in which organometallic precursors are swiftly injected into a hot solvent, resulting in instantaneous nucleation followed by controlled development. Another method is the heat-up method, which gradually heats precursors to cause nucleation, and the next method is the cluster-assisted method, which uses uniform seed clusters to control nanocrystal formation. The microwave-assisted approach allows for rapid heating and fine control over synthesis parameters [42]. Carbon quantum dots (CQDs) are synthesized by the bottom-up technique using controlled conditions to carbonize tiny molecule precursors. One extensively used method is to pyrolyze the mixture of urea and citric acid and autoclave it at 200 °C under high pressure [43]. Ekimov and Efros reported in 1985 that microcrystals have been synthesized in a dielectric glassy matrix [44].

8.6 Making QDs Biocompatible

For applications in bioimaging, biosensing, drug monitoring, and other biological applications, QDs must be turned into a biocompatible form so that they can be absorbed and reach the targeted site. Solubilization and bioconjugation of QDs are very crucial steps for making the QDs biocompatible. QDs are capped with functional groups such as amino, carboxyl, or thiol for being recognized by biological interaction mediators. There are some methods to make QDs sensitive to biological receptors [45]. First is hydrophilization, where the quantum crystals are solubilized by water. Second, post-synthesis ligand exchange, where hydrophobic coating is modified. Third, encapsulating hydrophobic quantum dots within various carrier vehicles. Bioconjugation is the process of making the QDs ideal for biological effect. For this, covalent and non-covalent binding methods are available to facilitate the selectivity towards specific conjugation with receptors. The use of covalent linkages is more suitable than electrostatic interactions when quantum dots are used, though both methods are good for the process of bioconjugation of QDs [46].

Covalent binding methods, such as EDC/NHS (1-ethyl-3-(3-dimethylaminopropyl) carbodiimide/*N*-hydroxysuccinimide) coupling, electrostatic interactions, metal-affinity coordination (e.g. binding of His-tagged proteins to QD surfaces), and ligand-receptor interactions like biotin–avidin coupling, facilitate the conjugation of quantum dots (QDs) with biomolecules. These approaches enable the attachment of proteins, peptides, nucleic acids, or antibodies to QDs, enhancing their functionality for targeted imaging, biosensing, and drug delivery applications. By employing these diverse bioconjugation strategies, researchers can tailor QD properties to meet specific requirements in various biomedical contexts [47].

Non-covalent bioconjugation methods include electrostatic interactions, where negatively charged QDs bind to positively charged proteins; metal-affinity coordination using nickel-nitrilotriacetic acid (Ni-NTA) and His-tagged proteins; and ligand-receptor interactions, such as biotin–avidin binding [48]. Unlike covalent bonding, this strategy preserves protein function, prevents aggregation, and maintains high photoluminescence quantum yield (PLQY). QDs are initially capped with trioctylphosphine (TOP) and trioctylphosphine oxide (TOPO), making them soluble in organic solvents. For biological applications, the cap exchanged TOP/TOPO with hydrophilic ligands such as dihydrolipoic acid (DHLA), mercaptoacetic acid (MPA), or polyethylene glycol (PEG), ensuring water solubility and biocompatibility. The consequent QD-bioconjugates are stable at multiple pH levels and ionic strengths, making them appropriate for bioimaging, biosensing, and drug administration. Their capacity to reversibly attach biomolecules broadens their use in fluorescence tracking, diagnostics, and targeted therapies [49] (Figure 8.3).

A one-step synthetic process of protein-conjugated quantum dots (QDs) is a rapid and efficient method conducted at room temperature in an aqueous solution. This process involves using ZnHg-Se QDs, where proteins directly bind to the QD surface through interactions with Zn^{2+} ions *via* thiol (–SH), carboxyl (–COOH), amine (–NH_2), and imidazole groups. The synthesis begins with the preparation of $Zn(OAc)_2$, $Hg(ClO_4)_2$, and NaHSe precursors in an NH_4HCO_3 solution (pH 12.3), followed by the rapid formation of QDs upon the injection of NaHSe. Proteins such as bovine serum albumin (BSA) and mercaptopropionic acid (MPA) are used as co-ligands. This method eliminates the need for multi-step conjugation, preserving protein activity while achieving high quantum yield, photostability, and colloidal stability [50].

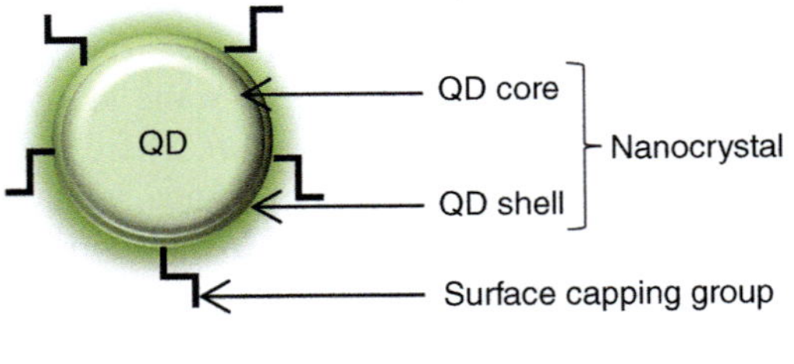

Figure 8.3 Modified QD-shell complex.

8.7 Detection of QDs (Bioimaging)

By the fluorescence microscopy technique, the bioimaging has been done on sentinel organs in mice [51]. Using a Leica DM RA2 photomicroscope attached to a 100 W halogen light source and a SPOT RT-SE slider high-resolution charge-coupled device camera (CCD), the images are captured. Tissue structures were visualized under bright-field illumination, while QD fluorescence (621 nm emission) was detected using a 415WB/100 excitation filter, a 475DCLP dichroic filter, and a 620WB20 emission filter. Deep tissue imaging in mice has been conducted by multiphoton fluorescence microscopy; a femtosecond laser with 900 nm excitation was used, delivering 40 mW power through a 20 × water-immersion objective (0.95 NA). The QD-labeled blood vessels beneath the skin with high clarity, and the dorsal skin clamp stabilized the sample. This setup allowed high-resolution, low-photobleaching imaging of capillaries and blood flow, demonstrating superior brightness and photostability of QDs [52]. A NIR fluorescence imaging technique has been used to detect micrometastases using the QDs [53]. QDs' high brightness and photostability allow the detection of low-abundant cancer biomarkers at the single-cell level. Magnetic nanoparticles (MNPs) coupled with pan-cytokeratin (pan-CK) antibodies were used for initial cell enrichment, and then the QDs were applied for specific detection of micrometastatic cells using fluorescence imaging. QDs emit fluorescence when excited by a shorter wavelength light source. A single-quantum-dot-based nanosensor for detection was conducted of RNA-peptide interactions using fluorescence resonance energy transfer (FRET). A 444-nm laser excites 605QD (FRET donor; 605-nm-emitting quantum dot), which transfers energy to Cy5-labeled Rev peptide (FRET acceptor), enabling fluorescence detection. Emitted photons pass through dichroic mirrors and band-pass filters, then are detected *via* avalanche photodiodes. This method quantifies Rev peptide-RRE RNA binding, assesses the inhibitory effect of proflavin, and minimizes background fluorescence. This provides real-time, high-sensitivity biomolecular interaction analysis, aiding drug screening and HIV research [54].

8.8 Conventional Neuroimaging Techniques

To diagnose the patient and assessment in emergency situations, neuroimaging and other radiological imaging techniques play the key role in the medical field. Conventional techniques include structural and functional methods that are employed as neurodiagnostics till date [55]. Newly emerging hybrid techniques were also introduced some decades back. Though they come with some limitations, they are still in use at the present time. Neuroimaging has evolved from early physiological observations to advanced imaging techniques that evolved neuroscience. In the 1970s, Computed Tomography (CT), developed by Hounsfield and Cormack, provided the first high-resolution, non-invasive brain scans, enabling structural analysis of brain injuries and diseases. Shortly after, PET, introduced by Ter-Pogossian, allowed researchers to measure brain metabolism and blood flow, significantly advancing the study of cognitive functions and neurological disorders. In the 1980s, PET became instrumental in mapping brain activity related to language, memory, and attention. A major breakthrough came in 1973 with MRI by Lauterbur, initially for structural imaging but later refined into functional MRI (fMRI) in 1990 by Ogawa, utilizing the blood oxygenation level dependent (BOLD) signal to track brain activity in real-time. The 2000s saw the rise of multimodal imaging, combining fMRI, PET, and electroencephalograph (EEG) for a comprehensive understanding of brain function. Recent advancements, including 7- Tesla MRI and AI-driven neuroimaging, continue to refine diagnostic accuracy and personalized treatments for neurological disorders. These innovations have transformed neuroscience, providing critical insights into brain function and paving the way for future breakthroughs in research and clinical applications [56–58].

8.8.1 Computed Tomography (CT)

CT is an advanced neuroimaging technique that utilizes rotating X-rays, emitting narrow radiation of X-rays that penetrates the subject from various angles. Modern CT scanners take images in a single spiral acquisition, which are compiled and integrated into 2D or 3D images based on tissue density, where high-density structures like bone appear white and lower-density tissues appear darker [59]. Advanced improvement in image contrast for better viewing of the brain, bones, and soft tissues; it is a very good tool to detect acute hemorrhage, ischemic stroke, malignancies, hydrocephalus, and infections. Iodine-containing contrast media is in high practice to yield high-resolution sliced tomography reported for ischemic stroke and paranasal sinus diagnosis [60, 61]. But frequent exposure to radiation from CT accumulates over time, increasing the risk of cancer, particularly in children and young adults. High-dose radiation exposure can lead to cellular mutations and DNA damage, increasing the risk of genetic disorders over time [62].

8.8.2 Positron Emission Tomography (PET)

A PET scan begins with the injection of a radiotracer, which binds to specific biological targets. Generally, fluorine-18 (^{18}F) or carbon-11 (^{11}C) has been used as a radioactive isotope. As the tracer decays, it emits positrons, which collide with electrons, producing gamma photons. PET scanners use scintillation detectors to detect gamma photons produced during positron annihilation. The most common detectors include bismuth germanate (BGO), lutetium oxyorthosilicate (LSO), lutetium-yttrium oxyorthosilicate (LYSO), and gadolinium oxyorthosilicate (GSO), each with varying properties like light yield and stopping power. These scintillators convert gamma rays into visible light, which is then processed by photomultiplier tubes (PMTs) or silicon photomultipliers (SiPMs) to generate electrical signals. Modern PET systems use SiPMs for higher resolution and sensitivity. The collected signals are processed using image reconstruction algorithms, creating detailed 3D functional images for neurological, oncological, and cardiac diagnostics [57, 63]. But PET scans involve the injection of radioactive tracers, which may likely expose patients to ionizing radiation, therefore increasing long-term cancer risks, especially with repeated scans [64].

8.8.3 Magnetic Resonance Imaging (MRI)

Magnetic resonance imaging (MRI) is another advanced and well-used technique used for brain imaging. MRI signals are based on fat and water content, with T1 (for fatty infiltration, indicating chronic muscle disease) and T2 (for muscle edema, which may suggest inflammatory changes) relaxation times determining tissue contrast [65]. MRI plays a crucial role in the diagnosis, monitoring, and differential diagnosis of Alzheimer's disease (AD). Structural MRI is used to diagnose brain atrophy, particularly in the medial temporal lobe, hippocampus, and entorhinal cortex, which are important markers of AD. MRI can predict progression from mild cognitive impairment (MCI) to AD and distinguish AD from other dementias such as vascular dementia, frontotemporal degeneration, and Lewy body dementia [66]. But MRI machines generate a high range of noise (near around 120 dB), which can cause temporary or permanent hearing damage. Additionally, MRI contrast agents (mostly gadolinium) can cause severe allergic reactions, and, in nephropathy patients, nephrogenic systemic fibrosis (NSF) can be seen rarely [67].

8.8.4 Electroencephalograph (EEG)

Electroencephalograph (EEG) is a widely used method used to detect neurophysiological changes. Quantitative EEG (qEEG) is an advanced method that utilizes mathematical phenomena to enhance the detection of specific waveform components and provide valuable diagnostic insights. EEG is used in diagnosing and monitoring cognitive disorders, learning disabilities, encephalopathies, dementia, delirium, attention disorders, and mood disorders [68]. The standard 10–20 system is used for placing electrodes on the scalp to record brain activity, sometimes used some more to acquire better spatial resolution. High-density EEG utilizes more electrodes (nearly 50–250 channels) for enhanced spatial resolution, mainly for research purposes [69]. But sometimes people cannot tolerate long-term application of this method.

8.8.5 Functional Magnetic Resonance Imaging (fMRI)

Another technique for the neuroimaging is fMRI. fMRI can measure brain activity by detecting the changes in blood oxygenation levels that are associated with a brain function. BOLD fMRI detects the metabolic rate of oxygen consumption by the brain during any neural activity [70]. In particular, this method visualizes the dynamic alterations brought about by changes in neural metabolism brought about by variations in brain tissue oxygenation levels as a result of neuronal responses to certain cognitive processes or spontaneous variations in resting states. Thus, a highly useful therapeutic application of this method is the differentiation between the vegetative state and the minimally conscious state [71]. Despite many contributions in this field, fMRI has some drawbacks that need to be addressed, *e.g.* low temporal resolution and signal-to-noise ratio [72].

8.8.6 Functional Near-Infrared Spectroscopic Imaging (fNIRS)

A non-invasive neuroimaging technique that measures the brain functions. Variations in the regional cerebral blood volume are thought to be indicated by changes in oxygenated and deoxygenated hemoglobin. fNIRS uses NIR light for the measurement of cerebral blood flow and oxygenation [73]. Besides the advantages like reduced cost, portability, and robustness to head movement, it has some disadvantages also, including lower spatial resolution than fMRI and restricted brain regions that may be addressed, particularly the cortical areas, as well as the interference of extracortical components and superficial veins and arteries that reduce signal specificity [74].

8.9 Application of QDs in Brain Health

Because of their special optical and physical characteristics, such as their tunable emission spectra and high quantum yield, quantum dots are excellent candidates for use as a tool in materials research, molecular biology, and chemical analysis [75]. Sometimes, using the Trojan horse approach, QDs enter tumor cells by being phagocytosed by macrophage cells and then release fluorescent signals [76]. The prolonged photostability of quantum dots makes them traceable for longer, which makes them preferable to traditional fluorescent dyes like fluorescein isothiocyanate (FITC) [77].

QDs are perfect for observing and monitoring chemical events within cells since they may generate iridescent fluorescence and target distinct cells by altering different molecules and proteins [78]. Being the most complex organ of a human body, the brain plays the most crucial role in controlling the whole-body processes through various neurological applications [79]. So, maintaining and improving brain health is a much-needed target we should focus on. Various neurological problems, like cerebrovascular diseases, brain tumors, and neurodegenerative diseases, e.g. Alzheimer's disease (AD), Parkinson's disease (PD), and Huntington's disease, affect brain health. But the diagnosis and therapy lack effectiveness due to its complex structure and the presence of BBB. So, the demand for developing novel techniques for early diagnosis and real-time prognostic monitoring of CNS (central nervous system) illnesses remains unmet [78].

8.9.1 Brain Tumors

Early detection of brain tumors can lead to a better prognosis for the disease. The ability of QDs to cross the BBB allows for real-time tracking of tumor growth and neovascularization [80]. The potential of QDs in the diagnosis and treatment of tumors has been confirmed by several research studies. Ganganboina et al. have developed an electrochemically impedimetric biosensor generated by an ultrasensitive pulse that has been successfully created for the identification of glioma cells. They prepared a dual functional probe by homogeneously depositing 4–11 nm sulfur-doped graphene QDs (S-GQDs) onto gold nanoparticles-decorated carbon nanospheres (Au-CNS) to form S-GQDs@Au-CNS nanocomposite through a gold-thiol linkage conjugated with angioprep-2 (Ang-2) for the detection of glioma cells. These nanocomposites provide detection of a wide range of 100–100,000 cells/mL with an LOD of 40 cells/mL [81]. Chunyan et al. have developed a unique gadolinium-Ag_2S-based nanoprobe known as Gd-DOTA-Ag_2S, or simply Gd-Ag_2S, to combine two imaging modalities, e.g. MRI and the second NIR window (NIR-II, 1000–1700 nm), with a three-step process. In a mouse model, a brain tumor (U87MG) was identified *in situ* using Gd-assisted T1 MR imaging. The tumor was then precisely resected using NIR-II fluorescence imaging of Ag_2S QDs after an intravenous injection of Gd-Ag_2S nanoprobe [82]. Wang et al. reported a single-layered boron-doped

GQD (SL-BGQD) that shows better T_1 contrast than Gd diethylene penta-acetic acid (Gd-DTPA) with lower toxicity concerns of releasing metal ions when used *in vivo* [83]. Yi et al. have developed nitrogen-doped carbon nanodots known as pN-CNDs with a size range of 5–15 nm, coated with 1-methyl-2-pyrrolidinone (NMP)-derived polymer for the targeted fluorescence imaging of glioma. pN-CNDs can enter glioma cells and produce fluorescence images with good contrast via elevated passive targeting. These QDs also bear desirable characteristics like water solubility and low cytotoxicity [84]. To create a multifunctional platform for tissue imaging and targeting to glioma cells for tissue imaging, Gao et al. functionalized carbon QDs with maleimide-polyethylene glycol-amino succinimide succinate (Mal-PEG-NHS) and the tumor-penetrating peptide (RGERPPR) [85].

8.9.2 Neurodegenerative Disorders (NDs)

The loss of synapses and neurons is a hallmark of a family of diseases known as neurodegenerative disorders (NDs) that affect the brain and spinal cord [86]. There are various reasons to cause NDs, including oxidative stress, metal ion disorder, immune inflammation, etc. [87]. Quantum dots, being a very useful nanotechnology, have an impactful effect on the diagnosis of various NDs [88]. Some types of NDs are discussed below along with the use of QDs to diagnose them.

8.9.2.1 Alzheimer's Disease (AD)

AD is a progressive brain disorder that is characterized by cognitive decline, memory loss, and behavioral changes [89]. Some of the most important histopathology of AD includes the presence of β-amyloid plaques, aggregation of tau protein, neurofibrillary tangles, etc. [90]. The complex pathology and the multifactorial nature of AD make it difficult to diagnose and treat. Tang et al. have reported developing a sensitive sandwich immunoassay using QDs for the detection of amyloid-beta peptide 1-42 ($A\beta_{1\text{-}42}$), a well-known component of amyloid plaques in AD. By measuring the fluorescence intensity in the supernatant, the concentration of $A\beta_{1\text{-}42}$ was ascertained. These QDs have a detection limit of 1.7 pM (7.6 pg/mL) and the linear range of 5.0–100 pM (0.023–0.45 ng/mL) under ideal circumstances [91]. Liu et al. have synthesized graphene QDs of an average size of 8 nm that have shown to have inhibitory activity of $A\beta_{1\text{-}42}$ aggregation with non-toxicity to PC-12 cells [92]. A microfluidic electrochemical immunoassay was developed by a group of researchers for the detection of apolipoprotein E (ApoE) using CdSe@ZnS quantum dots (QDs) as labeling carriers. The assay utilized tosyl-activated magnetic beads for preconcentration within a PDMS-PC microfluidic chip integrated with screen-printed electrodes (SPEs). Detection was performed via square-wave anodic stripping voltammetry (SWASV), achieving a limit of detection (LOD) of 12.5 ng/mL with a linear range of 10–200 ng/mL and high accuracy in diluted human plasma, demonstrating its potential for Alzheimer's disease biomarker detection [93].

8.9.2.2 Parkinson's Disease (PD)

Parkinson's is the second most prevalent neurodegenerative disease after AD. PD is mostly characterized by movement problems, stiffness of the body, and balance issues [94].

The loss of dopaminergic neurons in the substantia nigra is the most prevalent pathological characteristic of Parkinson's disease (PD), and it is followed by the appearance of cytoplasmic fibrillary inclusion bodies that contain ubiquitin and α-synuclein (α-syn) [95]. One of the key factors for the pathogenesis of PD is the aggregation of an unstructured presynaptic protein known as α-syn [96]. Various therapeutic modalities have been developed to alleviate Parkinson's disease symptoms, including medication interventions, deep brain stimulation, and physical therapy. Though some symptomatic treatment, including dopamine agonist, e.g. catechol methyl transferase and monoamine oxidase, are available for the treatment of PD, prolonged administration of medications induces various complications by disrupting neurotransmitter homeostasis and impairing hormonal regulation. These adverse effects include gastrointestinal dysfunction, cardiovascular pathologies, and neuropsychiatric disorders. Since the 1960s, levodopa has been a common medication used to treat Parkinson's disease; nevertheless, it has a number of adverse effects, including dystonia, motor fluctuation, and wearing-off [97]. So, identifying an effective therapeutic strategy to modify disease progression represents a significant challenge in PD treatment. Mohammadi *et al.* have investigated the pathogenesis of PD caused by mutation of A53T-α-syn aggregation by using nanomaterials as a whole-cell recombinant biosensor. Mutation of A53T is a missense mutation of the α-syn gene where alanine at the 53rd position is replaced by threonine, and it is associated with early onset of PD. In their research they have reported for the first time that accelerating the nucleation process, the primary rate-determining stage of fibrillation, GQDs at a concentration of 0.5 μg/mL enhance A53T-α-syn aggregation, creating a signal-on biosensor [96]. A multi-functional brain-targeted ZnO QDs composite was developed for PD treatment. Its multifunction includes gene silencing, neuroprotection, and biolabeling. To treat Parkinson's disease (PD) models, glutathione (GSH)-modified ZnO QDs composites loading gene and concurrently grafting nerve growth factor (NGF) were created. According to the results of the fluorescence tracking, ZnO QD composites may transfer genes into the brain through the BBB and then release them through lysosomal escape [98]. Kim *et al.* have shown in a report that GQDs can inhibit the fibrillization of α-syn, which is highly correlated with the pathogenesis of PD, without any severe toxicity *in vitro* and in a long-term *in vivo* study. GQDs possess unique neuroprotective characteristics against the α-syn aggregates that make them better than previously reported nano-graphene oxides (GOs). In their research, they have also proved that GQDs cross the BBB and guard against behavioral abnormalities, Lewy body/Lewy neurite pathology, and dopamine neuron loss brought on by α-syn preformed fibrils. For the visualization of GQDs and α-syn fibrils interaction, the GQDs were functionalized with PEGylated biotin [99]. A novel fluorescent glutathione-conjugated CQDs (GSH-CQDs) was developed by Park et al. for the sensitive and selective detection of levodopa, a standard drug for the treatment of PD. GSH-CQDs were prepared by the coupling of GSH and amine-functionalized CQDs using meta-phenylenediamine and ethylenediamine. These sensors show bright green fluorescence. But the fluorescent property is quenched when levodopa is tagged to it under alkaline conditions. In alkaline condition, dopaquinone is formed from the conversion of levodopa. The thiol groups of GHS-CQDs are thought to establish covalent connections with dopaquinone, which results in photoinduced electron transfer and fluorescence quenching. The intensity of fluorescence is inversely proportional to the concentration of levodopa in the range of 0.05–1 μM with an LOD of 0.057 μM [97].

8.10 Conclusion

Quantum dots are a revolutionary development in neurodiagnostics, providing remarkable imaging, biosensing, and therapeutic monitoring capabilities. It may be possible to improve early diagnosis, treatment approaches, and patient outcomes by incorporating them into neurological research and clinical practice. Drug delivery and treatment monitoring have also been incorporated into the use of QDs in neurodiagnostics. Drug molecules and QDs can be conjugated to provide more precise and less harmful targeted therapy for neurological disorders. Drug efficacy and individualized treatment strategies can be better understood because of QDs' real-time tracking of drug distribution in the brain. Furthermore, QD-based theranostic platforms, which include therapeutic and diagnostic capabilities, have enormous potential to transform the treatment of neurodegenerative diseases. However, the successful clinical translation of QD-based neurodiagnostic devices will depend on resolving regulatory obstacles and biocompatibility issues. Ongoing advances in nanotechnology and biomedical research position quantum dots as a promising tool in neurodiagnostics, enabling more precise, effective, and personalized approaches to the management of neurological disorders.

References

1 Jiang, X. and Nardelli, J. (2016). Cellular and molecular introduction to brain development. *Neurobiology of Disease* 92: 3–17.

2 Przedborski, S., Vila, M., and Jackson-Lewis, V. (2003). Series introduction: neurodegeneration: what is it and where are we? *The Journal of Clinical Investigation* 111 (1): 3–10.

3 Meliboyev, T.T.U. (2024). Epidemiology of infectious diseases of the brain. *Journal of Intellectual Prperty and Human Rights* 3 (10): 172–177.

4 McEwen, B.S., Bowles, N.P., Gray, J.D. et al. (2015). Mechanisms of stress in the brain. *Nature Neuroscience* 18 (10): 1353–1363.

5 Di Liegro, C.M., Schiera, G., Proia, P., and Di Liegro, I. (2019). Physical activity and brain health. *Genes* 10 (9): 720.

6 Ekstrand, B., Scheers, N., Rasmussen, M.K. et al. (2020). Brain foods - the role of diet in brain performance and health. *Nutrition Reviews* 79 (6): 693–708.

7 Galván, A. (2020). The need for sleep in the adolescent brain. *Trends in Cognitive Sciences* 24 (1): 79–89.

8 Singh, G., Sharma, M., Kumar, G.A. et al. (2021). The burden of neurological disorders across the states of India: the Global Burden of Disease Study 1990–2019. *The Lancet Global Health* 9 (8): e1129–e1144.

9 Ehman, E.C., Johnson, G.B., Villanueva-Meyer, J.E. et al. (2017). PET/MRI: where might it replace PET/CT? *Journal of Magnetic Resonance Imaging* 46 (5): 1247–1262.

10 Hawkins, B.T. and Egleton, R.D. (2006). Fluorescence imaging of blood–brain barrier disruption. *Journal of Neuroscience Methods* 151 (2): 262–267.

11 Tigari, G. and Manjunatha, J. (2020). Poly(glutamine) film-coated carbon nanotube paste electrode for the determination of curcumin with vanillin: an electroanalytical approach. *Monatshefte für Chemie-Chemical Monthly* 151: 1681–1688.

12 Pushpanjali, P., Manjunatha, J., Hareesha, N. et al. (2021). Voltammetric analysis of antihistamine drug cetirizine and paracetamol at poly(L-leucine) layered carbon nanotube paste electrode. *Surfaces and Interfaces* 24: 101154.

13 Raril, C., Manjunatha, J.G., Ravishankar, D.K. et al. (2020). Validated electrochemical method for simultaneous resolution of tyrosine, uric acid, and ascorbic acid at polymer modified nano-composite paste electrode. *Surface Engineering and Applied Electrochemistry* 56: 415–426.

14 Dudani, T., Saini, S.M., Chandrashekharappa, S., and Manjunatha, J.G. (2024). Biosensors for the quantification of flavonoids. In: *Biosensing Technology for Human Health*, 386–401. Royal Society of Chemistry.

15 Mahajan, P.G., Yadav, A., Tiwari, P., and Chandrashekharappa, S. (2025). Brief overview of voltammetry for biosensing applications. In: *Advancements in Voltammetry for Biosensing Applications. Springer Nature;* (ed. J.G. Manjunatha), 45–62.

16 Liz-Marzán, L.M., Artzi, N., Bals, S. et al. (2023). Celebrating a Nobel prize to the "discovery of quantum dots, an essential milestone in nanoscience". *ACS Publications* 17: 19474–19475.

17 Wang, Z., Hu, T., Liang, R., and Wei, M. (2020). Application of zero-dimensional nanomaterials in biosensing. *Frontiers in Chemistry* 8: 320.

18 Probst, C.E., Zrazhevskiy, P., Bagalkot, V., and Gao, X. (2013). Quantum dots as a platform for nanoparticle drug delivery vehicle design. *Advanced Drug Delivery Reviews* 65 (5): 703–718.

19 Devi, S., Kumar, M., Tiwari, A. et al. (2022). Quantum dots: an emerging approach for cancer therapy. *Frontiers in Materials* 8: 798440.

20 Samantara, A.K., Maji, S., Ghosh, A. et al. (2016). Good's buffer derived highly emissive carbon quantum dots: excellent biocompatible anticancer drug carrier. *Journal of Materials Chemistry B* 4 (14): 2412–2420.

21 Valizadeh, A., Mikaeili, H., Samiei, M. et al. (2012). Quantum dots: synthesis, bioapplications, and toxicity. *Nanoscale Research Letters* 7: 1–14.

22 Yoffe, A. (2002). Low-dimensional systems: quantum size effects and electronic properties of semiconductor microcrystallites (zero-dimensional systems) and some quasi-two-dimensional systems. *Advances in Physics* 51 (2): 799–890.

23 Nozik, A.J., Beard, M.C., Luther, J.M. et al. (2010). Semiconductor quantum dots and quantum dot arrays and applications of multiple exciton generation to third-generation photovoltaic solar cells. *Chemical Reviews* 110 (11): 6873–6890.

24 Li, Q., Wu, K., Zhu, H. et al. (2024). Charge transfer from quantum-confined 0D, 1D, and 2D nanocrystals. *Chemical Reviews* 124 (9): 5695–5763.

25 Vasudevan, D., Gaddam, R.R., Trinchi, A., and Cole, I. (2015). Core–shell quantum dots: properties and applications. *Journal of Alloys and Compounds* 636: 395–404.

26 Kalsoom, U.E., Yi, R., Qu, J., and Liu, L. (2021). Nonlinear optical properties of CdSe and CdTe core-shell quantum dots and their applications. *Frontiers in Physics* 9: 612070.

27 Yi, R., Qu, J., and Liu, L. (2021). Nonlinear optical (NLO) properties of CdSe and CdTe core-shell quantum dots (QDs) and their applications. *Frontiers in Physics* 9: 78.

28 Sahu, A. and Kumar, D. (2022). Core-shell quantum dots: a review on classification, materials, application, and theoretical modeling. *Journal of Alloys and Compounds* 924: 166508.

29 Jang, Y., Shapiro, A., Isarov, M. et al. (2017). Interface control of electronic and optical properties in IV–VI and II–VI core/shell colloidal quantum dots: a review. *Chemical Communications* 53 (6): 1002–1024.

30 Bailey, R.E. and Nie, S. (2003). Alloyed semiconductor quantum dots: tuning the optical properties without changing the particle size. *Journal of the American Chemical Society* 125 (23): 7100–7106.

31 Mittelstädt, A., Schliwa, A., and Klenovský, P. (2022). Modeling electronic and optical properties of III–V quantum dots—selected recent developments. *Light: Science & Applications* 11 (1): 17.

32 Osypiw, A.R., Lee, S., Jung, S.-M. et al. (2022). Solution-processed colloidal quantum dots for light emission. *Materials Advances* 3 (17): 6773–6790.

33 Kang, H., Kim, D.Y., and Cho, J. (2023). Top-down fabrication of luminescent graphene quantum dots using self-assembled Au nanoparticles. *ACS Omega* 8 (6): 5885–5892.

34 Li, X., Xu, Q., and Zhang, Z. (2023). Molecular beam epitaxy growth of quantum wires and quantum dots. *MDPI* 13: 960.

35 Shi, M., Dong, L., Zheng, S. et al. (2019). "Bottom-up" preparation of MoS_2 quantum dots for tumor imaging and their *in vivo* behavior study. *Biochemical and Biophysical Research Communications* 516 (4): 1090–1096.

36 Zuo, P., Lu, X., Sun, Z. et al. (2016). A review on syntheses, properties, characterization and bioanalytical applications of fluorescent carbon dots. *Microchimica Acta* 183: 519–542.

37 Cheng, X., Lowe, S.B., Reece, P.J., and Gooding, J.J. (2014). Colloidal silicon quantum dots: from preparation to the modification of self-assembled monolayers (SAMs) for bio-applications. *Chemical Society Reviews* 43 (8): 2680–2700.

38 Cao, Y., Gong, X., Li, L. et al. (2023). Xylenol orange-modified CdTe quantum dots as a fluorescent/colorimetric dual-modal probe for anthrax biomarker based on competitive coordination. *Talanta* 261: 124664.

39 Andreev, A. and Lipovskii, A. (1999). Anisotropy-induced optical transitions in PbSe and PbS spherical quantum dots. *Physical Review B* 59 (23): 15402.

40 Mattoussi, H., Palui, G., and Na, H.B. (2012). Luminescent quantum dots as platforms for probing *in vitro* and *in vivo* biological processes. *Advanced Drug Delivery Reviews* 64 (2): 138–166.

41 Nakata, Y., Mukai, K., Sugawara, M. et al. (2000). Molecular beam epitaxial growth of InAs self-assembled quantum dots with light-emission at 1.3 μm. *Journal of Crystal Growth* 208 (1–4): 93–99.

42 Sinatra, L., Pan, J., and Bakr, O.M. (2017). Methods of synthesizing monodisperse colloidal quantum dots. *Mater Matters* 12 (1): 3–7.

43 Bruno, F., Sciortino, A., Buscarino, G. et al. (2021). A comparative study of top-down and bottom-up carbon nanodots and their interaction with mercury ions. *Nanomaterials* 11 (5): 1265.

44 Ekimov, A.I., Efros, A.L., and Onushchenko, A.A. (1985). Quantum size effect in semiconductor microcrystals. *Solid State Communications* 56 (11): 921–924.

45 Martynenko, I., Litvin, A., Purcell-Milton, F. et al. (2017). Application of semiconductor quantum dots in bioimaging and biosensing. *Journal of Materials Chemistry B* 5 (33): 6701–6727.

46 Bilan, R., Fleury, F., Nabiev, I., and Sukhanova, A. (2015). Quantum dot surface chemistry and functionalization for cell targeting and imaging. *Bioconjugate Chemistry* 26 (4): 609–624.

47 Zhou, J., Yang, Y., and Zhang, C.-y. (2015). Toward biocompatible semiconductor quantum dots: from biosynthesis and bioconjugation to biomedical application. *Chemical Reviews* 115 (21): 11669–11717.

48 Mattoussi, H., Mauro, J.M., Goldman, E.R. et al. (2000). Self-assembly of CdSe–ZnS quantum dot bioconjugates using an engineered recombinant protein. *Journal of the American Chemical Society* 122 (49): 12142–12150.

49 Hines, M.A. and Guyot-Sionnest, P. (1996). Synthesis and characterization of strongly luminescing ZnS-capped CdSe nanocrystals. *The Journal of Physical Chemistry* 100 (2): 468–471.

50 He, X., Gao, L., and Ma, N. (2013). One-step instant synthesis of protein-conjugated quantum dots at room temperature. *Scientific Reports* 3 (1): 2825.

51 Gopee, N.V., Roberts, D.W., Webb, P. et al. (2007). Migration of intradermally injected quantum dots to sentinel organs in mice. *Toxicological Sciences* 98 (1): 249–257.

52 Larson, D.R., Zipfel, W.R., Williams, R.M. et al. (2003). Water-soluble quantum dots for multiphoton fluorescence imaging *in vivo*. *Science* 300 (5624): 1434–1436.

53 Alam, F. and Yadav, N. (2013). Potential applications of quantum dots in mapping sentinel lymph node and detection of micrometastases in breast carcinoma. *Journal of Breast Cancer* 16 (1): 1–11.

54 Zhang, C.-y. and Johnson, L.W. (2007). Quantifying RNA–peptide interaction by single-quantum dot-based nanosensor: an approach for drug screening. *Analytical Chemistry* 79 (20): 7775–7781.

55 Hammoud, D.A., Hoffman, J.M., and Pomper, M.G. (2007). Molecular neuroimaging: from conventional to emerging techniques. *Radiology* 245 (1): 21–42.

56 Raichle, M.E. (1998). Behind the scenes of functional brain imaging: a historical and physiological perspective. *National Academy of Sciences of the United States of America* 95 (3): 765–772.

57 Raichle, M.E. (ed.) (1998). Imaging the mind. In: *Seminars in Nuclear Medicine*, 278–289. Elsevier.

58 Raichle, M.E. and Snyder, A.Z. (2007). A default mode of brain function: a brief history of an evolving idea. *NeuroImage* 37 (4): 1083–1090.

59 Skipper, N., Igra, M., and Davidson, A. (2018). Brain imaging for anaesthetists and intensivists: part 1—Computed tomography. *BJA Education* 18 (10): 300–309.

60 Zidan, M.M., Altayeb, A., Altayeb, A. et al. (2024). Revealing the surprising: computed tomography-guided exploration of paranasal sinuses incidental findings. *Journal of Radiation Research and Applied Science* 17 (4): 101101.

61 Copen, W.A., Lev, M.H., and Rapalino, O. (2016). Brain perfusion: computed tomography and magnetic resonance techniques. *Handbook of Clinical Neurology* 135: 117–135.

62 Lubell, D.L. (2005). Drawbacks and limitations of computed tomography. *Texas Heart Institute Journal* 32 (2): 250.

63 Giovacchini, G., Squitieri, F., Esmaeilzadeh, M. et al. (2011). PET translates neurophysiology into images: a review to stimulate a network between neuroimaging and basic research. *Journal of Cellular Physiology* 226 (4): 948–961.

64 Anton-Rodriguez, J.M. (2017). *Assessment of the Benefits and Drawbacks of High Resolution PET for the Imaging of Cancer in the Head*. The University of Manchester (United Kingdom).

65 Lovitt, S., Moore, S.L., and Marden, F.A. (2006). The use of MRI in the evaluation of myopathy. *Clinical Neurophysiology* 117 (3): 486–495.

66 Frisoni, G.B., Fox, N.C., Jack, C.R. Jr., et al. (2010). The clinical use of structural MRI in Alzheimer disease. *Nature Reviews Neurology* 6 (2): 67–77.

67 Al-Shwaiheen, F.A., Wang, S.J., Uzelac, A. et al. (2015). The advantages and drawbacks of routine magnetic resonance imaging for long-term post-treatment locoregional surveillance of oral cavity squamous cell carcinoma. *American Journal of Otolaryngology* 36 (3): 415–423.

68 Kanda, P.A.M., Anghinah, R., Smidth, M.T., and Silva, J.M. (2009). The clinical use of quantitative EEG in cognitive disorders. *Dementia & Neuropsychologia* 3 (3): 195–203.

69 Fowle, A.J. and Binnie, C.D. (2000). Uses and abuses of the EEG in epilepsy. *Epilepsia* 41: S10–S18.

70 Kim, S.-G. and Ogawa, S. (2012). Biophysical and physiological origins of blood oxygenation level-dependent fMRI signals. *Journal of Cerebral Blood Flow and Metabolism* 32 (7): 1188–1206.

71 Owen, A.M. and Coleman, M.R. (2008). Functional neuroimaging of the vegetative state. *Nature Reviews Neuroscience* 9 (3): 235–243.

72 Haller, S. and Bartsch, A.J. (2009). Pitfalls in fMRI. *European Radiology* 19: 2689–2706.

73 Ferrari, M. and Quaresima, V. (2012). A brief review on the history of human functional near-infrared spectroscopy (fNIRS) development and fields of application. *NeuroImage* 63 (2): 921–935.

74 Irani, F., Platek, S.M., Bunce, S. et al. (2007). Functional near infrared spectroscopy (fNIRS): an emerging neuroimaging technology with important applications for the study of brain disorders. *The Clinical Neuropsychologist* 21 (1): 9–37.

75 Teleanu, D.M., Chircov, C., Grumezescu, A.M. et al. (2019). Contrast agents delivery: an up-to-date review of nanodiagnostics in neuroimaging. *Nanomaterials* 9 (4): 542.

76 Jackson, H., Muhammad, O., Daneshvar, H. et al. (2007). Quantum dots are phagocytized by macrophages and colocalize with experimental gliomas. *Neurosurgery* 60 (3): 524–530.

77 Kairdolf, B.A., Smith, A.M., Stokes, T.H. et al. (2013). Semiconductor quantum dots for bioimaging and biodiagnostic applications. *Annual Review of Analytical Chemistry* 6 (1): 143–162.

78 Hu, Y., Wang, X., Niu, Y. et al. (2024). Application of quantum dots in brain diseases and their neurotoxic mechanism. *Nanoscale Advances* 6 (15): 3733–3746.

79 Liu, X.-A., Li, X., Shen, P. et al. (2024). Fundamental role of brain-organ interaction in behavior-driven holistic homeostasis. *Fundamental Research* 1–13.

80 Guan, S. and Tang, M. (2024). Exposure of quantum dots in the nervous system: central nervous system risks and the blood–brain barrier interface. *Journal of Applied Toxicology* 44 (7): 936–952.

81 Ganganboina, A.B., Dega, N.K., Tran, H.L. et al. (2021). Application of sulfur-doped graphene quantum dots@ gold-carbon nanosphere for electrical pulse-induced impedimetric detection of glioma cells. *Biosensors & Bioelectronics* 181: 113151.

82 Li, C., Cao, L., Zhang, Y. et al. (2015). Preoperative detection and intraoperative visualization of brain tumors for more precise surgery: a new dual-modality MRI and NIR nanoprobe. *Small* 11 (35): 4517–4525.

83 Wang, H., Revia, R., Mu, Q. et al. (2020). Single-layer boron-doped graphene quantum dots for contrast-enhanced *in vivo* T 1-weighted MRI. *Nanoscale Horizons* 5 (3): 573–579.

84 Wang, Y., Meng, Y., Wang, S. et al. (2015). Direct solvent-derived polymer-coated nitrogen-doped carbon nanodots with high water solubility for targeted fluorescence imaging of glioma. *Small* 11 (29): 3575–3581.

85 Gao, L., Zhao, X., Wang, J. et al. (2018). Multiple functionalized carbon quantum dots for targeting glioma and tissue imaging. *Optical Materials* 75: 764–769.

86 Brettschneider, J., Tredici, K.D., Lee, V.M.-Y., and Trojanowski, J.Q. (2015). Spreading of pathology in neurodegenerative diseases: a focus on human studies. *Nature Reviews Neuroscience* 16 (2): 109–120.

87 Teleanu, D.M., Niculescu, A.-G., Lungu, I.I. et al. (2022). An overview of oxidative stress, neuroinflammation, and neurodegenerative diseases. *International Journal of Molecular Sciences* 23 (11): 5938.

88 Hasannejadasl, B., Janbaz, F.P., Choupani, E. et al. (2020). Quantum dots application in neurodegenerative diseases. *Thrita* 9 (2): 100105.

89 Kirova, A.-M., Bays, R.B., and Lagalwar, S. (2015). Working memory and executive function decline across normal aging, mild cognitive impairment, and Alzheimer's disease. *BioMed Research International* 2015 (1): 748212.

90 Sonawane, S.K., Dubey, T., Balmik, A.A. et al. (2022, 2022). *Alzheimer's Disease Pathology: A Tau Perspective*, 55–88. Royal Society of Chemistry.

91 Tang, M., Pi, J., Long, Y. et al. (2018). Quantum dots-based sandwich immunoassay for sensitive detection of Alzheimer's disease-related Aβ1–42. *Spectrochimica Acta Part A: Molecular and Biomolecular Spectroscopy* 201: 82–87.

92 Liu, Y., Xu, L.-P., Dai, W. et al. (2015). Graphene quantum dots for the inhibition of β amyloid aggregation. *Nanoscale* 7 (45): 19060–19065.

93 Medina-Sánchez, M., Miserere, S., Morales-Narvaez, E., and Merkoçi, A. (2014). On-chip magneto-immunoassay for Alzheimer's biomarker electrochemical detection by using quantum dots as labels. *Biosensors & Bioelectronics* 54: 279–284.

94 Balestrino, R. and Schapira, A.H. (2020). Parkinson disease. *European Journal of Neurology* 27 (1): 27–42.

95 Dickson, D.W. (2012). Parkinson's disease and parkinsonism: neuropathology. *Cold Spring Harbor Perspectives in Medicine* 2 (8): a009258.

96 Mohammadi, S., Nikkhah, M., and Hosseinkhani, S. (2017). Investigation of the effects of carbon-based nanomaterials on A53T alpha-synuclein aggregation using a whole-cell recombinant biosensor. *International Journal of Nanomedicine* 8831–8840.

97 Park, S.W., Kim, T.E., and Jung, Y.K. (2021). Glutathione-decorated fluorescent carbon quantum dots for sensitive and selective detection of levodopa. *Analytica Chimica Acta* 1165: 338513.

98 Lin, D., Li, M., Gao, Y. et al. (2022). Brain-targeted gene delivery of ZnO quantum dots nanoplatform for the treatment of Parkinson disease. *Chemical Engineering Journal* 429: 132210.

99 Kim, D., Yoo, J.M., Hwang, H. et al. (2018). Graphene quantum dots prevent α-synucleinopathy in Parkinson's disease. *Nature Nanotechnology* 13 (9): 812–818.

9

Quantum Biosensing in Infectious Disease Diagnostics

Begüm Sarac[1], *Seydanur Yücer*[1], *and Fatih Ciftci*[1,2]

[1]*Faculty of Engineering, Department of Biomedical Engineering, Fatih Sultan Mehmet Vakıf University, Istanbul, Turkey*
[2]*Biomedical Electronic Design Application and Research Center (BETAM), Fatih Sultan Mehmet Vakıf University, Istanbul, Turkey*

9.1 Introduction

Infectious diseases continue to pose significant global health challenges, necessitating the development of rapid, accurate, and sensitive diagnostic technologies. Traditional diagnostic methods, including culture-based techniques, enzyme-linked immunosorbent assays (ELISA), and polymerase chain reaction (PCR), have been widely employed for pathogen detection. However, these methods often suffer from limitations such as long processing times, high costs, and the requirement for specialized laboratory infrastructure. In recent years, biosensing technologies have emerged as powerful tools for disease diagnostics, offering improved sensitivity, specificity, and real-time detection capabilities. Among these, quantum biosensing has gained increasing attention due to its ability to exploit quantum mechanical properties for enhanced biomolecular detection.

Quantum biosensors leverage phenomena such as quantum entanglement, superposition, and spin properties to achieve superior sensitivity compared to classical biosensors. These sensors can detect minute biological changes at the molecular level, making them highly suitable for the early and precise detection of infectious agents. The integration of quantum dots, nitrogen-vacancy (NV) centers in diamonds, and quantum-enhanced imaging techniques has further expanded the potential of quantum biosensors in medical diagnostics. Despite these advantages, the application of quantum biosensing in infectious disease diagnostics remains in its early stages, with several technical and practical challenges to overcome, including stability, reproducibility, and the need for scalable fabrication methods.

This review explores the advancements in quantum biosensing technologies and their potential to revolutionize infectious disease diagnostics. By addressing current limitations in conventional diagnostic techniques, quantum biosensors offer a promising avenue for the development of next generation, ultrasensitive, and rapid disease detection platforms.

Quantum Biosensing in Medical Diagnostics, First Edition. Edited by J. G. Manjunatha.

9.2 Fundamentals of Quantum Biosensors

Quantum biosensors, which leverage quantum mechanical principles, have emerged as a revolutionary technology in the detection and diagnosis of infectious diseases. Their unparalleled sensitivity and specificity enable the identification of pathogens and biomarkers at extremely low concentrations, facilitating early diagnosis and timely medical intervention. Among these biosensors, quantum dots (QDs) have gained significant attention due to their exceptional optical and electronic properties. These nanomaterials, derived from semiconductor and carbon-based structures, exhibit size-dependent quantum confinement effects, which profoundly influence their photoluminescence and energy transfer characteristics.

Quantum dots, particularly those composed of semiconductor materials, share fundamental electronic properties with their bulk counterparts, including capacitance, conductivity, and bandgap characteristics. However, a crucial distinction arises in the behavior of electron transfer. In bulk materials, electrons move freely within the lattice, whereas in QDs, electron mobility is restricted within a critical spatial range known as the Bohr radius. When the size of QDs approaches this limit, quantum confinement effects become dominant, altering the electronic and optical properties significantly. These properties underpin their potential applications in biosensing and disease diagnostics.

Among quantum dots, carbon quantum dots (CQDs) and graphene quantum dots (GQDs) have gained prominence as viable alternatives to semiconductor QDs. Their biocompatibility, low toxicity, and environmentally friendly synthesis make them attractive for biomedical applications. Despite these advantages, challenges such as broad size distribution, synthesis reproducibility, and relatively low quantum efficiency remain critical hurdles in their widespread adoption. Additionally, variations in fabrication methods lead to inconsistencies in optical properties, affecting the reliability of biosensing applications.

9.2.1 Graphene Quantum Dots (GQDs) in Pathogen Detection

Graphene quantum dots, which are nanoscale fragments of graphene, exhibit remarkable photoluminescence, stability, and electrochemical properties, making them ideal candidates for biomedical sensing [1]. Their unique characteristics have been exploited in various biosensing platforms, including electrochemical and fluorescence-based detection systems. For instance, GQDs have been utilized in the development of an immunosensor for detecting *Yersinia enterocolitica*, a pathogen responsible for yersiniosis. This sensor demonstrated exceptional sensitivity, detecting bacterial concentrations as low as 5 colony-forming units per milliliter (CFU/mL) in milk and 30 CFU/mL in human serum without requiring pre-sample treatment. Such advancements highlight the potential of GQDs in rapid and precise infectious disease diagnostics [2].

Beyond pathogen detection, GQDs have been widely investigated in biosensors targeting biomarkers for cancer, cardiovascular diseases, and neurodegenerative disorders. Their water solubility, nontoxicity, and excellent electrocatalytic properties enable the development of highly sensitive optical, electrochemical, and photoelectrochemical biosensors [3].

9.2.2 Nitrogen-Vacancy (NV) Centers in Diamond for Viral RNA Sensing

Another promising quantum biosensing approach involves nitrogen-vacancy (NV) centers in diamond, which have been proposed for detecting viral RNA, such as that of SARS-CoV-2. These NV centers can transduce the presence of specific RNA sequences into detectable magnetic noise signals, providing a rapid and highly sensitive diagnostic method. Theoretical analyses suggest that NV-based quantum sensors could detect viral RNA at concentrations as low as a few hundred copies, with a false negative rate below 1%. This capability offers a substantial advantage over conventional diagnostic techniques in terms of accuracy and speed [4].

A comparative overview of the advantages and limitations of semiconductor, carbon, and GQDs in biosensing applications is provided in Table 9.1 [9].

The integration of quantum biosensors, particularly those based on GQDs and NV centers in diamond, represents a significant leap in the field of infectious disease diagnostics. Their ability to detect minute pathogen and biomarker concentrations with high specificity and sensitivity holds promise for advancing early detection, improving patient outcomes, and informing public health strategies [4]. Despite challenges related to fabrication, reproducibility, and large-scale implementation, ongoing research and technological advancements continue to enhance their feasibility for clinical applications. Further studies aimed at improving quantum efficiency, standardizing synthesis methods, and optimizing biosensing platforms will be crucial in translating these novel quantum technologies into real-world diagnostic tools [10, 11].

Table 9.1 Advantages and disadvantages of semiconductor, carbon, and graphene quantum dots in biosensing applications.

Quantum dot type	Advantages	Disadvantages	References
Semiconductor QDs	High photostability, resistance to photobleaching, narrow and symmetric emission peaks, high quantum yield, long lifetime, stable under chemical/biological conditions	Fluorescence blinking, high background signals, potential cytotoxicity, incomplete clearance from biological systems	[5]
Carbon QDs	Nontoxic, cost-effective, environmentally friendly, rapid and scalable synthesis, size and nanostructure controllability	Use of toxic acid/base reagents, broad size distribution, long synthesis duration	[6]
Graphene QDs	Simple synthesis, high stability, environmentally friendly, uniform size distribution, improved production yield	Use of strong oxidizers (potential safety hazards), requirement for oxidation pretreatment, challenges in large-scale production	[7, 8]

9.3 Application of Quantum Biosensors in Infectious Disease Detection

9.3.1 Viral Infections

The use of quantum dots (QDs) for the detection and elimination of SARS-CoV-2 has garnered considerable attention from researchers. This interest stems primarily from the unique optical properties of QDs, which allow them to be tracked under specific wavelengths of light. QDs have demonstrated the ability to effectively target and penetrate SARS-CoV-2, which has a particle size ranging from 60 to 140 nm. In particular, carbon-based QDs, which carry a positive surface charge, can facilitate the separation and inactivation of the spike (S) protein of the virus. Additionally, these QDs can interact with the virus's negatively charged RNA strand, leading to the generation of reactive oxygen species (ROS) that contribute to viral inactivation. Previous studies have also highlighted the antiviral potential of carbon dots (CDs) against other viruses, such as the porcine reproductive and respiratory syndrome virus and pseudorabies virus. CDs have been found to induce interferon-stimulated genes, including interferon-α, which plays a role in suppressing viral replication. Moreover, functionalized QDs can effectively interact with SARS-CoV-2 entry receptors, thereby influencing the virus's genomic replication. In the realm of biosensor technology, carbon nanotube field-effect transistor (CNT-FET) electronic biosensors have been utilized for detecting various biomolecules, including metal ions, whole cells, and viruses [12]. Scientists have leveraged CNT-FET technology to develop an electrochemical biosensor specifically designed for SARS-CoV-2 detection. This approach allows for digital detection of the virus's S1 antigen in saliva samples. The biosensor operates by leveraging noncovalent interactions between single-walled carbon nanotubes (SWCNTs) and a stabilizing binder, 1-pyrenbutanoic acid succinimidyl ester. The antibody specific to SARS-CoV-2 S1 is immobilized on the SiO_2 surface, positioned between the source and drain channels. The biosensor integrates a nanotube-based field-effect transistor with a liquid gate, using RNA hybridization as a signal transducer. To validate this sensor, researchers employed a commercially available SARS-CoV-2 S1 antigen, demonstrating that the technology can accurately differentiate SARS-CoV-2 from other coronaviruses, such as SARS-CoV-1 and MERS-CoV. This rapid and sensitive detection method allows for efficient identification of viral particles in saliva samples. QDs also hold promise as biomarkers and targeted antiviral agents. Research has shown that graphene-based QDs and their derivatives can be utilized for coronavirus detection, with potential applications that could significantly advance efforts to combat COVID-19. In an innovative study, scientists developed a method to target the spike protein of SARS-CoV-2 by integrating graphene sheets—each thinner than a postage stamp (approximately 0.34 nm thick)—with specific antibodies. These graphene sheets were then exposed to artificial saliva samples containing either COVID-positive or COVID-negative specimens. Atomic surface vibrations of the graphene sheets were measured to determine viral presence. The study also assessed the response of these sheets when exposed to other coronaviruses, such as MERS-CoV. The results revealed that graphene sheets bound to SARS-CoV-2 exhibited distinct vibrational changes, while those exposed to other coronaviruses or COVID-negative samples showed no significant alterations. This detection process, which relies on Raman spectrometry, was completed in under five minutes, offering a fast and reliable diagnostic tool. Additionally, researchers have explored the use of electrochemical biosensors to detect SARS-CoV-2 and other pathogenic viruses. One such method involves the design of a rapid electrochemical detection kit featuring screen-printed electrodes. To facilitate virus detection in biological samples like blood and saliva, scientists coated a graphene oxide (GO) layer with chemically sensitive compounds and gold nanostars (Au NS) to activate the sensor's working electrode. This approach enables the detection of viral traces within approximately one minute, eliminating the need for biomarker extraction or complex sample preparation. The developed nanosensor demonstrated a limit of detection (LOD) of 1.68×10^{-22} μg/mL and a sensitivity of 0.0048 μAμg/mL cm^{-1}. These findings highlight the potential of QD-based biosensors in providing highly sensitive, rapid, and reliable diagnostics for infectious diseases like COVID-19 (Figure 9.1a) [13].

Middle East respiratory syndrome (MERS) is a viral respiratory illness caused by the MERS coronavirus (MERS-CoV), first identified in Saudi Arabia in 2012. Common symptoms include fever, cough, and difficulty breathing, with pneumonia frequently observed, although not in all cases. Additionally, gastrointestinal symptoms such as diarrhea have been reported. Some MERS-CoV infections have been classified as asymptomatic, meaning affected individuals show no clinical symptoms, with many cases detected through thorough contact tracing of confirmed cases. Despite this, MERS-CoV has been identified as the cause of death in approximately 35% of confirmed cases. Numerous biosensors have been developed for MERS detection, as reported in the literature. One method for detecting MERS-CoV involves a label-free colorimetric assay utilizing gold nanoparticles (AuNPs), which enables the identification of target analytes based on color changes in an indicator. Kim et al. introduced a colorimetric test that relies on an expanded form of double-stranded DNA (dsDNA) self-assembly to protect AuNPs in an electrolyte-rich environment, facilitating MERS-CoV detection. This method detects viral molecules through localized surface plasmon resonance (LSPR) shifts and variations in AuNP color within the UV–vis wavelength range. The researchers designed thiol-modified probes complementary to specific regions of the MERS-CoV genome, namely, the upE and ORF1a genes, at either the 5′ or 3′ end. The target sequence and probes formed a disulfide-induced self-assembled complex, preventing salt-induced AuNP aggregation and changes in optical properties. This assay successfully detected 30 bp MERS-CoV sequences at a concentration of 1.0 pmol/L. Another approach employed a multiplexed paper-based colorimetric DNA sensor utilizing pyrrolidinyl peptide nucleic acid (PNA)-induced silver nanoparticle (AgNP) aggregation for MERS-CoV detection. The presence of the oligonucleotide target was determined by monitoring AgNP color changes, achieving a detection limit of 1.53 nM for MERS-CoV. The acpcPNA probe exhibited high selectivity for complementary oligonucleotides, effectively distinguishing them from single- and two-base mismatches, as well as noncomplementary DNA sequences [17].

Influenza viruses are categorized into three genera: influenza A, B, and C. Among them, influenza A is the most prevalent and pathogenic, primarily due to its high genetic variability and rapid mutation rate, making it a significant threat to human health. Conventional diagnostic approaches for influenza viruses face several limitations, necessitating the development of more advanced detection techniques. Fluorescent semiconductor quantum dot (QD) nanocrystals have emerged as valuable fluorescence reporters in biosensor applications. Consequently, an ultrasensitive and highly specific LSPR-induced immunofluorescence nanobiosensor has been developed for influenza virus detection. This biosensor leverages a gold nanoparticle (AuNP)-induced quantum dot (QD) fluorescence signal. The detection mechanism is based on conjugating anti-neuraminidase (NA) antibodies (anti-NA Ab) with thiolated AuNPs and anti-hemagglutinin (HA) antibodies (anti-HA Ab) with alloyed quaternary L-cysteine-capped CdSeTeS QDs. The binding of these antibodies to the antigens present on the surface of the influenza virus triggers an LSPR signal from nearby AuNPs, enhancing QD fluorescence in correlation with the viral concentration. This biosensor has been utilized for detecting the influenza virus (H1N1) in both deionized water and human serum. The detection sensitivity for H1N1 was determined to be 0.03 pg/mL in deionized water, while for clinically isolated H3N2, the detection limit was 10 PFU/mL in human serum [18].

Biomarkers offer a promising alternative to ELISA kits for detecting infectious diseases; however, their sensitivity remains a key limitation. A major challenge in bioanalytical research is the development of biosensors capable of detecting multiple analytes simultaneously. Quantum dot (QD)-based biomarkers provide an innovative approach for HIV detection by enabling real-time visualization of interactions between the virus and human immune cells. One effective strategy involves using boronic acid-functionalized QDs, which can be selectively activated during the viral life cycle. This functionalization enhances QD interactions through covalent, noncovalent, and electrostatic forces, making them highly effective for HIV diagnostics. Research by Iannazzo et al. [19] demonstrated that modifying GQDs or traditional QDs with non-nucleoside reverse transcriptase inhibitors (NNRTIs) significantly improves their diagnostic capability for HIV-1. A notable advancement in this field is the development of fluorescence-based biosensors incorporating CQDs and cadmium telluride quantum dots (CdTe QDs) for detecting HIV dsDNA in serum samples. In one study, CdTe QDs were coated with mercapropionic acid and linked with mitoxantrone (MTX), an anthraquinone derivative that binds to DNA. This interaction led to fluorescence quenching at 599 nm due to electron transfer between CdTe QDs and MTX. In this state, only fluorescence at 435 nm was observed due to green QDs. However, when HIV-dsDNA was present, MTX preferentially bound to the dsDNA, causing CdTe QDs to move away from the MTX-ssDNA complex due to electrostatic repulsion, thereby altering fluorescence emission. This innovative approach showcases the potential of QD-based biosensors in enhancing HIV diagnostics through high specificity and sensitivity. In the presence of HIV-dsDNA, specific binding of MTX to dsDNA led to discrimination between MTX and CdTe QDs (Figure 9.1b) [14].

A dual-color quantum dot (QD) detection method has been developed to simultaneously identify co-infections of human papillomavirus (HPV) and HIV. Xue et al. [20] utilized QD in situ hybridization (QD-ISH) to investigate the relationship between oral squamous cell carcinoma (OSCC) and HPV. Their study revealed that QD-ISH was more effective than conventional ISH, successfully detecting high-risk HPV16/18. These findings indicate that QD technology could serve as a reliable tool for diagnosing HPV infections and HPV-related OSCC. Further research by Piao et al. [21] explored the use of fluorescent nanomaterials, including metal nanoparticles, QDs, and fluorophore-modified silica nanoparticles, for detecting HPV DNA. Additionally, Nejdl et al. studied the interactions between

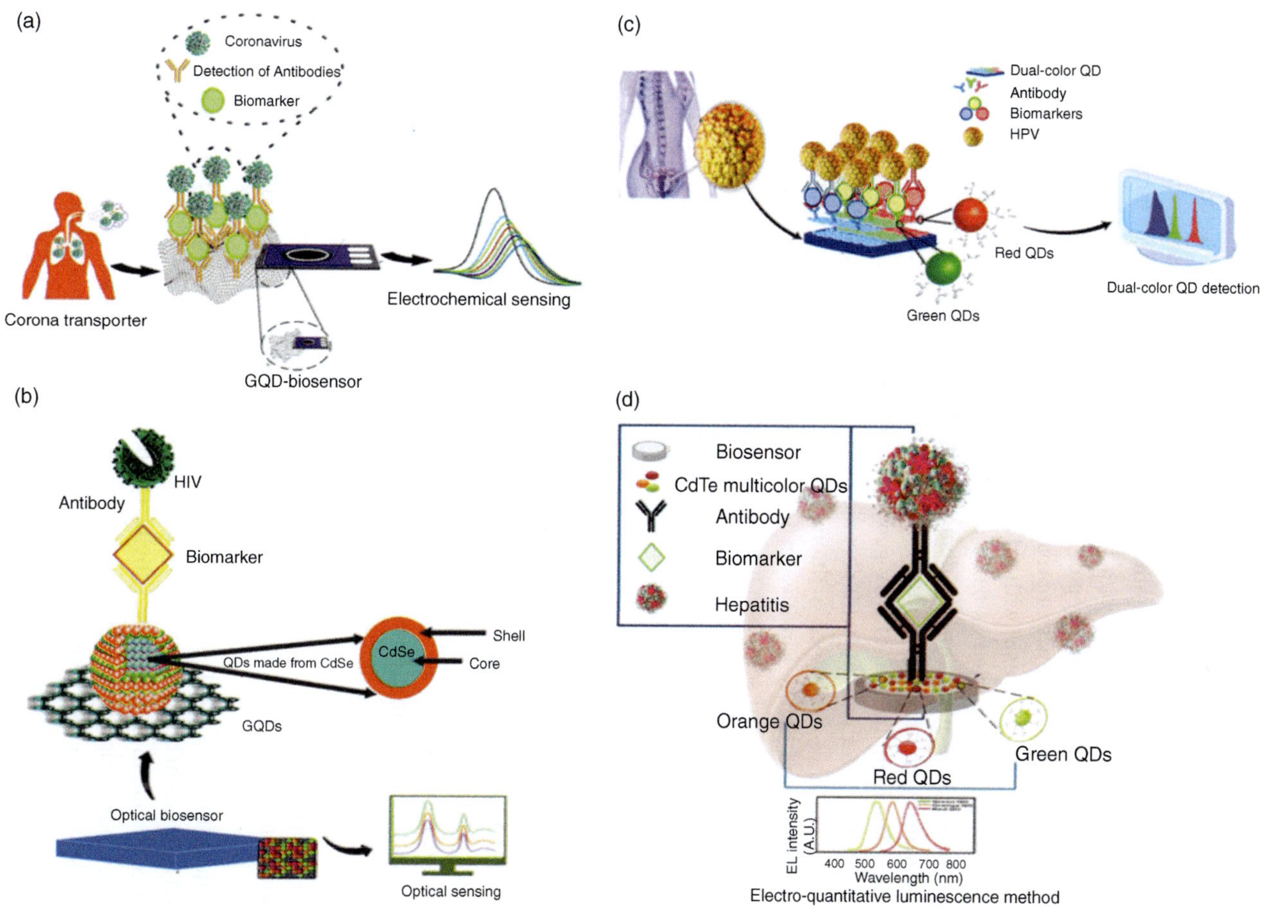

Figure 9.1 (a) Schematic summary shows the detection of coronaviruses using QD-based graphene biomarkers. Adapted from Hashemi et al. [13]. (b) Schematic summary shows the detection of HIV using QD-based biomarkers made from CdSe with optical sensors. Liang et al. [14] / with permission of Elsevier. (c) Schematic summary shows the detection of HPV infection using the dual-color QD detection method. Adapted from Nejdl et al. [15]. (d) Schematic summary shows the detection of hepatitis C and hepatitis B viruses using CdTe multicolored QDs and gold nanoparticles in an electro-quantitative luminescence method. Wang et al. [16] / with permission of American Chemical Society.

blue and yellow fluorescent CdS QDs to analyze the HPV-16 oncogene E6. Their results demonstrated that yellow-low fluorescent CdS QDs exhibited a stronger affinity for the E6 DNA sequence compared to blue fluorescent dyes (Figure 9.1c) [15].

Recent advancements have led to the development of quantum dot (QD)-based biomarkers, which offer a promising approach for detecting hepatitis viruses. Unlike conventional methods, these biomarkers do not require extensive sample preparation, making the detection process more efficient. The identification of specific hepatitis virus targets using QD-based biomarkers is achieved through interactions with biological molecules, such as antigen-antibody binding, ligand-receptor interactions, or enzyme-substrate reactions. An electro-quantitative luminescence approach has been utilized to identify hepatitis B (HBV) and hepatitis C (HCV) viruses with CdTe multicolor QDs. The detection limits for HBV and HCV have been reported as 0.082 and 1.341 picomoles per liter, respectively. The DNA sensor used in this technique demonstrates high specificity, stability, reproducibility, and sensitivity, making it suitable for detecting HBV DNA and HCV RNA in human samples. The identification of HBV gene mutations plays a crucial role in clinical diagnostics, particularly for patients with chronic hepatitis B (CHB) undergoing antiviral therapy. Researchers have developed a fluorescence-based QD detection method that is both cost-effective and highly sensitive in identifying these mutations. Additionally, QD-enhanced electrochemical sensors have been explored for detecting the hepatitis E virus (HEV), showing significant potential for accurate and efficient diagnostics. Further advancements include the use of QD-encoded protein arrays for hepatitis C virus detection. A comparative analysis between this method and the recombinant immunoblot assay (RIBA), a standard diagnostic technique, has confirmed its effectiveness in identifying HCV-positive and HCV-negative cases. Additionally, the development of QD barcodes has enabled the simultaneous and selective detection of multiple HBV biomarkers, demonstrating high efficiency in multiplex hepatitis diagnostics. These findings highlight the potential of QD-based biosensors in enhancing the accuracy and efficiency of hepatitis virus detection, paving the way for improved diagnostic applications in the field of infectious disease monitoring (Figure 9.1d) [16].

Quantum dots have been applied in detecting viruses such as dengue, influenza, Zika, and norovirus. Traditional diagnostic techniques for these infections typically require between two to six days, which can delay timely treatment. The lengthy process of current diagnostic methods, which may take up to five days to confirm the presence of a virus and additional time to identify the specific strain, highlights the need for faster alternatives.

Graphene quantum dots (GQDs) and CQDs offer a more efficient approach, utilizing multi-colored fluorescent beads that bind to distinct molecular structures on viral envelopes and infected cells. Due to their high sensitivity, QD-based systems can detect viral particles within a few hours, significantly reducing the time required compared to conventional testing methods. Additionally, quantum dots are both cost-effective and capable of simultaneously identifying multiple respiratory viruses, including dengue, influenza, Zika, and norovirus. This is achieved by attaching QDs of different colors to linker molecules that specifically bind to surface structures of various viruses. Since quantum dots come in a wide range of colors, they can be paired with antibodies specific to these viruses, enhancing detection accuracy and efficiency.

Mosquitoes serve as vectors for transmitting dengue virus (DENV) infections. DENV is a single-stranded, positive-sense RNA virus enclosed within a lipid envelope and is classified into five distinct serotypes. Nearly half of the global population is at risk of DENV infection, affecting approximately 3 billion individuals worldwide. Annually, around 400 million people contract DENV, with 100 million developing symptomatic infections, and dengue-related complications result in approximately 22,000 deaths each year. Avian influenza viruses primarily infect birds but can also be transmitted to humans. The disease spreads through direct contact with infected birds and can further propagate through human-to-human transmission. The symptoms closely resemble those of the common cold, including fever, cough, sore throat, headache, muscle aches, and respiratory difficulties. If administered within the first two days of symptom onset, certain antiviral medications may help alleviate the severity of the illness. To enhance DENV detection, a biofunctionalized tapered optical fiber (TOF)-based sensor incorporating a polyamidoamine (PAMAM) dendrimer was developed for detecting the DENV E protein. The TOF structure created an evanescent field responsive to external medium variations, while PAMAM facilitated bio-recognition molecule attachment and anti-DENV II E protein antibody binding. This design resulted in an increased number of active sites for DENV II E protein absorption along the tapering region. The sensor demonstrated a dissociation constant (K_d) of 1.02×10^{-10} M, with a detection resolution of 19.53 nm/nM and a detection limit of 1.0 pM, indicating significant potential for improving dengue diagnostics. Another method for detecting DENV E protein involved immobilizing a monoclonal antibody (IgM) onto a gold/Fe-MPA-NCC-CTAB/EDC-NHS thin film via spin coating. Upon interaction with DENV E protein, the IgM-functionalized thin film generated a surface plasmon resonance (SPR) signal within a concentration range of 0.0001–10 nM. The method demonstrated a linear correlation between SPR angle and DENV E protein concentration up to 0.01 nM, achieving a sensitivity of 39.96°/nM.

Various biomarkers, including DNA, RNA, peptides, antibodies, glycoproteins, and antigens, serve as key analytes for detecting viral infections. These biomarkers can be broadly categorized into two main groups: antigens and antibodies. Many viruses contain either RNA or DNA as their genetic material, which encodes proteins essential for viral replication and function. Structurally, viruses typically consist of three core components: genetic material (either DNA or RNA), nucleocapsid proteins, and capsid proteins. The genetic material is enclosed by nucleocapsid proteins, which are further surrounded by envelope proteins, such as capsid proteins [10]. Viral entry into host cells is initiated through interactions with cell-surface receptors, followed by the release of the viral genome into the cytoplasm. This process occurs via two main mechanisms: endocytic and non-endocytic pathways. Upon successful infection, viral components, including RNA, DNA, and nucleocapsid/capsid proteins, become immediately available within the host cell. In response to viral antigens, B lymphocytes secrete immunoglobulins, facilitating immune recognition. The presence of viral antigens (RNA, DNA, and proteins) and their corresponding antibodies in the cellular microenvironment enables their detection through optical biosensing techniques utilizing specific biomolecules. In addition to viral diagnostics, assessing the severity of viral infections is crucial for effective disease management. A major challenge in clinical diagnostics involves differentiating viral infections from bacterial ones due to overlapping symptoms. Misdiagnosis occurs in approximately 50% of cases, often leading to unnecessary antibiotic prescriptions intended to prevent secondary bacterial infections. Ideally, biomarkers for infection diagnosis should be independent of factors such as illness duration and comorbidities while also being detectable from minimally invasive samples. Disease severity can be assessed through biomarkers such as interleukins, C-reactive protein (CRP), tumor necrosis factor-alpha (TNF-α), glutamate, interferons, D-dimer, and hematological markers. These biomarkers, classified into therapeutic and pathogenetic categories, play essential roles in clinical diagnostics and prognosis. Optical biosensing techniques can be employed for their detection. Most studies in this field focus on identifying host-derived biomarkers that are produced as part of the immune system's response to infection. The immune system releases various biomolecules into circulation, reflecting real-time pathological changes. These molecules serve as critical biomarkers due to their biological relevance in disease progression. However, not all biomarkers are suitable for diagnostic applications and must meet specific criteria. For instance, hematological markers such as polymorphonuclear leukocytes (PMNs), human neutrophil lipocalin (HNL), neutrophil counts, white blood cells (WBCs), and erythrocyte sedimentation rate (ESR) have been identified as highly effective biomarkers, with significant statistical correlations. PMNs, a subset of phagocytes, exhibit increased levels in the bloodstream during infections and play a central role in host immune defense. The immune response to different pathogens leads to distinct biomarker profiles, which provide insights into disease etiology. Inflammatory markers such as CRP and procalcitonin (PCT) are widely used to differentiate bacterial from nonbacterial infections. CRP, a calcium-dependent plasma protein composed of cyclic homo-pentameric non-glycosylated polypeptide subunits, is a key indicator of inflammation in humans. PCT, a precursor of the hormone calcitonin synthesized by C cells of the thyroid gland, is also recognized as a crucial biomarker in acute and severe viral infections. Cytokines, which mediate inflammatory responses to pathogens, represent another important class of host protein biomarkers. These include pro-inflammatory cytokines such as IL-1, IL-2, IL-6, IL-8, and TNF, as well as anti-inflammatory cytokines like IL-4, IL-10, and IL-13. Among them, IL-6, a 21 kDa glycoprotein primarily produced by macrophages and lymphocytes, plays a significant role in immune regulation. It facilitates the activation and recruitment of immune cells, along with the release of monocyte chemoattractant protein-1 (MCP-1). Elevated IL-6 levels are associated with severe infections, making it a valuable disease marker [22].

Table 9.2 provides a structured overview of relevant literature in the field, showcasing various biosensing techniques and their contributions to infectious disease diagnostics.

9.3.2 Bacterial Infections

A highly sensitive electrochemical (EC) aptasensor was developed for detecting *Staphylococcus aureus* using gold nanoparticles (AuNPs), nitrogen-doped carbon nano-onions, and thiol-modified aptamers in combination with screen-printed carbon electrodes. The nitrogen-doped carbon significantly enhanced the sensor's analytical performance, while the thiol-functionalized aptamers were securely attached via covalent bonding with AuNPs. This aptasensor demonstrated excellent sensitivity and specificity for *S. aureus* detection, with a thorough evaluation of its reproducibility, repeatability, and

Table 9.2 Overview of the literature on viral infections.

Study	Main topic	Method/technique	Findings	References
Yazirlioğlu. L.,	Functional activity analysis of polymorphonuclear leukocytes (PMNs) using optical chemiluminescence-based methods	Chemiluminescence (CL) measurements for assessing the oxidative potential of PMNs	Detection of viral and bacterial infections from whole blood samples of individuals with acute infections	[23]
Chen et al.	Use of SPR-based biosensor for morphological analysis of SARS-CoV	Surface plasmon resonance (SPR) technology	Enabled the structural analysis of SARS-CoV proteins	[24]
Qui et al.	COVID-19 diagnosis using plasmonic photothermal (PPT) and localized surface plasmon resonance (LSPR)	2D gold nano-islands (AuNIs) for cDNA-based biosensing	Introduced a novel optical biosensor approach for COVID-19 detection	[25]
Shafiee et al.	Label-free optical biosensors for detection using nanostructured photonic crystals	Nanostructured photonic crystals with rough surfaces	Developed a high-resolution method for detecting viruses and biomolecules at low surface concentrations	[26]

storage stability. In a separate study, another EC aptasensor was engineered using a nanocomposite composed of silver nanoparticles, chitosan, GQDs, and nitrogen-doped titanium dioxide (Ag–Cs-Gr QDs/$NTiO_2$). This composite was utilized to modify the surface of a screen-printed carbon electrode, not only improving the sensor's overall performance but also providing an optimal immobilization site for aptamers due to the presence of silver nanoparticles. The resulting aptasensor exhibited superior sensitivity, reproducibility, repeatability, and stability compared to previously developed platforms. Additionally, *Acinetobacter baumannii*, an opportunistic gram-negative aerobic pathogen, has been recognized for its role in hospital-acquired infections such as pneumonia, bacteremia, and urinary tract infections. Recent studies have also highlighted its contribution to antibiotic-resistant infections, making its rapid and precise detection crucial for effective treatment and management. To meet this need, a label-free EC aptasensor was constructed using glyoxal-modified magnetic silica nanoparticles (Fe_3O_4@SiO_2@Gly) for highly sensitive detection of *A. baumannii*. Experimental findings confirmed that this aptasensor could selectively and effectively identify *A. baumannii* even in complex biological samples such as human serum [27].

GQDs produced through various synthesis techniques have been extensively applied in biosensing due to their large surface area, which allows functionalization. This characteristic enables the direct detection of biomolecules such as DNA, enzymes, proteins, antigens, and antibodies, facilitated by oxide components formed during synthesis. For instance, Safardoust et al. [28] developed GQDs from citric acid and ethylene diamine to create a photoluminescent biosensor for identifying *S. aureus* and *E. coli*. Their biosensor exhibited a linear correlation between fluorescence intensity and bacterial concentration up to 9×10^7 CFU/mL. In another study, Hazani et al. [29] designed a highly sensitive electrochemical biosensor using functionalized graphene oxide combined with cadmium sulfide quantum dots (CdS QDs). This PNA-based sensor was tailored for detecting *Mycobacterium tuberculosis* and achieved a LOD of 8.948×10^{-13} M. Integrating GQDs into biosensors enhances key performance metrics, including selectivity, sensitivity, and reproducibility, further advancing their potential in diagnostic applications [30].

A highly selective biosensing platform was also developed based on the electropolymerization of polydopamine (PDA) molecularly imprinted polymer (MIP) combined with nitrogen-doped GQDs (N-GQDs) for the detection of Escherichia coli. The sensor fabrication involved the direct electropolymerization of PDA with E. coli to form a PDA-MIP film on a glassy carbon electrode (GCE). To remove the bacterial template, the PDA-MIP/GCE was incubated overnight in a solution containing sodium dodecyl sulfate (SDS) and acetic acid. A polyclonal antibody labeled with N-GQDs was subsequently used for target recognition. The characterization of the sensor was performed via cyclic voltammetry (CV) and electrochemical impedance spectroscopy (EIS), confirming its sensitivity and specificity. This method enabled the detection of E. coli in a wide concentration range (10–10^7 CFU/mL) with a detection limit of 8 CFU/mL in environmental water samples.

Furthermore, Ju et al. [31] synthesized N-GQDs via a hydrothermal method using dicyandiamide and citric acid, achieving a quantum yield (QY) of 32.4% at 350 nm. These GQDs were employed for glutathione (GSH) detection in HeLa cells, utilizing a fluorescence quenching mechanism mediated by Hg^{2+} ions. The quenching of GQD photoluminescence (PL) resulted from electron transfer and electrostatic interactions between N-GQDs and Hg^{2+}, which was subsequently reversed upon the addition of GSH. This "turn-on" fluorescence sensor demonstrated high selectivity and sensitivity for GSH, with a detection limit of 87 nM. In a similar approach, Safardoust-Hojaghan et al. [28] synthesized GQDs with a QY of 25% using citric acid and ethylenediamine. These GQDs were utilized for the fluorescence-based detection of bacterial pathogens, specifically Escherichia coli and Staphylococcus aureus, showing a linear fluorescence intensity response up to 9×10^7 CFU/mL.

Additionally, Ye et al. [32] developed an impedimetric immunosensor for the rapid and sensitive detection of *Salmonella typhimurium*. The biosensor was based on a nanoporous alumina membrane coupled with a polydimethylsiloxane (PDMS) chamber, where bacterial detection was influenced by impedance changes across the membrane. As *S. typhimurium* accumulated, impedance increased, while the introduction of antibiotics resulted in a reduction in impedance due to bacterial lysis. This biosensor leveraged modified GQDs to enhance the membrane's surface-to-volume ratio. Anti-*Salmonella* antibodies were immobilized onto the GQD-enriched membrane using glutaraldehyde as a cross-linking agent, enabling the detection of *S. typhimurium* within 10 minutes, with a detection limit of 1 pM. The sensor demonstrated high specificity for *S. typhimurium* with negligible interference from other bacteria, highlighting its potential for clinical applications in bacterial infection diagnostics.

Collectively, these studies underscore the advancements in quantum dot-based biosensing strategies for bacterial infections. By integrating nanomaterials such as GQDs, AuNPs, and functionalized polymers, these biosensors offer high sensitivity, rapid response times, and specificity, making them promising tools for real-time infectious disease diagnostics. The continued development and optimization of these biosensing platforms may facilitate more effective pathogen detection, contributing to improved disease management and public health outcomes [9] (Table 9.3).

CDots-based biosensors offer significant advantages in terms of rapid detection and high sensitivity. By minimizing detection times and improving specificity, these biosensors present a promising approach for diagnosing bacterial infections, particularly in resource-limited settings. Further research and investment in their development can enhance global healthcare capabilities in combating infectious diseases [44].

With the rapid emergence of antibiotic-resistant pathogens, the swift and accurate detection of antibiotic resistance genes (ARGs) is essential to curbing their spread. Engineered DNA-binding zinc finger proteins (ZFPs) can be tailored to identify specific double-stranded (ds) DNA sequences within ARGs. In this study, a rapid and straightforward approach was developed for ARG detection in bacteria using engineered ZFPs and a two-dimensional graphene oxide (GO) nanosheet-based sensing platform. The detection mechanism is based on fluorescence signal modulation, where the presence or absence of the target ARG dictates signal activation or quenching, respectively. Due to GO's inherent electronic properties, quantum dot (QD)-labeled ZFPs are adsorbed onto its surface, leading to fluorescence quenching via fluorescence resonance energy transfer (FRET). Upon interaction with the target DNA, ZFPs selectively bind to the ARG sequence, causing their dissociation from the GO surface and restoring fluorescence. This system enables the detection of target DNA by monitoring the re-emergence of QD fluorescence, which directly correlates with DNA concentration. The engineered ZFPs demonstrated high specificity for the tetracycline resistance gene tetM, achieving detection within just 10 minutes. By integrating FRET-based ZFP technology with GO, this platform provides a single-step, quantitative method for ARG identification, offering a detection limit as low as 1 nM. The findings highlight the potential of GO-based biosensors in combination with engineered DNA-binding domains as an effective and rapid screening method for antibiotic resistance, addressing the escalating challenge posed by drug-resistant bacteria [45].

9.3.3 Parasitic and Fungal Infections

In recent years, nanotechnology has significantly facilitated the identification and diagnosis of various parasitic organisms responsible for infections. Parasitic pathogens that can be detected using nanotechnology include protozoa, helminths, and ectoparasites. Protozoan parasites, being unicellular, replicate rapidly within the host and exhibit high virulence, leading to severe infections. In contrast, multicellular helminths are generally macroscopically observable in their adult stages. This group includes worm-like parasites such as *Ascaris lumbricoides*, trematodes, cestodes, and nematodes. Various studies have explored the applications of nanotechnology in the detection and treatment of helminth infections.

The diagnosis of parasitic infections is often based on indirect methods, including microscopic graphic analysis, PCR-based techniques, and flow cytometry for the detection of

Table 9.3 Overview of the literature on bacterial infections.

Study	Target pathogen/ biomarker	Biosensing mechanism	Key findings	References
Li, Wang et al.	Zearalenone	Fluorescence-based lateral flow immunochromatographic assay using Förster resonance energy transfer (FRET)	Developed a biosensor using CDots conjugated with ovalbumin (OVA) and AgNPs linked with a monoclonal antibody for detection	[33]
Bhattacharyya et al.	Tuberculosis (VOBs)	Fluorescence quenching	VOBs reduced CDots fluorescence, indicating a potential tuberculosis detection method	[34]
Weng et al.	*Escherichia coli*	Mannose-modified carbon quantum dots (Man-CQDots)	Selective binding to FimH lectin in bacterial flagella, achieving a detection limit of 450 CFU/mL	[35]
Yang et al.	*Escherichia coli* O157:H7	Carboxyl-functionalized graphene quantum dots (cf-CGDots) conjugated with anti-*E. coli* O157:H7 antibody	High specificity detection with a threshold of 100 CFU/mL	[36]
Ahmadian-Fard-Fini et al.	*Escherichia coli*	Fluorescence quenching using CDots from lemon extract/citric acid	Fluorescence reduction is proportional to increasing bacterial concentration	[37]
Gao et al.	*Pseudomonas aeruginosa*	FRET between CGDots and 5-carboxyfluorescein (FAM)	Fluorescence restoration in the presence of complementary dDNA from P. aeruginosa	[38]
Wang et al.	*Pseudomonas aeruginosa*	CDots-DNA probe fluorescence quenching with graphene oxide (GO)	High specificity and wide detection range	[39]
Huang et al.	*Escherichia coli* toxin (Colitoxin)	Electrochemical DNA biosensor using CDots synthesized from banana peels	DNA probe immobilized on Pd–Au-CD nanocomposite for colitoxin detection in human serum	[40]
Nandi et al.	*E. coli*, *S. typhimurium*, *P. aeruginosa*, *B. cereus*	CDots with long hydrocarbon chains	Fluorescence variations based on bacterial species	[41]
Yang et al.	*Staphylococcus aureus*	CDots encapsulated in breakable organosilicate nanocapsules (BONs)	Fluorescence-based pathogen detection after capsule breakdown using NaBH4	[42]
Liang et al.	HIV	On-and-off biosensor using CDots, CdTeQDs, mitoxantrone (MTX), and dsDNA	Fluorescence recovery in the presence of dsDNA	[14]
Singh et al.	*Plasmodium falciparum* (Malaria)	ssDNA aptamer targeting PfGDH, linked to CDots	Highly sensitive and selective malaria detection method	[43]

parasite eggs in water sources. However, these methods are costly and time-consuming. Laboratory protocols have been developed for detecting helminths in wastewater using PCR techniques. More recently, a recombinase polymerase amplification (RPA) mechanism integrated into lateral flow strips has offered a faster and more practical diagnostic approach. Additionally, studies have emerged in recent years focusing on the direct detection of parasites through nanozymology. For instance, an electrochemical sensor based on nanogold was developed for the identification of *Leishmania* parasites. This method utilizes the catalytic capacity of gold nanoparticles to convert hydrogen ions into hydrogen gas, enabling specific detection through faradaic current measurements. Moreover, the zinc metalloprotease leishmanolysin, found on the *Leishmania* surface, was evaluated for its ability to specifically bind to casein, and gold nanoparticles conjugated with casein were used for highly sensitive detection.

Furthermore, a nanozyme-based enzyme-linked immunosorbent assay (ELISA) was employed for the detection of *Trichinella spiralis*. Instead of the conventional peroxidase-based ELISA test, a system utilizing gold nanoparticles as nanozymes instead of horse peroxidase enzymes was developed. Polyclonal immunoglobulin antibodies facilitated the capture and immobilization of *T. spiralis*, enhancing diagnostic sensitivity.

Among nanotechnology-based biosensors, electrochemical biosensors have gained significant attention. For instance, a magnetic immunosensor-based method was designed for the detection of *Plasmodium falciparum* lactate dehydrogenase (PfLDH). This system utilized capture antibodies immobilized on magnetic beads and enzyme-labeled detection antibodies to develop a microfluidic paper-based electrochemical sensor. The amperometric detection of PfLDH enabled rapid diagnosis with low detection limits in clinical samples.

Additionally, *Toxoplasma gondii*, an intracellular parasite primarily hosted by felines, is transmitted through cat feces, infecting various animals and eventually humans via multiple routes. Due to its widespread prevalence, controlling its transmission remains a major concern. To address this issue, an antibody-based electrochemical biosensor was developed for the direct detection of *T. gondii*, providing potential applications in disease monitoring and management. This biosensor consists of a screen-printed electrode modified with chitosan and gold nanoparticles, where anti-*T. gondii* antibodies are immobilized using glutaraldehyde as a cross-linking agent. The biosensor's surface properties were characterized using differential pulse voltammetry, cyclic voltammetry, and EIS, while scanning electron microscopy was employed to examine its surface morphology. The biosensor's linear response range and detection limit were determined, and its effectiveness was validated for detecting *T. gondii* in synthetic serum samples. Such innovations pave the way for the development of sensitive and specific diagnostic tools crucial for effective disease prevention [46].

Apart from *Toxoplasma gondii*, biosensors have demonstrated promising potential in detecting other parasitic pathogens, such as *Trichinella spiralis* and *Trypanosoma* species. *T. spiralis*, a zoonotic nematode transmitted through the consumption of undercooked meat, has been effectively targeted using quantum dot-based fluorescent biosensors. These biosensors employ fluorescent quantum dots conjugated with specific antibodies to detect *T. spiralis* antigens with high sensitivity and specificity, providing a rapid and accurate diagnostic tool [47].

On the other hand, semiconductor nanoparticles known as quantum dots (QDs) have been used in both in vivo and ex vivo studies on *Trypanosoma cruzi*. However, concerns regarding the potential toxicity of QDs on living cells, microorganisms, and tissues persist. A study examining the nanotoxicity effects of QDs on *T. cruzi* investigated the impact of different CdTe QD concentrations on parasite development, morphology, and viability. Findings indicated that exposure to a 200 μM CdTe QD solution for 72 hours resulted in morphological changes such as DNA damage, plasma membrane blebbing, and mitochondrial swelling. In contrast, parasites incubated with 2 μM CdTe QDs continued to proliferate for seven days. These results suggest that while low QD concentrations may be suitable for bioimaging applications, higher concentrations could exert toxic effects [48].

In conclusion, nanozyme-based methods, electrochemical sensors, and quantum dot-based bioimaging approaches offer promising technologies for the diagnosis of parasitic infections. However, further research is required to integrate these systems into clinical applications and evaluate their safety profiles [49].

9.4 Advantages of Quantum Biosensing in Disease Diagnostics

The rapid and accurate detection of infectious diseases remains a cornerstone of effective disease management and public health interventions. Traditional diagnostic methods, such as PCR and enzyme-linked immunosorbent assay (ELISA), often require complex sample preparation, extended processing times, and specialized laboratory infrastructure. Quantum biosensing (QB), leveraging the principles of quantum mechanics, presents a transformative approach that addresses these limitations by offering unprecedented sensitivity, specificity, and real-time detection capabilities.

Quantum biosensors utilize quantum phenomena such as superposition and entanglement to enhance molecular detection accuracy. Unlike classical biosensors, which rely on bulk signal measurements, quantum biosensors can detect individual biomolecules, such as viral RNA, bacterial proteins, or disease-specific biomarkers, at ultralow concentrations. For instance, quantum dots (QDs), semiconductor nanocrystals exhibiting quantum confinement effects, have demonstrated remarkable sensitivity in detecting infectious agents like HIV, malaria, and SARS-CoV-2. By functionalizing QDs with antibodies or DNA strands, their fluorescence properties change upon binding with target biomolecules, enabling highly specific detection with minimal false positives or negatives.

One of the most significant advantages of quantum biosensors is their capability for real-time and non-invasive diagnostics [50]. Traditional diagnostic tools often involve laborious sample processing, leading to delayed results. In contrast, quantum-based sensors, such as nitrogen-vacancy (NV) centers in diamonds, allow for instantaneous detection of pathogen-induced changes in biological samples. NV centers are highly sensitive to variations in magnetic and electric fields, making them powerful tools for monitoring biochemical reactions in real time. This capability is particularly beneficial in epidemic scenarios, where timely detection is crucial for outbreak containment and management.

The ability to detect diseases at their earliest stages is pivotal for effective treatment and improved patient outcomes. Quantum biosensors can detect minute biochemical changes associated with infections before the onset of symptoms. For example, studies have shown that GQDs integrated with electrochemical aptasensors can successfully identify early-stage biomarkers of viral and bacterial infections. This early detection advantage enables prompt medical interventions, reducing disease progression and transmission [51, 52].

Advances in nanotechnology and microfluidics have facilitated the development of portable and cost-effective quantum biosensors suitable for point-of-care (POC) diagnostics. Unlike conventional laboratory-based methods, which necessitate extensive equipment and trained personnel, miniaturized quantum biosensors provide on-site testing solutions with rapid turnaround times. These devices are particularly valuable in resource-limited settings and remote areas where access to centralized diagnostic facilities is limited. Moreover, smartphone-integrated biosensors using quantum-enhanced fluorescence or electrochemical detection mechanisms are emerging as accessible and scalable solutions for infectious disease screening.

Despite their promising advantages, quantum biosensors still face technical and practical challenges that must be addressed before widespread clinical adoption. Issues such as stability, reproducibility, and integration with existing diagnostic platforms require further research and development. Innovations in material science, fabrication techniques, and data standardization protocols will be essential in optimizing QB performance and ensuring regulatory compliance. Furthermore, interdisciplinary collaborations between quantum physicists, biomedical engineers, and clinicians will be crucial in translating quantum biosensing technologies from the laboratory to real-world healthcare applications [53].

Quantum biosensing represents a paradigm shift in infectious disease diagnostics by offering superior sensitivity, rapid detection, and POC feasibility. As research in this field progresses, quantum biosensors hold the potential to revolutionize medical diagnostics, enabling early disease detection and real-time monitoring with unprecedented accuracy. The continued advancement of quantum biosensing technologies will pave the way for a new era of precision medicine, ultimately improving patient outcomes and global health security [54].

9.5 Current Research and Practical Implementations

Quantum biosensing represents a cutting-edge approach in the detection and management of infectious diseases, leveraging quantum mechanics principles to achieve unprecedented sensitivity and specificity. This section delves into the latest breakthroughs, commercial and clinical applications, and the integration of quantum sensors into medical devices [55].

Recent advancements have showcased the potential of quantum biosensors in revolutionizing infectious disease diagnostics. A notable development is the theoretical design of a sensor based on nitrogen-vacancy (NV) centers in diamonds, proposed by researchers at the Massachusetts Institute of Technology (MIT). This sensor aims to detect the SARS-CoV-2 virus by exploiting atomic-scale defects in diamonds that are highly sensitive to minute perturbations. The proposed method involves coating nanodiamonds containing NV centers with a material magnetically coupled to them and engineered to bind specifically to the viral RNA sequence. Upon binding, disruptions in the magnetic coupling occur, leading to detectable changes in the diamond's fluorescence, which can be measured using a laser-based optical sensor. Mathematical simulations suggest that this approach could result in faster, cheaper, and more accurate detection, with false-negative rates below 1% and the ability to detect just a few hundred strands of viral RNA within seconds [56].

While many quantum biosensing technologies are in the experimental or theoretical stages, there is a growing interest in translating these innovations into commercial and clinical settings. Companies like Quantum Diamond Technologies Inc. are actively exploring the use of diamond-based quantum sensors for biomedical diagnostics, recognizing their potential to provide powerful tools for disease detection. The adaptability of quantum sensors to detect various viruses by modifying the compounds attached to the nanodiamond sensors underscores their versatility in addressing both current and emerging infectious diseases.

Integrating quantum sensors into medical devices necessitates interdisciplinary collaboration, combining expertise in quantum physics, engineering, chemistry, and biology. The development process involves:

- *Quantum sensor fabrication*: Creating detectors with atomic-scale precision, such as NV centers in diamonds, to ensure high sensitivity to biological interactions.
- *Biochemical functionalization*: Engineering molecules that can specifically bind to target viral RNA sequences and attaching these to the surface of quantum sensors to facilitate selective detection.
- *Optical detection systems*: Utilizing laser-based optical sensors to monitor changes in fluorescence or other optical properties of the quantum sensors upon interaction with target molecules.

The potential for miniaturization and integration of these components into portable devices could lead to POC diagnostic tools capable of rapid and accurate detection of infectious agents. Such advancements would be particularly valuable in resource-limited settings or during outbreaks where timely diagnostics are crucial.

In summary, quantum biosensing is poised to significantly enhance infectious disease diagnostics through its superior sensitivity, specificity, and rapid detection capabilities. Ongoing research and development efforts are crucial to transition these technologies from theoretical constructs to practical medical applications, ultimately improving disease management and patient outcomes [57].

9.6 Challenges and Future Perspectives

The development of quantum biosensors depends heavily on advances in materials science and nanotechnology. However, key technical challenges include the stability and shelf life of these sensors. For instance, certain quantum dots tend to degrade over time, which can pose issues for long-term storage and usage. Additionally, inconsistencies in manufacturing processes can negatively impact the reproducibility and reliability of these sensors. Moreover, the effective immobilization of biomolecules on quantum materials requires precise control over surface chemistry, adding another layer of complexity to the fabrication process.

The integration of quantum biosensors into clinical applications is hindered by regulatory approval processes and clinical validation requirements. These sensors must undergo extensive clinical trials to prove their safety, efficacy, and compliance with quality standards. However, these validation processes are often time-consuming and costly, slowing down the commercialization of innovative technologies. Furthermore, the lack of standardization makes it difficult to compare results across different laboratories and manufacturers, further complicating widespread adoption.

Future research aims to overcome the limitations of existing quantum biosensors and enhance their clinical applicability. A particularly promising direction is the development of multiplexed sensors capable of detecting multiple pathogens simultaneously. Additionally, integrating artificial intelligence and machine learning algorithms can improve sensor accuracy and speed by facilitating the analysis of large datasets. Wearable and implantable microsensors may also revolutionize personalized medicine by enabling real-time, non-invasive health monitoring. Finally, reducing production costs and simplifying fabrication processes will make quantum biosensors more accessible in low- and middle-income countries, allowing them to be incorporated into global health initiatives.

The role of quantum biosensors in infectious disease diagnostics will continue to expand as these challenges are addressed. Future advancements in this field have the potential to significantly enhance healthcare quality by enabling early disease detection and improving treatment outcomes.

9.7 Conclusion

Quantum biosensing has emerged as a groundbreaking technology with the potential to transform the landscape of infectious disease diagnostics. By harnessing the quantum mechanical properties of novel nanomaterials, such as GQDs and nitrogen-vacancy centers in diamonds, researchers have developed highly sensitive and selective biosensors capable of detecting pathogens at an unprecedented level of accuracy. These quantum-based approaches offer numerous advantages over conventional diagnostic methods, including real-time detection, enhanced specificity, and the ability to detect multiple pathogens simultaneously.

In this chapter, we have explored the fundamental principles and various applications of quantum biosensors in detecting viral, bacterial, and fungal infections. Compared to traditional diagnostic techniques, quantum biosensing provides significant improvements in terms of detection limits, processing time, and operational simplicity. The ability to rapidly and accurately identify infectious agents is crucial for timely medical intervention, especially in the case of emerging pandemics and antimicrobial-resistant infections.

Despite its potential, several challenges must be addressed before quantum biosensing can be fully integrated into clinical settings. Key obstacles include the complexity of large-scale manufacturing, the high cost of certain quantum materials, and the need for standardized protocols to ensure reproducibility and reliability. Furthermore, regulatory approvals and clinical validation are essential steps that must be undertaken to ensure the safe and effective deployment of quantum biosensors in healthcare settings.

Looking ahead, future advancements in quantum biosensing will likely focus on enhancing sensor miniaturization, improving biocompatibility, and integrating quantum technologies with digital health platforms for remote diagnostics. The convergence of quantum biosensors with artificial intelligence and machine learning may further enhance their diagnostic capabilities, allowing for more accurate disease predictions and automated pathogen identification. Additionally, the development of portable, wearable, and smartphone-compatible quantum biosensors could pave the way for real-time, on-the-go infectious disease monitoring, particularly in resource-limited regions.

In conclusion, quantum biosensing represents a transformative approach in medical diagnostics, offering new possibilities for rapid, precise, and cost-effective infectious disease detection. While challenges remain, ongoing research and technological advancements are steadily bringing quantum biosensors closer to clinical application. As this field continues to evolve, quantum biosensing is poised to play a crucial role in shaping the future of global healthcare, improving disease surveillance, and ultimately enhancing patient care.

References

1 Hareesha, N. and Manjunatha, J.G. (2021). Electro-oxidation of formoterol fumarate on the surface of novel poly(thiazole yellow-G) layered multi-walled carbon nanotube paste electrode. *Scientific Reports* 11 (1): 12797.

2 Wei, S.C., Chang, C.C., Chuang, T.L. et al. (2022). Rapid detection of virus nucleic acid via isothermal amplification on plasmonic enhanced digitizing biosensor. *Biosensors* 12: 75.

3 Savas, S. and Altintas, Z. (2019). Graphene quantum dots as nanozymes for electrochemical sensing of *Yersinia enterocolitica* in milk and human serum. *Materials* 12: 2189.

4 Li, C., Soleyman, R., Kohandel, M., and Cappellaro, P. (2022). SARS-CoV-2 Quantum sensor based on nitrogen-vacancy centers in diamond. *Nano Letters* 22: 43–49.

5 Wang, L., Xu, D., Gao, J. et al. (2020). Semiconducting quantum dots: modification and applications in biomedical science. *Science China Materials* 63: 1631–1650.

6 Farshbaf, M., Davaran, S., Rahimi, F. et al. (2018). Carbon quantum dots: recent progresses on synthesis, surface modification and applications. *Artificial Cells, Nanomedicine, and Biotechnology* 46 (7): 1331–1348.

7 Chen, W., Lv, G., Hu, W. et al. (2018). Synthesis and applications of graphene quantum dots: a review. *Nanotechnology Reviews* 7: 157–185.

8 Kundu, S. and Pillai, V.K. (2020). Synthesis and characterization of graphene quantum dots. *Physical Sciences Reviews* 5: 20190013.

9 Mansuriya, B.D. and Altintas, Z. (2020). Applications of graphene quantum dots in biomedical sensors. *Sensors (Basel)* 20: 1072.

10 Mousavi, S.M., Hashemi, S.A., Yari Kalashgrani, M. et al. (2022). The pivotal role of quantum dots-based biomarkers integrated with ultra-sensitive probes for multiplex detection of human viral infections. *Pharmaceuticals* 15: 880.

11 Pushpanjali, P.A., Manjunatha, J.G., Hareesha, N. et al. (2021). Voltammetric analysis of antihistamine drug cetirizine and paracetamol at poly(L-Leucine) layered carbon nanotube paste electrode. *Surfaces and Interfaces* 24: 101154.

12 Zamzami, M.A., Rabbani, G., Ahmad, A. et al. (2022). Carbon nanotube field-effect transistor (CNT-FET)-based biosensor for rapid detection of SARS-CoV-2 (COVID-19) surface spike protein S1. *Bioelectrochemistry* 143: 107982.

13 Hashemi, S.A., Golab Behbahan, N.G., Bahrani, S. et al. (2021). Ultra-sensitive viral glycoprotein detection NanoSystem toward accurate tracing SARS-CoV-2 in biological/non-biological media. *Biosensors and Bioelectronics* 171: 112731.

14 Liang, S.S., Qi, L., Zhang, R.L. et al. (2017). Ratiometric fluorescence biosensor based on CdTe quantum and carbon dots for double strand DNA detection. *Sensors and Actuators B: Chemical* 244: 585–590.

15 Nejdl, L., Skalickova, S., Kudr, J. et al. (2014). Interaction of E6 gene from human papilloma virus 16 (HPV-16) with CdS quantum dots. *Chromatographia* 77: 1433–1439.

16 Wang, G., Leng, Y., Dou, H. et al. (2013). Highly efficient preparation of multiscaled quantum dot barcodes for multiplexed hepatitis b detection. *ACS Nano* 7: 471–481.

17 Teengam, P., Siangproh, W., Tuantranont, A. et al. (2017). Multiplex paper-based colorimetric DNA sensor using pyrrolidinyl peptide nucleic acid-induced AgNPs aggregation for detecting MERS-CoV, MTB, and HPV oligonucleotides. *Analytical Chemistry* 89: 5428–5435.

18 Kumar, P.R. (2023). Quantum dots in diagnostics of infectious and non- infectious diseases: current scenario and future prospectus. *Open Access Journal of Microbiology & Biotechnology* 8: 1–7.

19 Iannazzo, D., Pistone, A., Ferro, S. et al. (2018). Graphene quantum dots based systems as HIV inhibitors. *Bioconjugate Chemistry* 29: 3084–3093.

20 Xue, J., Chen, H., Fan, M. et al. (2009). Use of quantum dots to detect human papillomavirus in oral squamous cell carcinoma. *Journal of Oral Pathology and Medicine* 38: 668–671.

21 Piao, J.Y., Park, E.H., Choi, K. et al. (2009). Direct visual detection of DNA based on the light scattering of silica nanoparticles on a human papillomavirus DNA chip. *Talanta* 80: 967–973.

22 Sharma, A., Mishra, R.K., Goud, K.Y. et al. (2021). Optical biosensors for diagnostics of infectious viral disease: a recent update. *Diagnostics* 11: 2083.

23 Yazirlioğlu, L. (2021). Sustainable design considerations for emotional durability and product longevity through product care activities by repair enthusiasts. Frontiers in Neuroscience.

24 Chen, S., Chen, L., Tan, J. et al. (2005). Severe acute respiratory syndrome coronavirus 3C-like proteinase N terminus is indispensable for proteolytic activity but not for enzyme dimerization: biochemical and thermodynamic investigation in conjunction with molecular dynamics simulations. *The Journal of Biological Chemistry* 280: 164–173.

25 Qiu, G., Gai, Z., Tao, Y. et al. (2020). Dual-functional plasmonic photothermal biosensors for highly accurate severe acute respiratory syndrome coronavirus 2 detection. *ACS Nano* 14: 5268–5277.

26 Shafiee, H., Lidstone, E.A., Jahangir, M. et al. (2014). Nanostructured optical photonic crystal biosensor for HIV viral load measurement. *Scientific Reports* 4: 4116.

27 El-Safty, S.A. and Shenashen, M.A. (2020). Nanoscale dynamic chemical, biological sensor material designs for control monitoring and early detection of advanced diseases. *Materials Today Bio* 5: 100044.

28 Safardoust-Hojaghan, H., Salavati-Niasari, M., Amiri, O., and Hassanpour, M. (2017). Preparation of highly luminescent nitrogen doped graphene quantum dots and their application as a probe for detection of *Staphylococcus aureus* and *E. coli*. *Journal of Molecular Liquids* 241: 1114–1119.

29 Mat Zaid, M.H., Abdullah, J., Yusof, N.A. et al. (2017). PNA biosensor based on reduced graphene oxide/water soluble quantum dots for the detection of *Mycobacterium tuberculosis*. *Sensors and Actuators B: Chemical* 241: 1024–1034.

30 Castillo-Henríquez, L., Brenes-Acuña, M., Castro-Rojas, A. et al. (2020). Biosensors for the detection of bacterial and viral clinical pathogens. *Sensors (Basel)* 20: 6926.

31 Ju, J. and Chen, W. (2015). In situ growth of surfactant-free gold nanoparticles on nitrogen-doped graphene quantum dots for electrochemical detection of hydrogen peroxide in biological environments. *Analytical Chemistry* 87: 1903–1910.

32 Ye, W., Guo, J., Bao, X. et al. (2017). Rapid and sensitive detection of bacteria response to antibiotics using nanoporous membrane and graphene quantum dot (GQDs)-based electrochemical biosensors. *Materials* 10: 603.

33 Li, S., Wang, J., Sheng, W. et al. (2018). Fluorometric lateral flow immunochromatographic zearalenone assay by exploiting a quencher system composed of carbon dots and silver nanoparticles. *Microchimica Acta* 185: 388.

34 Bhattacharyya, D., Sarswat, P.K., and Free, M.L. (2017). Quantum dots and carbon dots based fluorescent sensors for TB biomarkers detection. *Vacuum* 146: 606–613.

35 Weng, C.I., Chang, H.T., Lin, C.H. et al. (2015). One-step synthesis of biofunctional carbon quantum dots for bacterial labeling. *Biosensors and Bioelectronics* 68: 1–6.

36 Yang, X., Feng, L., and Qin, X. (2018). Preparation of the Cf-GQDs-*Escherichia coli* O157: H7 bioprobe and its application in optical imaging and sensing of *Escherichia coli* O157: H7. *Food Analytical Methods* 11: 2280–2286.

37 Ahmadian-Fard-Fini, S., Salavati-Niasari, M., and Ghanbari, D. (2018). Hydrothermal green synthesis of magnetic Fe_3O_4-carbon dots by lemon and grape fruit extracts and as a photoluminescence sensor for detecting of *E. coli* bacteria. *Spectrochimica Acta - Part A: Molecular and Biomolecular Spectroscopy* 203: 481–493.

38 Gao, R., Zhong, Z., Gao, X., and Jia, L. (2018). Graphene oxide quantum dots assisted construction of fluorescent aptasensor for rapid detection of pseudomonas aeruginosa in food samples. *Journal of Agricultural and Food Chemistry* 66: 10898–10905.

39 Wang, H., Chi, Z., Cong, Y. et al. (2018). Development of a fluorescence assay for highly sensitive detection of: *Pseudomonas aeruginosa* based on an aptamer-carbon dots/graphene oxide system. *RSC Advances* 8: 32454–32460.

40 Huang, Q., Zhang, H., Hu, S. et al. (2014). A sensitive and reliable dopamine biosensor was developed based on the Au@carbon dots-chitosan composite film. *Biosensors and Bioelectronics* 52: 277–280.

41 Nandi, S., Ritenberg, M., and Jelinek, R. (2015). Bacterial detection with amphiphilic carbon dots. *The Analyst* 140: 4232–4237.

42 Yang, L., Deng, W., Cheng, C. et al. (2018). Fluorescent immunoassay for the detection of pathogenic bacteria at the single-cell level using carbon dots-encapsulated breakable organosilica nanocapsule as labels. *ACS Applied Materials & Interfaces* 10 (4): 3441–3448.

43 Singh, N.K., Chakma, B., Jain, P., and Goswami, P. (2018). Protein-induced fluorescence enhancement based detection of *Plasmodium falciparum* glutamate dehydrogenase using carbon dot coupled specific aptamer. *ACS Combinatorial Science* 20 (6): 350–357.

44 Pasinszki, T., Krebsz, M., Tung, T.T., and Losic, D. (2017). Carbon nanomaterial based biosensors for non-invasive detection of cancer and disease biomarkers for clinical diagnosis. *Sensors* 17 (8): 1919.

45 Ha, D.T., Nguyen, V.-T., and Kim, M.-S. (2021). Graphene oxide-based simple and rapid detection of antibiotic resistance gene via quantum dot-labeled zinc finger proteins. *Analytical Chemistry* 93: 8459–8466.

46 Neethirajan, S. (2017). Recent advances in wearable sensors for animal health management. *Sensing and Bio-Sensing Research* 12: 15–29.

47 Vieira, C.S., Almeida, D.B., de Thomaz, A.A. et al. (2011). Studying nanotoxic effects of CDTE quantum dots in *Trypanosoma cruzi*. *Memórias do Instituto Oswaldo Cruz* 106: 158–165.

48 Parihar, A., Yadav, S., Sadique, M.A. et al. (2023). Internet-of-medical-things integrated point-of-care biosensing devices for infectious diseases: toward better preparedness for futuristic pandemics. *Bioengineering & Translational Medicine* 8: e10481.

49 Songca, S.P. (2022). Applications of nanozymology in the detection and identification of viral, bacterial and fungal pathogens. *International Journal of Molecular Sciences* 23: 4638.

50 Raril, C. and Manjunatha, J.G. (2018). Sensitive electrochemical analysis of resorcinol using polymer modified carbon paste electrode: a cyclic voltammetric study. *Analytical and Bioanalytical Electrochemistry* 10 (4): 488–498.

51 Zhou, X., Schuh, D.A., Castle, L.M., and Furst, A.L. (2022). Recent advances in signal amplification to improve electrochemical biosensing for infectious diseases. *Frontiers in Chemistry* 10: 911678.

52 Monnappa, A.B., Manjunatha, J.G., and Bhatt, A.S. (2020). Design of a sensitive and selective voltammetric sensor based on a cationic surfactant-modified carbon paste electrode for the determination of alloxan. *ACS Omega* 5 (36): 23481–23490.

53 Wang, C., Liu, M., Wang, Z. et al. (2021). Point-of-care diagnostics for infectious diseases: from methods to devices. *Nano Today* 37: 101092.

54 Farid, I. and Khondakar, K.R. (2024). Chapter 6 - Smart sensors for infectious disease diagnosis. In: *Next-Generation Smart Biosensing* (ed. K. Reza Khondakar and A.K. Kaushik), 149–187. Academic Press.

55 Raril, C., Manjunatha, J.G., Ravishankar, D.K. et al. (2020). Validated electrochemical method for simultaneous resolution of tyrosine, uric acid, and ascorbic acid at polymer modified nano-composite paste electrode. *Surface Engineering and Applied Electrochemistry* 56 (4): 415–426.

56 Ramesh, M., Janani, R., Deepa, C., and Rajeshkumar, L. (2023). Nanotechnology-enabled biosensors: a review of fundamentals, design principles, materials, and applications. *Biosensors* 13: 40.

57 Kaushik, A. and Mostafavi, E. (2022). To manage long COVID by selective SARS-CoV-2 infection biosensing. *Innovations* 3: 100303.

10

Quantum Biosensors in Drug Discovery and Development

Santosh Nandi[1], *Vinayak Adimule*[2], *Shankramma S. Nesargi*[3], *Savita Hanaji*[4], *Rangappa Keri*[5], *and Praveen Barmavatu*[1]

[1]*Department of Mechanical Engineering, Faculty of Engineering, Universidad Tecnológica Metropolitana, Santiago, Chile*
[2]*Department of Chemistry, Angadi Institute of Technology and Management (AITM), Belagavi, Karnataka, India*
[3]*Department of Chemistry, KLE Technological University, Dr. M. S. Sheshgiri Campus, Belagavi, Karnataka, India*
[4]*Department of Engineering Science and Humanities, KLE Technological University, Dr. M. S. Sheshgiri Campus, Belagavi, Karnataka, India*
[5]*Centre for Nano and Material Sciences (CNMS), Jain (Deemed-to-be) University, Bangalore, Karnataka, India*

10.1 Introduction

Quantum biosensing (QB) is the use of quantum mechanics or quantum phenomena to biological sensing and detection [1, 2]. Identifying biological molecules, infections, or other biological substances involves taking advantage of the unique properties of quantum systems in order to form extremely sensitive, accurate methods. Quantum biosensors (QB) provide important knowledge about a wide variety of biological processes through the ability to detect and measure the motions of small molecules, or even single atoms, in biological samples. The concept of biosensing began to emerge in the 1950s and matured in the 1960s with the development of the first commercially viable enzyme-type sensors, like the glucose biosensor [3]. The applications were similar to classical ideas from chemistry and physics and created an invaluable foundation of possibilities for environmental monitoring, food safety, and medical diagnostics. The speed of development in biosensing technologies advanced rapidly as it responded to demands for sensitivity, specificity, and up-to-date detection methods. Prominent turning points in the discipline were the advent of piezoelectric, optical, and electrochemical sensors [4]. By the late twentieth century, developments in materials science and nanotechnology had significantly enhanced biosensing capabilities, allowing for the more selective and lower concentration detection of analytes. Quantum biology (QB) is an intriguing research direction that applies principles of quantum mechanics to detect and analyze biological molecules, with a high level of specificity and sensitivity. A biosensor that utilizes the principles of quantum mechanics to enhance its sensitivity and capabilities will be called a QB. A biosensor is defined as an analytic tool utilizing some type of physicochemical detector and a biological component (e.g. enzymes, antibodies, or DNA) for the purpose of identifying and measuring a

Quantum Biosensing in Medical Diagnostics, First Edition. Edited by J. G. Manjunatha.

specific biological agent [5]. The sensitivity and precision that can be attained by classical biosensors are fundamentally constrained by the shot-noise limit [6, 7] of classical physics, notwithstanding these developments, which have greatly increased biosensor sensitivity. This is where quantum biosensing comes into play, which shifts the way we think about biosensing problems [8, 9]. Superposition, entanglement, and quantum coherence are some of the special qualities of quantum mechanics that are used in quantum biosensing to get beyond the drawbacks of traditional sensors. QBs exploit quantum phenomena to improve biological entity identification's sensitivity, accuracy, and selectivity. Several potential types of sensing materials have been engineered to sense various material properties, including temperature, strain, thermal conductivity, electric and magnetic fields, and pH [10]. Leveraging quantum mechanics principles, similar examinations can be applied to biosensors in the form of quantum dots (QDs), other nanoparticles, and quantum materials. For example, quantum confinement properties are displayed by semiconductor nanocrystals called QDs. QDs can be functionalized in biosensing using biomolecules such as DNA strands or antibodies [11].

Quantum biosensing addresses the limitations of classical sensors through the unique aspects of quantum mechanics, such as entanglement, quantum coherence, and superposition [12]. Even when detecting single molecules, this methodology takes advantage of these quantum phenomena to demonstrate ultrasensitive detection and monitoring of very small changes in biological systems. Not only does quantum biosensing expand the potential of biosensing to previously unachievable levels, but it also builds upon the foundation established by classical biosensors. Research on QBs is still in its infancy and is developing quickly. QBs should discover more useful uses and make a substantial contribution to the development of biotechnology and healthcare as technology develops and our comprehension of quantum phenomena deepens [13]. QBs are different from traditional biosensors that utilize visible, macroscopic means; QBs utilize quantum mechanical phenomena (e.g. quantum entanglement or quantum interference) for detecting and measuring biomolecules from extremely low concentrations—for example, at single-molecule concentrations as shown in Figure 10.1 [14]. A vast array of physical properties, including magnetic, electric, and thermal fields, can be measured by QB with exceptional sensitivity [15]. The data that QBs can gather is not possible for present sensors because of their small size. Small sensing devices are not even sufficient to justify the needs that can be met by applying QBs. The sensitivity and subcellular precision of QBs allow for a descriptive investigation of contemporary biological components and interactions, as well as the discovery of their applications in numerous fields. Systems known as QBs use quantum interference, entanglement, and coherence to measure specific physical attributes [16]. They may consist of individual systems or collections of systems. But fresh developments in physics and quantum technology point to the possibility of a much-improved generation of biological sensors in the future. Furthermore, QBs can provide higher accuracy in the identification and quantification of biological analytes. The quantum characteristics of these biosensors allow for the adjustment of detection parameters, resulting in more accurate results. QBs can be designed at the nanoscale, which reduces their size. Quantum-based biosensors employ quantum dots (QDs), nitrogen-vacancy (NV) centers in diamond, and other quantum-confined materials to provide a detection level that is unmatched [17]. These sensors can detect extremely small temperature and pH changes,

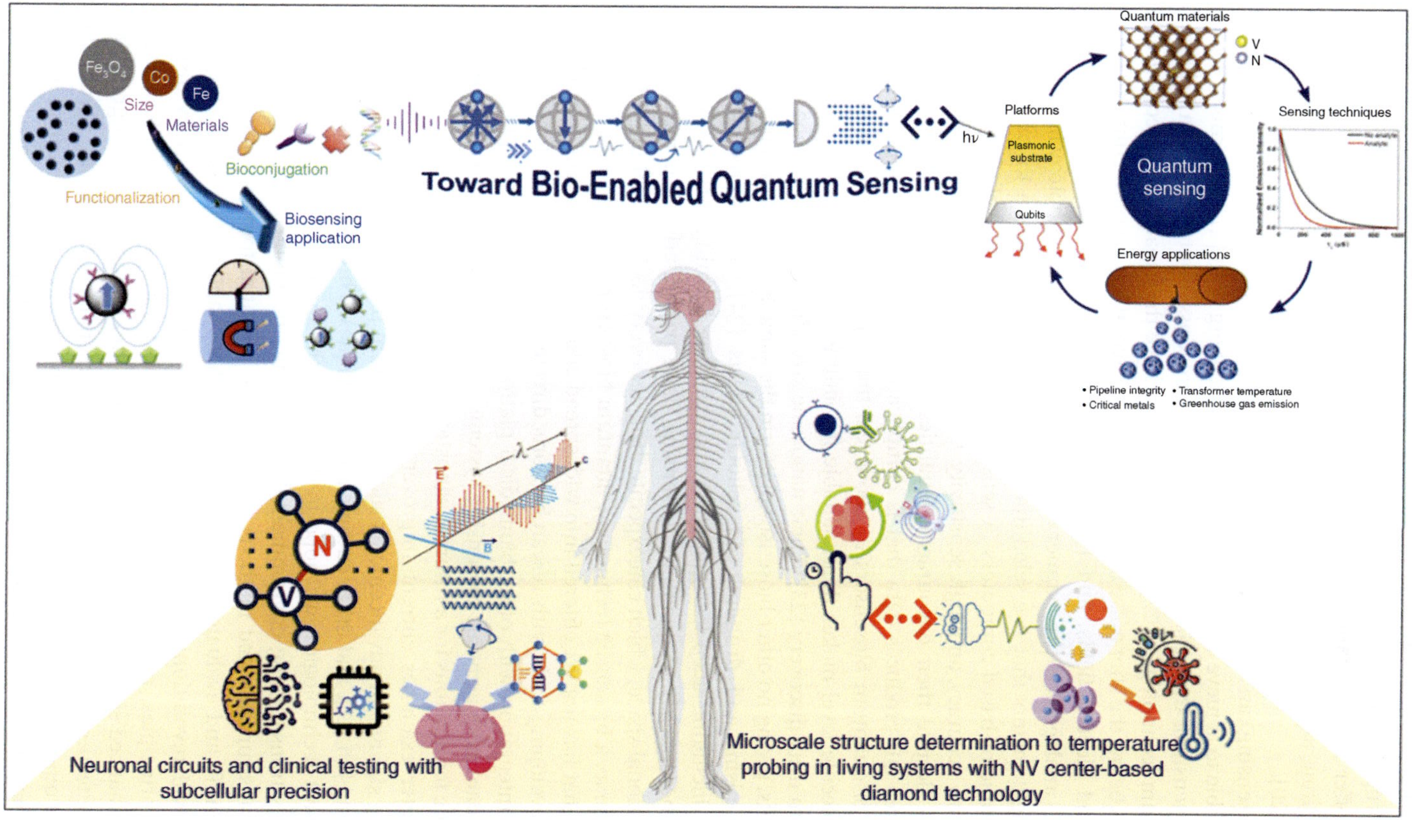

Figure 10.1 Graphical representation of quantum biosensing. Suparna et al. [14] / IOP Publishing / CC BY 3.0.

electric and magnetic fields, and molecular binding events, making them extremely useful for drug screening, pharmacokinetics, and biomarker detection [18]. Quantum biosensors offer a new realm of new opportunities, a greater sense, approaching the Heisenberg limit, and exceeding the classical shot-noise limit of chemical and biological sensors [19, 20].

Quantum sensing can achieve measurement accuracies below the shot noise limit, and it may be possible to achieve accuracy beyond the Heisenberg limit [21]. In terms of sensing, being more sensitive than the shot noise level is a great asset but would be a liability for other quantum technologies [22]. Quantum sensors have the capacity to detect extremely tiny changes below the shot noise limit in rotation, temperature, pressure, time, frequency, magnetic and electric fields, and rotation. This recent positive advance in quantum technology is analogous to the manner in which semiconductor technology has developed. There are light meters that use selenium photocells with no relation to, or applications for, semiconductor-based computers. A sensing device that works well should have high specificity, low accuracy, and high sensitivity. QBs are still a long way from being utilized in hospitals, despite their encouraging qualities and the paucity of ongoing scientific investigations. QB produces data that is complicated and frequently high-dimensional. The ability of quantum biosensors to allow real-time drug-target interaction studies at a single-molecule level makes them suitable for aiding the identification of therapeutic candidates in drug discovery [23]. Their nanoscale size allows them to probe subcellular environments, giving no other biosensor platform the same level of information regarding drug mechanisms and drug-induced cellular effects [24]. AI-powered data analysis techniques, including deep learning (DL) and machine learning (ML), will be able to help with the analysis of quantum biosensing data, particularly as it moves through high-dimensional space, which can expedite drug discovery decision-making [25]. Despite their promise as a biosensor technology, quantum biosensors are a developing emerging technology that has not yet entered into widespread clinical and industrial use. These are currently developing tools with research efforts directed toward scaling, longevity of the signal, and integration with other biomedical platforms. New and developing platforms, including quantum plasmonic biosensing systems and diamond defect-based sensors, are providing the foundations for a future diagnostic tool.

This chapter examines current improvements in quantum biosensing technologies, emphasizing significant advancements such as quantum plasmonic biosensing, diamond defect sensing [26, 27], and other cutting-edge techniques. Through our analysis of these state-of-the-art technologies, we hope to present a thorough picture of how quantum biosensing is transforming environmental monitoring and biomedical diagnostics, providing new instruments and methods that have the potential to completely change the way we identify and evaluate biological systems. This chapter examines how quantum biosensors will revolutionize drug discovery and development. We start by discussing underlying principles, recent advances, and real-world applications, emphasizing how quantum-enhanced sensing can reinvent diagnostics and therapeutics through improved biomarker detection, increased drug efficacy, and personalized medicines. By translating principles of quantum physics to biotechnology, quantum biosensors can open new paths in the pharmaceutical realm by generating speedier, more accurate, and cheaper solutions for challenges in modern healthcare.

10.1.1 Fundamentals of Quantum Biosensing

Biosensors have been one of the most fortunate tools to have developed in the biomedical and global health sectors, with respect to development. Some of the issues that biosensors are addressing are disease diagnostics, detection of drug-resistant organisms, identification of emerging epidemics, and detection of minute levels of harmful toxins/microorganisms in food and drinking water. However, there are still significant challenges that scientists and engineers whose work is devoted to addressing those problems will have to overcome. For example, if biosensors are to enhance the accuracy and efficiency of diagnoses in medicine, they need to be able to detect extremely small quantities of a virus/bacteria in a blood or other biological sample. Quantum sensing takes advantage of unique quantum mechanical properties like superposition, entanglement, and coherence to provide ultrasensitive measurements that exceed classical limits. Another fundamental principle concerns spin states, where spin acts like an environmental indicator. Any environmental changes, such as magnetic and electric fields, temperature, and pressure, act analogously on these spin states. Quantum sensors can maintain coherence (stable quantum state) long enough to detect perturbations that are extremely small [28]. Entanglement in this context relates to a situation in which one system causes other systems to be coupled together. If a sensor could harness this coupling, it could measure some phenomenon even more precisely than the standard quantum limit. In general, via the act of manipulating and "reading out" these quantum states, quantum sensors make for robust platforms for sensing extremely small signals across a myriad of domains, including biomedical diagnostics and environment monitoring [29]. The capabilities of quantum sensing are contingent on the availability of new quantum materials that can fabricate stable and programmable quantum states. Some notable examples of platforms that fulfill these criteria are nitrogen-vacancy (NV) defects embedded in diamond crystals that are highly tunable nanoscale devices to probe for magnetic and electric fields, temperature, and strain that can be deployed at room temperature as solutions for bioimaging applications, or magnetometry at the nanoscale [30]. Quantum dots are also being deployed for photonic sensing and single-photon emission because they are nanocrystals of semiconductors that confine electrons in all three dimensions and can yield tunable optical properties. Nanopores, often incorporating quantum tunneling, are often employed for single-molecule detection and for DNA sequencing as part of nanopore sequencing methods. These materials remain diversely placed and provide the hardware necessary to enable the emergence of novel applications that leverage quantum sensing. In quantum sensors, signal transduction transforms changes in quantum systems (spin orientations, energy levels, etc.) into signals that can be measured, often optically, electrically, or microwave-related. NV center-based sensors rely upon the fact that the fluorescent intensity of the defect falls upon its functional spin state to provide optical readout of magnetic or thermal variation [31, 32]. Nanopore sensors utilize changes in quantum tunneling and quantum capacitance to measure events from molecular translocation. Other platforms will use resonance frequency, where a change in the target analyte causes the measurable change in the oscillating system's behavior. These platforms often measure with significantly improved precision due to quantum coherence, while often utilizing more complex control schemes for the coherent system, such as microwave, lasers, or manifold methods involved in spin-locking to amplify or stabilize the signal being measured. The QB may assist in monitoring a single ion as it passes through a cell membrane, measuring the electrical hooks at a synaptic junction, or tracking peptides as they travel across small organelles in the cell fluid, all of which are quite challenging to qualitatively evaluate without superior imaging approaches. Indeed, hardware and technology that integrates biological and quantum engineering could change everything we think we know about medicine [33]. QB provides the health sector with diagnostic scans with lower cost, greater accuracy, and lesser risk for negative clinical outcomes. The most impactful technologies in global health and biomedical technologies include biosensors to diagnose disease and detect multi-drug resistant organisms and emerging epidemics and to qualify low levels of toxins and microorganisms from food and beverage sources.

10.2 Examples of Quantum Biosensors

Quantum biosensing is an exciting new field that bridges quantum physics and biological sensing, giving it the potential to detect biologically important molecules with unattainable levels of sensitivity and accuracy. Quantum biosensing will facilitate a new, sensitive methodology for detecting biological analytes at extremely low concentrations and, potentially, at the single-molecule scale, which would be well beyond the limits of traditional biosensing technologies by accessing the properties of quantum systems (superposition, entanglement, and coherence). The additional sensitivity afforded by quantum biosensors is related to the degree to which quantum effects can be used in the measurement process to detect small changes to biological systems compared with traditional biosensing methods [34]. This additional sensitivity provides a continuous advantage. Quantum systems can detect changes to environmental variables, including temperature changes, changes in the local magnetic field, and changes to chemical concentrations. Quantum biosensing is an emerging disruptive technology in drug discovery and development that provides new levels of precision, sensitivity, and speed far beyond classical biosensing. It exploits quantum phenomena such as superposition, coherence, and entanglement in ways that allow detection of effectively infinitesimal concentrations of biomarkers to the single-molecule level, which is helpful in early-stage drug discovery and diagnostics.

10.2.1 Optical Biosensors

Optical biosensors utilize light-based detection mechanisms to detect biological analytes at high sensitivity and specificity. Fluorescence-based sensors are especially popular because of their ease of use and speed of detection [35]. These sensors often utilize fluorescent dyes or, more commonly, proteins. In these cases, the fluorescent signaling occurs when the probe binds to target molecules, along with a simple detection chamber that allows the signaling light to be read. These sensors can be adapted to a range of biological assays, including detecting pathogens and *in vitro* biosensing, and throughout cellular research they can even inform studies of cellular activity during a live experiment. Förster resonance energy transfer (FRET) sensors, which indicate the transfer of energy

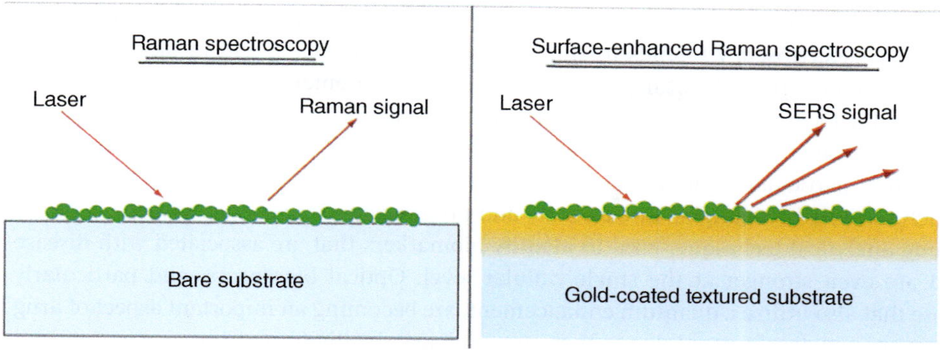

Figure 10.2 An example to demonstrate the different modalities of Raman spectroscopy and surface-enhanced Raman spectroscopy (SERS). In SERS, the incident laser creates an electric field that enhances the probability of Raman scattering, which translates to an enhanced Raman signal. Thorlabs [37] / Thorlabs, Inc.

between two fluorophores separated by a finite distance, act as a popular method for detecting different types of molecular interactions, changes in a protein's conformation, or detection of various signaling pathways in cells. Surface-enhanced Raman scattering (SERS) sensors enhance the scattering signals of Raman scattering by several orders of magnitude using plasmonic nanoparticles (gold or silver) to enhance light and enable ultrasensitive, label-free detection of target analytes, including DNA, proteins, and small molecules, often including detection at the single-molecule level [36]. Raman scattering is a rare process, with around 1 in every 1 million incident photons to the sample generating actual light scattering (see Figure 10.2). With low rates of these outcomes, higher-power light sources are often used in excitation of the sample, and this variety mostly alludes to the fact that with very low occurrence rates, it does not matter as much if we use greater levels of light intensity. Recently there has been progress with some of the initial studies on quantum biosensing and several other sensing technologies that are showing some really exciting outcomes. Coherent Raman Microscopy has been shown to be able to image below the shot noise limit by utilizing a bright squeezed state [37]. Similar advances have been made in stimulated emission microscopy based on stimulated emission from intensity-squeezed light for sub-shot noise limited microscopy. Surface enhanced Raman scattering (SERS) was developed similarly to how Raman scattering was enhanced using some metallic surface as a substrate to enhance Raman scattering. The enhanced scattering is indeed due to an enhancement of the electric field caused by the exciting of localized surface plasmons. Portable SERS platforms are being developed, with many being founded on easy-to-use, simple disposable end-user accessories. This new generation of sensor is creating opportunities for direct use in point-of-care (POC) diagnostics and in-person environmental monitoring. The second mechanism, chemical enhancement, describes charge transfer from the metal to the target molecule and from the molecule to the metal, or, in general, intra- and intermolecular charge transfer. In this way, SERS provides a non-destructive, very sensitive way to detect very small concentrations of a sample, and it uses much less laser power than classical Raman methods [38]. Classical optical detection methods suffer from shot noise, which imposes a fundamental limit on what can be achieved in thermodynamic terms in spectroscopic measurements. Quantum Raman (qRaman) spectroscopy intends to surpass these classical limits, and in fact, it behaves vastly different from classical (or non-quantum) measurements, utilizing non-classical light and achieving high-precision spectrographic fingerprinting. This divergence marks the opportunity of discovery where accuracy in medical diagnostics is dramatically improved, particularly for the identification of biomarkers at the single cellular level, which essentially highlights disease state and progression. Raman spectroscopy is a very strong analytical technique, used to identify biomarkers that are associated with disease and are even stronger at the single cellular level. Optical biosensors, and particularly those that also utilize a quantum enhancement, are becoming an important aspect of drug discovery and development due to their unparalleled sensitivity, non-invasive nature, and ability to follow biological events as they happen. Quantum technologies are driving new generations of biosensing [39]. Classical optical activities, including fluorescence, FRET, and SERS, are incorporated into multiple quantum technologies to push the horizons of traditional biosensors to affects you can measure and biomolecular events detected with exquisite accuracy to detect drug-target interactions, follow disease biomarkers, and screen for therapeutic efficacy. Whereas nanoquantum sensors serve as both luminescent probes and nanocarriers, quantum dots (QDs) bound to therapeutics and sensitizers have come leaps and bounds as theranostic platforms. These multifunctional nanoscale materials have the purpose of offering diagnostic and therapeutic support simultaneously, especially in cancer therapy [40]. In this work, we present four exemplified formats of theranostics from nanoquantum sensors: imaging-assisted computed tomography (CT); photothermal therapy (PTT); sonodynamic therapy (SDT); and radiotherapy (RT). In the first example, Jiangbing Zhou et al. used surface-engineered carbon quantum dots (CQDs) as imaging-assisted CT, and the CQDs also aided as they were used for drug loading, targeted tumor delivery, and fluorescence-guided imaging, which were used multifacetedly in order to assist the diagnosis and treatment of cancer (Figure 10.3a–c) [41].

10.2.2 Quantum Plasmonic Biosensors

Quantum plasmonic biosensors are a type of quantum biosensing that utilize quantum light interaction with plasmon through the coherent oscillators of free electrons at the surface of a metal to enhance detection limits. A good example of a biosensor utilizing surface plasmon resonance (SPR) (Figure 10.4), where biosensors can act in a label-free manner and provide inherently high sensitivity to biomolecules [42]. Quantum-enhanced plasmonic biosensing has the potential to achieve below shot-noise limits for sensing to detect single or smaller biological entities. While many biosensors shall operate on SPR methods based off quantum light (squeezed light and entangled photons), they are an enhancement for plasmonic sensing, as they increase the SNR(signal-to-noise ratio) and accuracy for detection. Quantum plasmonic biosensors have been used for a variety of functions, sensing cancer biomarkers and real-time monitoring of molecular interactions. Quantum plasmonic biosensors are members of the family of optical quantum biosensors and have gained significant interest as methods of drug discovery and development with high speed, high sensitivity, and label-free detection capabilities [43]. Quantum plasmonic biosensors consist of a combination of quantum optics and plasmonics (which is the

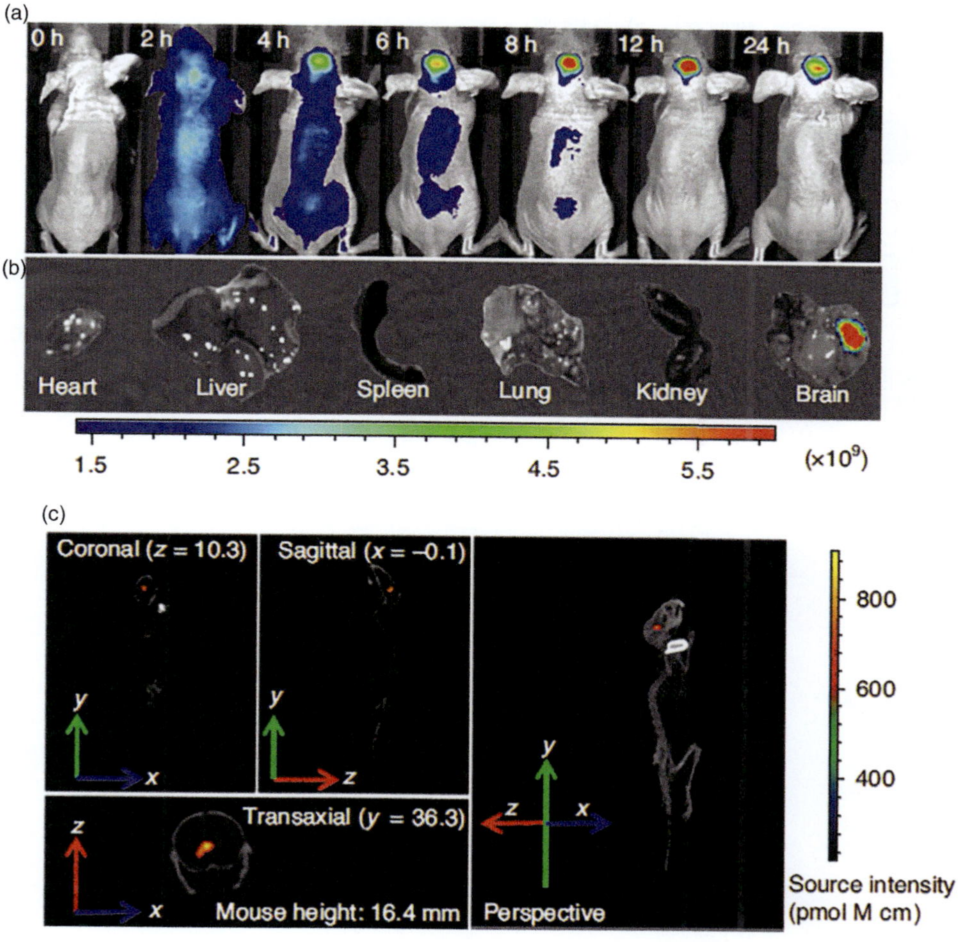

Figure 10.3 A summary of the application of LAAM TC-CQDs for imaging and therapy of brain tumors. (a) NIR fluorescence images from a U87-tumour bearing mouse received intravenously injected LAAM TC-CQDs at the indicated time points. All images are at the same scale. (b) Ex vivo NIR fluorescence imaging of the indicated organs and tumor at eight hours post LAAM TC-CQDs injection. (c) 3D reconstruction of LAAM TC-CQDs distribution in the mouse at eight hours post LAAM TC-CQDs injection. Li et al. [41] / Springer Nature.

coherent blown-out oscillations of electrons at metallic surfaces) to allow for examination of molecular interactions with resolutions that exceed classical limits.

10.2.3 Magnetic Biosensors

Magnetic biosensors measure biological signals based on changes in the magnetic field where the target analyte or magnetic tag has altered this field. Diamond NV (nitrogen-vacancy) centers represent the most significant advancement in nanoscale magnetometry. These quantum defects in diamond crystals allow for label-free, real-time

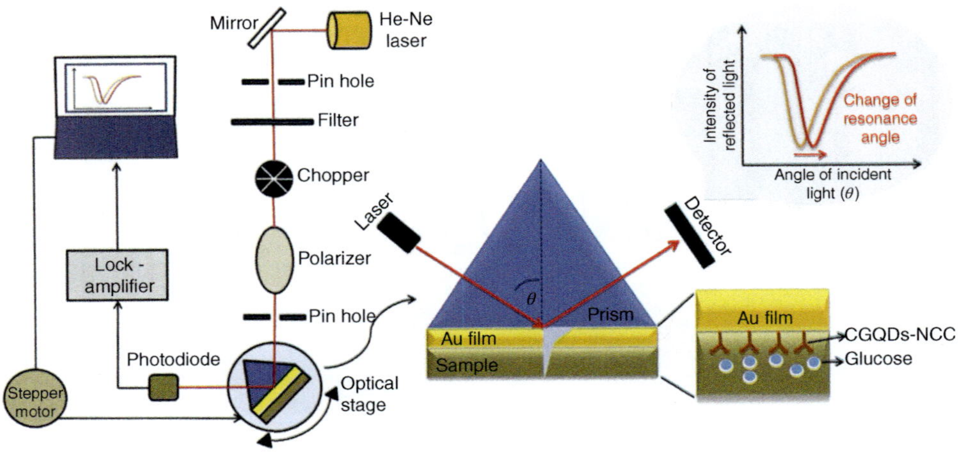

Figure 10.4 Schematic diagram of SPR setup and glucose detection under Krestchmann configuration. Rosddi et al. [42] / with permission of Elsevier.

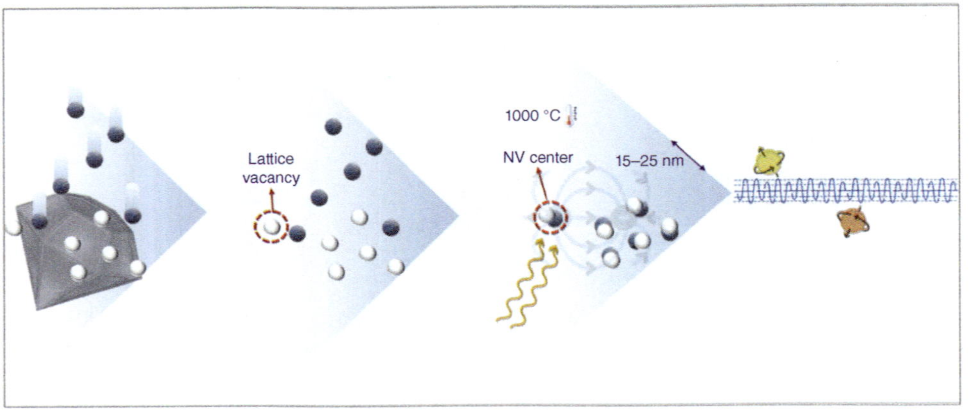

Figure 10.5 Schematic of Diamond NV quantum biosensor technologies. Suparna et al. [14] / IOP Publishing / CC BY 3.0.

detection of the magnetic field of biomolecules or neuronal activity in the room temperature regime (Figure 10.5) [14]. In particular, NV centers allow for *in vivo* imaging and monitoring of bioelectrical signals without direct contact with electrodes. On the extreme end of sensitivity, superconducting quantum interference devices (SQUIDs) are the most sensitive magnetic detectors capable of measuring minuscule magnetic fields (10^{-18} T). SQUIDs require cryogenic temperatures but can be applied in magnetoencephalography (MEG), biomagnetic imaging, early-stage cancer detection, and beyond. Exploratory work remains focused on further miniaturization and hybridization of these platforms for practical clinical application. The current flowing through the device corresponds to the phase difference of the superconducting wavefunctions on either side of the junction and depends on the external magnetic field. SQUIDs have a variety of systematic utility in a variety of fields, including medicine (to record variations of a brain's magnetic field in MEG), materials science, and fundamental measurements in physics [44]. NV centers are defects in the crystal structure of diamonds, which are very responsive to fluctuations in magnetic fields, temperature, and other environmental variations due to some of their interesting quantum properties. NV center biosensors have several specific advantages, which are primarily characterized by being able to operate at room temperature and the ability to monitor biological mechanisms noninvasively and in real-time. If NV centers are included in biosensors for biological applications, these sensors may be employed in a variety of applications, which include MRI with a spatial resolution of nanometers, single-molecule detection, and measurement of intracellular dynamics. These sensors can demonstrate remarkable spatial resolution for detecting changes in magnetic fields. This is helpful if, for example, the intent was to measure whether or not molecules inside cells were moving or to measure small deviations to the local environment in which biomolecules are located [45]. Magnetic quantum biosensors use quantum features like spin coherence and superposition to make ultrasensitive, label-free measurements of biomolecular interactions and physiological signals. In particular, diamond NV centers and SQUIDs have unique sensitivity, spatial resolution, and non-invasive functionality, presenting exciting opportunities in pharmaceutical research, diagnostics, and personalized therapy.

10.2.4 Quantum Dot-Based Quantum Biosensors

Quantum dot quantum biosensors are an emerging biosensing device that works with quantum dots (QDs) and principles of quantum mechanics to improve the detection and characterization of biological macromolecules. QDs are semiconductor nanocrystals of varying size that possess strong fluorescence and high photostability and emit narrow bandwidths, which allow them to be robust and be bound to biological moieties, therefore enhancing detection sensitivity and multiplexed detection. Combine the Qs with techniques that leverage quantum coherence, entanglement, and other quantum properties, and then quantum biosensors based on QDs can achieve greater overall sensitivity and specificity compared to conventional biosensors for detection of biomolecules. QDs are nanoparticles of semiconductors (Ag-In-Ga-Se, CdSe, and PbS), MXene (V_2C), and carbon, which have excellent optical properties [46, 47]. The typical nanometer size of QDs confines electron motion in three dimensions and the size and material format of QDs can be adjusted to engineer their energy levels (i.e. their optical properties). The optical properties of QDs may be better described by ultrahigh brightness and sharp emission; tunability of luminescent wavelength from visible to near-infrared (NIR) based on size and material; and extreme photostability. Additionally, QD nanoparticles have a high surface area and complex surface chemistry, and the biophysical and biochemical functionalities provided by their surfaces can be finely tuned utilizing QD surface modifications. Smith et al. made mercury cadmium chalcogenide QDs ($Hg_xCd_{1-x}Se/S$) that have tunable visible-to-SWIR emission while maintaining high quantum yield and photostability (Figure 10.6a). These QDs had a novel polyacrylamide (histamine-*co*-TEG) coating for aqueous stability and applicability to bioconjugation with click chemistry (via 30% azide-terminated TEG groups) (Figure 10.6b) [48]. These QDs maximized quantum confinement, biological compatibility, and target specificity in sensing. With quantum methodologies, these QDs

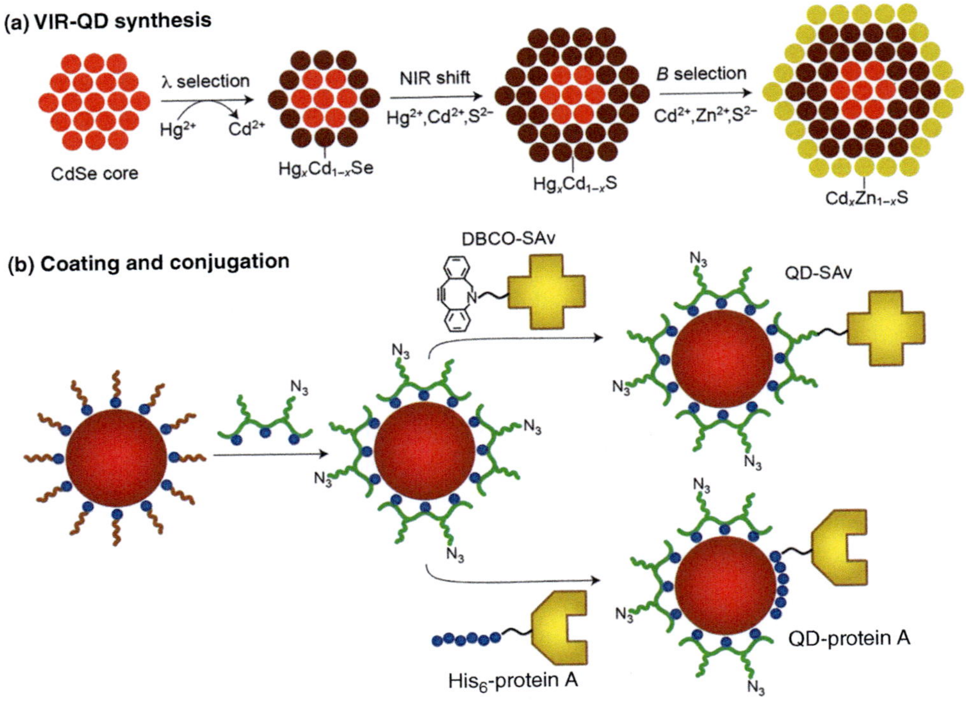

Figure 10.6 Schematic representation of three-domain quantum dots (QDs) that can emit visible-to-infrared (VIR) light and the steps necessary to synthesize them to be molecular probes. (a) Cartoon representation of the multi-step synthesis methods for producing those nanocrystals that are VIR-QDs. (b) The surface-coating of the VIR-QDs with a hydrophilic multidentate azide (N3)-functional polymer to displace the surface-bound hydrophobic ligands. Sarkar et al. [48] / with permission of American Chemical Society.

made it possible to monitor real-time drug-target interactions, perform HTS, perform single-molecule imaging, and catalyze drug development timelines. The challenges of scalability and quantum coherence still exist, but these customized nanomaterials are a significant breakthrough in the field of precision pharmacology research.

Quantum dot-based biosensors offer much promise in situations requiring ultrasensitive detection, such as medical diagnostics, where they can sense trace amounts of actual biomarkers for cancer, neurodegenerative disease, and infectious disease. In addition to medical diagnostics, biosensing platforms may be applicable to environmental monitoring applications. Quantum biosensors can detect pollutants or pathogens at a very low degree of concentration with exceptionally high specificity and accuracy. Another benefit of quantum dots is that you can have different sizes of quantum dots (related to the size of the molecules), giving you potential for multiplexing, therefore allowing you to measure multiple targeted molecules, cells, or pathogens simultaneously from the same sample. Furthermore, using quantum dots and quantum technologies opens the possibility to employ quantum imaging and sensing, where the benefits of quantum phenomena can be utilized to improve sensor resolution and signal-to-noise ratio. However, there are certain challenges that will need to be overcome, particularly biocompatibility issues and toxicity associated with molecular quantum dot materials and the challenges associated with using quantum principles to make the applications of practical biosensors.

Although facing challenges of operationalization and commercialization, quantum dot-based quantum biosensors (QDB-QBS) are on the edge of the biosensing technologies clade, and quantum dots provide additional opportunities for revolutionary effects in diagnostic biosensing applications and environmental applications due to quantum physics. Although they are experiencing challenges of operationalization and commercialization, QDB-QBS are on the edge of the biosensing technologies branch, and quantum dots provide even further opportunities for revolutionary effects in diagnostic biosensing applications and environmental applications because of quantum physics. QDB-QBS amalgamate nanotechnology and quantum physics into ultrasensitive multiplexed detection of biomolecules and molecular resolution point detection of biomolecules. A QDB-QBS is a specialized optical sensor that incorporates quantum dots (QDs), semiconductor nanocrystals with fluorescence properties that have their fluorescence properties tunable by size, and employs quantum phenomena such as coherence, superposition, and entanglement to detect biomolecules. QDs have come a long way from being suggested for use as markers of biologically active species in studies to being used in infectious diseases as drug screening assays and photo-induction for use in killing pathogens. Research shows that CdSe/CdS/ZnS QDs containing recombinant viral antigens could make excellent molecular probes to employ in biochemical and cellular assays to screen drug activities inhibiting viral antigens and their respective host cell surface receptors (Figure 10.7) [49]. An illustrative example of this can be seen for severe acute respiratory syndrome coronavirus 2019 (SARS-CoV-2), where this virus has a requirement on the binding of the spike receptor-binding domain (RBD) on the virus surface with the cell-surface angiotensin-converting enzyme 2 (ACE2) receptor to enter the intracellular space, and the recombinant spike RBD conjugated QDs could serve as a model for a pseudovirus.

10.2.5 Electrochemical and Mechanical Biosensors

Electrochemical biosensors work by transducing biochemical interactions to electrical signals by observing changes in current, voltage, or impedance. Electrochemical sensors are ideal for POC applications because they are inexpensive, can be miniaturized, and can connect with a smartphone [50]. Electrochemical biosensors have many applications, including glucose sensors, pathogen biosensors, and wearable health monitors, among many. Recently, advanced platforms have added nanomaterials, including graphene and MXenes, to facilitate higher sensitivity and evaluation of multiple analytes. Mechanical biosensors include microcantilevers (MCs) or piezoelectric sensors. They transduce physical change events—usually, mass loading or stress—into biosensor event measures that rely on the target binding to the sensor surface. The change in resonance frequency or quantifiable deflection provides a measurable result to allow for label-free, real-time detection of biomolecules. This promotes applications of the biosensors in drug discovery, environmental testing, and label-free viral and bacterial pathogen detection. Through the development of MEMS (microelectromechanical system) and AI integration, these biosensors have significantly improved responsiveness and multiplexing capability.

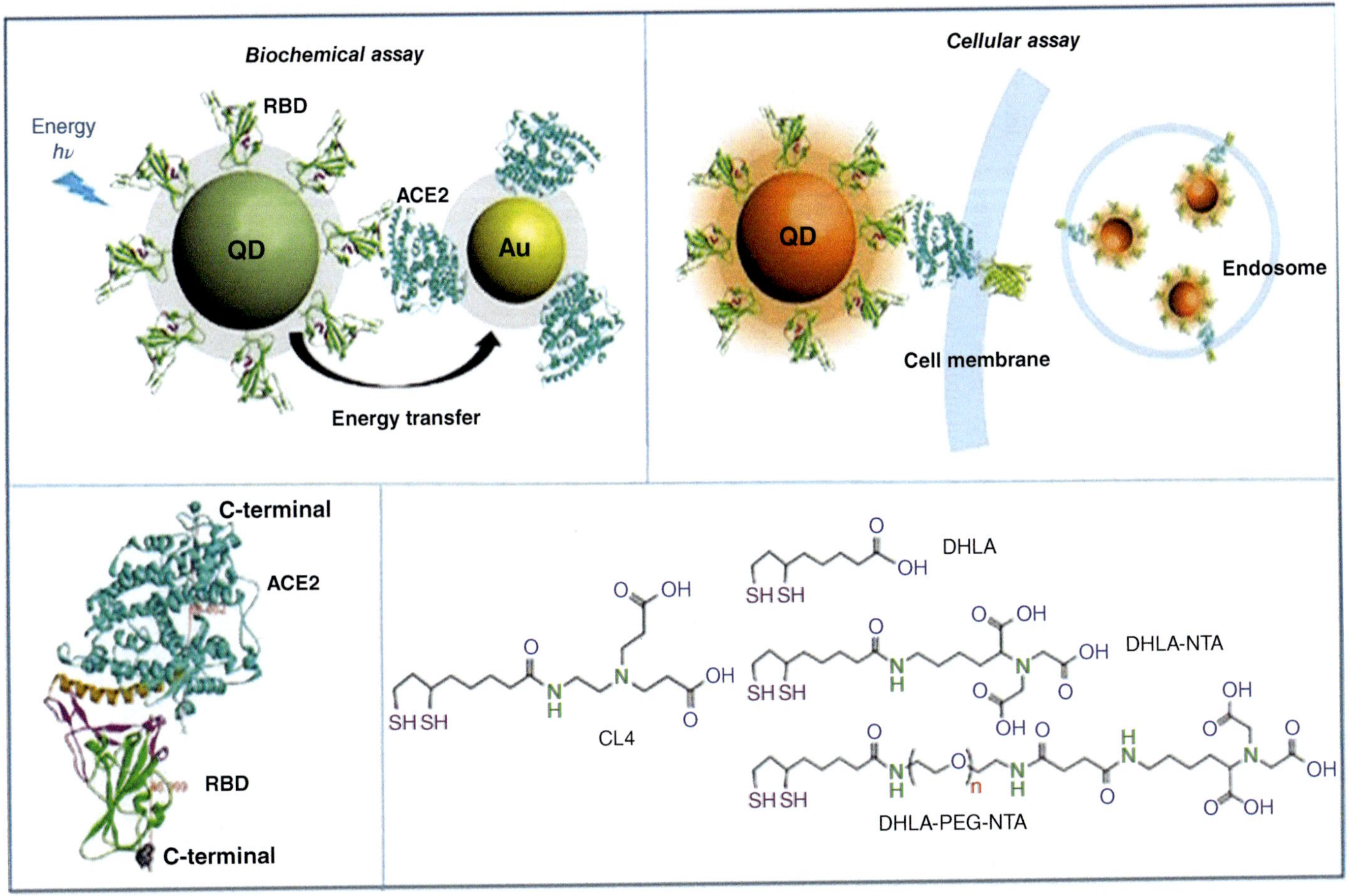

Figure 10.7 Schematic representation of the biochemical assay with energy transfer from QD-RBD to AuNP-ACE2 (top left) and the cellular assay with QD-RBD binding to ACE2 (with or without GFP modification at the end of the C-terminal at the cell membrane (top right). The image below shows the binding of ACE2, RBD and the structural schematic of the chemical structure of surface ligands for both QDs (CL4) and AuNPs (DHLA ligands) (bottom right). Kirill et al. [49] / with permission of American Chemical Society.

Electrochemical and mechanical biosensors serve as the gateway to the development of quantum biosensing platforms by providing real-time, label-free, and miniaturized diagnostics for researchers. The advances in these nano-biosensor platforms in tandem with numerous quantum nanomaterials (i.e. graphene, MXenes, quantum-dot materials, etc.) have vastly increased the utility of mass spectrometry approaches in drug discovery, pharmacokinetic studies, and HTS in research labs [51]. These new developments in quantum biosensing platforms provide several advantages as low-cost, scalable, and portable diagnostics platforms; this is especially true for diagnostic applications in the lab and decentralized health care (Table 10.1).

Table 10.1 The list illustrates common examples of nanoquantum sensors to be used in sensing and drug development.

Sensor materials	Drug discovery application	Detection principle	References
Nitrogen-vacancy (NV) nanodiamonds	Real-time tracking of drug-protein binding kinetics at single-molecule resolution.	Magnetic field sensing	[52]
CdSe/ZnS quantum dots (QDs)	High-throughput screening of drug candidates via multiplexed biomarker detection.	Fluorescence emission	[53]
RBD-conjugated CdSe/CdS/ZnS QDs	Pseudo-virus entry imaging and neutralizing antibody activity assays.	Targeted fluorescence	[54]
CdTe QDs	Photodynamic killing of drug-resistant bacteria (antimicrobial therapy development).	Reactive oxygen species (ROS)	[55]
D-penicillamine-modified CdSe/ZnS QDs	Intracellular pH mapping for drug metabolism and toxicity studies.	pH-dependent fluorescence	[56]
Drug-conjugated fluorescent nanodiamonds (FNDs)	Intracellular drug release monitoring via free radical generation sensing.	Spin-based sensing	[57]
Liposome-coated FNDs	Detection of drug-induced oxidative stress in immune cells (e.g. dendritic cells).	Free radical sensing	[58]
RGD-modified PbS/CdS QDs	Tumor-targeted drug delivery and personalized radiotherapy optimization.	NIR imaging	[59]
Nitrogen-vacancy (NV) nanodiamonds	Real-time tracking of drug-protein binding kinetics at single-molecule resolution.	Magnetic field sensing	[60]
CdSe QDs + Pd(II) porphyrin micelles	*in vivo* oxygen sensing for evaluating drug effects on tumor hypoxia.	Phosphorescence quenching	[61]

10.3 Transformative Applications of Quantum Biosensors in Drug Development

Using quantum biosensors, drug discovery and development is fundamentally changing by providing unmatched sensitivity and specificity with the ability to monitor in real time. With their quantum-induced detection, new applications can be developed that not only expedite research timelines but also improve clinical outcomes. To evaluate biological states of activity in a living subject, it is important to sense spatially local physicochemical quantities (e.g. oxygen, temperature, and pH) at the cellular and tissue levels. There is a vast scientific literature on scientific and technological achievements in developing biosensors; however, there is an imbalance between the level of research achievements and available commercial products based on QB [62]. QBs face the exceptional challenge of demonstrating both high levels of specificity and sensitivity to their target biomolecules in situ, with complex biological matrices. Future work will focus on novel surface functionalization strategies, signal processing algorithms, and ML strategies to combat non-specific binding and reduce the impact of background noise.

10.3.1 Accelerated High-Throughput Screening (HTS)

The use of quantum biosensors has the potential to fundamentally change HTS of biologically active compounds via highly multiplexed and ultrasensitive assays that can assess very large libraries of drug candidates quickly and effectively. Conventional HTS platforms tend to be challenged with decreased sensitivity, high background noise, slow readout environments, and increased false positive/negative results. Quantum dot-based and quantum plasmonic biosensors are the next generation for biosensing, with favorable optical properties such as high brightness, narrow emission spectra, and strong photobleaching resistance that remediate many of the disadvantages of traditional optical methods. These changes allow for improved signal-to-noise ratio and, consequently, much more accurate measurement of smaller molecular interactions, such as ligand-receptor binding or enzyme inhibition. Moreover, the ability for quantum biosensors to be individually mass-produced enables the inclusion of detection systems engineered for multiplexed detection (the ability to simultaneously assess different biological targets in the same assay), conserving reagents and time, and maximizing throughput. The high temporal resolution of quantum biosensors also enables real-time kinetic analysis of the interactions, greatly augmenting the value of screening information. Pharmaceutical scientists can leverage quantum biosensors on HTS platforms to expedite the early stages of drug discovery, effectively narrowing the pool of candidate molecules with specific activity or efficacy. This represents decreased timelines and costs, as well as improved odds of downstream clinical success, ultimately leading to safer and better drugs.

10.3.2 Single-Molecule Pharmacokinetic Monitoring

Quantum biosensing technologies are transforming pharmacokinetic studies using high-throughput technologies that provide a direct means to monitor the dynamics of individual drug molecules within complex biological systems. Conventional pharmacokinetic studies generally use averaged measurements over large populations of molecules, which eliminates significant variability in drug behavior *in vivo*. In contrast, quantum biosensors, including NV centers in diamonds or quantum dots, can detect single-molecule events with an unprecedented level of sensitivity and spatial resolution. This capability enables scientists to create remarkably detailed absorption, distribution, metabolism, and excretion (ADME) PK profiles where each of the key ADME parameters can be distinguished [63]. By observing the absorption of individual drug molecules into the bloodstream, the distribution of individual drug molecules in the tissues, the metabolism by enzymes, and the final excretion of drug molecules, scientists can access more subtle drug-related information that cannot be found using bulk measurements. This ability to study drug kinetics at the single drug molecule level is important for researchers studying the inter-patient variability of drugs, drug-drug interactions, and disease states and the effects of these state variables on drug efficacy. The bonus of a single molecule can be potentially useful for biological measurement feedback, which would allow fine-tuned and on-demand dosing and/or drug delivery, particularly when considering a precision medicine approach. For example, this means you could base the drug dosing regimen on how much of a specific drug a specific patient processes, minimizing side effects and maximizing therapeutic effectiveness. Single-molecule pharmacokinetic monitoring using quantum biosensing could prove to be a serious advancement in drug development, therapeutic monitoring, and personalized treatment design.

10.3.3 Target Engagement and Mechanism Studies

Quantum biosensors can detect small molecular interactions and conformational changes with high sensitivity, enabling detailed studies of the binding kinetics between drug targets and their binding partners. This information supports characterization of drug mechanisms of action as well as their off-target effects and resistance pathways, which are all important considerations for rational drug design and therapeutic refinement. Sulfonic-GQDs were developed by Sushil S. Burle et al. [59] that selectively target tumor cell nuclei based on cancer's increased interstitial fluid pressure. These anionic nanostructures (<10 nm) are capable of translocating into tumor membranes but, due to electrostatic charge, repel healthy cells. A self-cleaning translocation mechanism, transferring through the nuclear membrane from tumor cells. This ligand- and binding-free approach to tumor-targeted nanoparticles is a significant advancement in tumor-specific nanomedicine because it exploits the inherent biophysics of tumors to deliver agents specifically into the nuclear membrane from the tumor cell (Figure 10.8). Marco Raabe et al. [64] showed that functionalized CdSe/ZnS quantum dots (QDs) were created for pH sensing by changing the fluorescence lifetime (fluorescence lifetime modulation). QD surfaces were modified by ligand exchange with D-penicillamine and small peptides. Deprotonation of the amine groups on the ligands in response to a change in pH affected their sulfide interactions with the surface of the QD, changing the fluorescence lifetime as a result. The QDs produced pH-responsive fluorescence lifetimes within a physiologically relevant range (pH 3.9–8.0) in synthetic intracellular buffers. Following the internalization of the QDs by cells, they facilitated the mapping of pH units as distributions within the cells, noting differences between the pH profiles of co-cultured C3H10T1/2 and SK-MEL-2 cells (Figure 10.9).

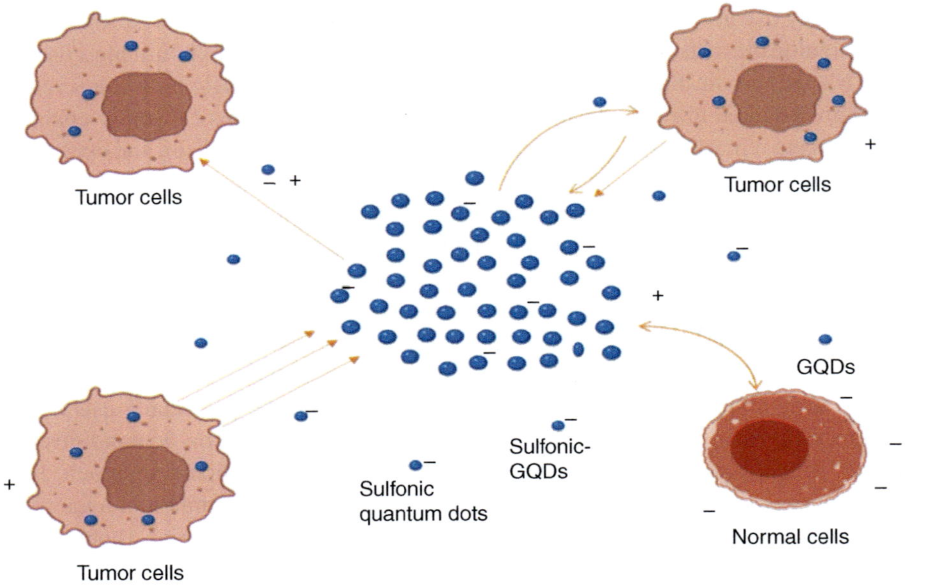

Figure 10.8 Mechanism of sulfonic-GCDs. Pratiksha et al. [59] / with permission of Elsevier / CC BY 4.0.

10.3.4 Early Disease Biomarker Detection

Quantum biosensors can identify disease biomarkers that are present at extremely low concentrations. This ultrasensitive detection is what makes quantum biosensors such powerful tools, especially for the pre-symptomatic detection of disease states. Early detection of disease states is particularly important in diseases like cancer and neurodegenerative diseases, where disease symptoms are not often presented until a more severe disease stage has been reached. In the case of cancer, if the disease can present a biomarker at an early disease state (a pre-symptomatic stage), then a detection at that stage would allow for an informed stratification of patients to be enrolled in clinical trials and an informed stratification of the appropriate patients for therapies developed for a specific biomarker. This pre-symptomatic detection of biomarkers accomplished with quantum biosensors may also provide real-time assessment of a patient's response to treatments. The insight gained from real-time feedback allows for more informed treatment decisions and ultimately a more timely and individualized patient experience. Given that quantum biosensors use non-invasive sampling methods such as blood or saliva, patient comfort is still maintained while continuing to have timely and periodic assessments of disease biomarkers (which also encourages compliance). Furthermore, given the sensitivity and specificity of quantum biosensors and their ability to detect biomarkers at very low concentrations, they hold the potential to make valuable contributions to making personalized medicine more feasible as well as to significantly impact the speed and efficiency of drug discovery and development activities.

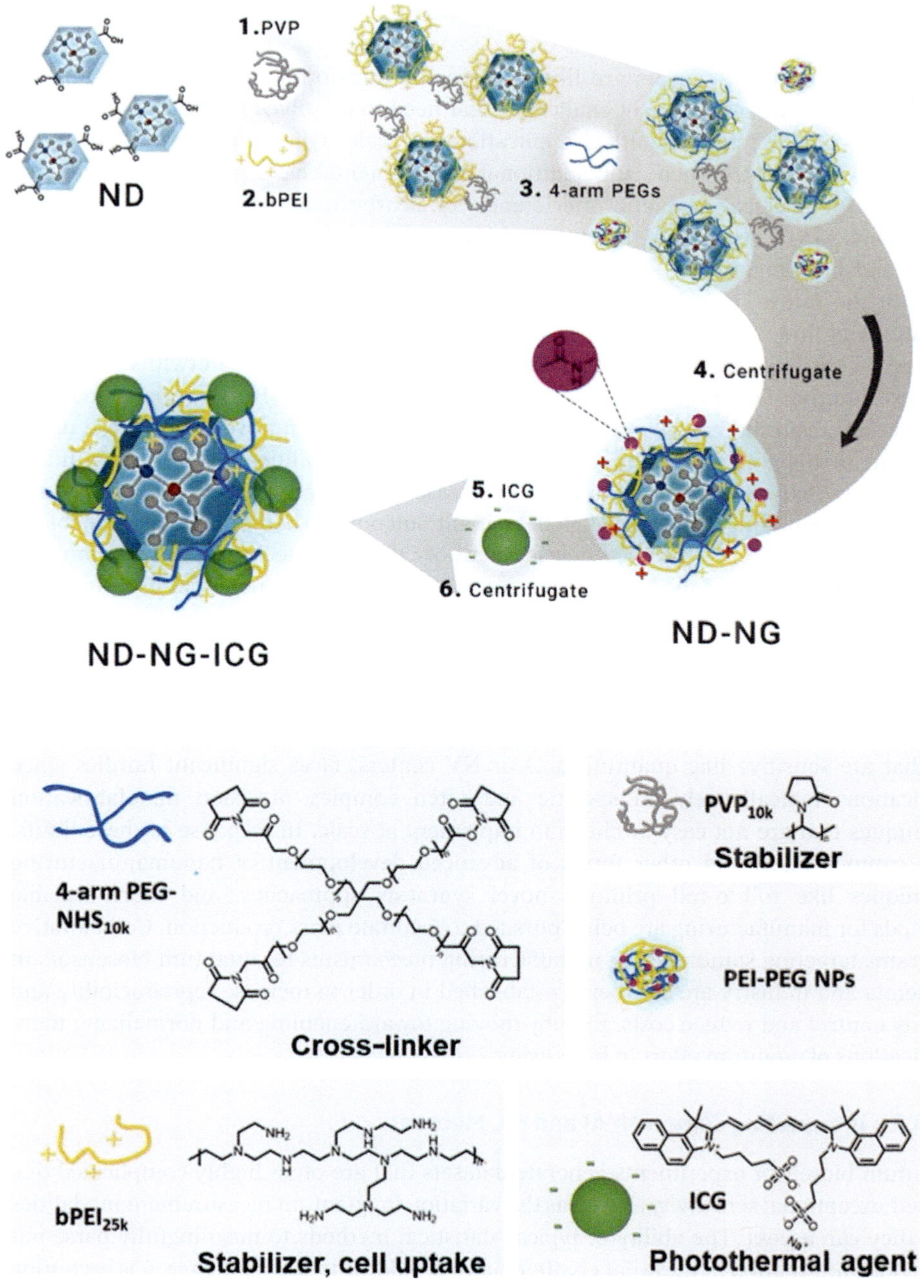

Figure 10.9 Schematic illustration of the synthesis route of nanodiamond–nanogel–indocyanine green. Wu et al. [64] / with permission of American Chemical Society.

10.4 Integration Issues and Challenges

Although quantum biosensors are likely to provide transformative capabilities in drug discovery, there are a number of challenges that need to be solved to determine effective implementation in pharmaceutical applications at scale. Quantum systems are easily affected by their environment, and additional environmental factors like any fluctuations in temperature, electromagnetic interference, or nearby materials can cause quantum decoherence [64]. Additional barriers encountered involve privacy and data security, safety and biocompatibility, and clinical verification. It may become essential at some point in the future for scientists, engineers, and health practitioners to collaboratively research and find ways to resolve obstacles such as these and truly assess QB technology as a real contribution to medicine (Figure 10.9) [14]. At present, notwithstanding all of these limiting factors as they relate to quality control for QBs, there is no denying that these transformative technologies could lead to radical improvements in the use of medicine as this area of development continues, with possibilities representing a future that produces physicians' ability to identify disease earlier and implement more tailored therapies, culminating in improved global patient outcomes. Therefore, it is reasonable to conclude that, through continued development of QBs, these technologies may prove to be a contributive part of medical diagnostics in the future and a potential ushering in of precision medicine into clinical practice (Figure 10.10).

10.4.1 Challenges of Scale and Manufacturing

The fabrication of quantum biosensors, especially those made using quantum materials that are sensitive like quantum dots or NV centers, faces significant hurdles since applications typically rely on specific and often complex precision nanofabrication techniques that are not easy or cheap to implement at scale. In response to these limits, both campus-based and other forms of advanced development of nanomanufacturing techniques like roll-to-roll printing, novel synthesis approaches, and more scalable methods for manufacturing are being pursued to facilitate mass production. Collaborative programs targeting standardizing manufacturing mechanisms for quantum biosensors in academia and industry are also being established in order to increase reproducibility and quality control and reduce costs, thereby moving toward enabling and normalizing more applications of quantum effort in biosensing.

10.4.2 Interpreting Data with AI and ML Methods

Quantum biosensor experiments generate datasets that are often highly complicated due to their exceptional sensitivity as well as the variation in quantum measurement modalities that they can access. The ability of typical statistical methods to meaningfully parse out biological information in these data is often limited. Adding to this challenge, it is becoming increasingly common for artificial intelligence (AI) and machine learning (ML) models to form part of the data analysis workflow [65]. AI and ML methods are advantageous as they are capable of identifying subtle patterns in data, removing noise, and improving signal. AI/design algorithms allow for real-time decision-making, optimization of calibration information, and improved predictive estimates, all of which have made these models

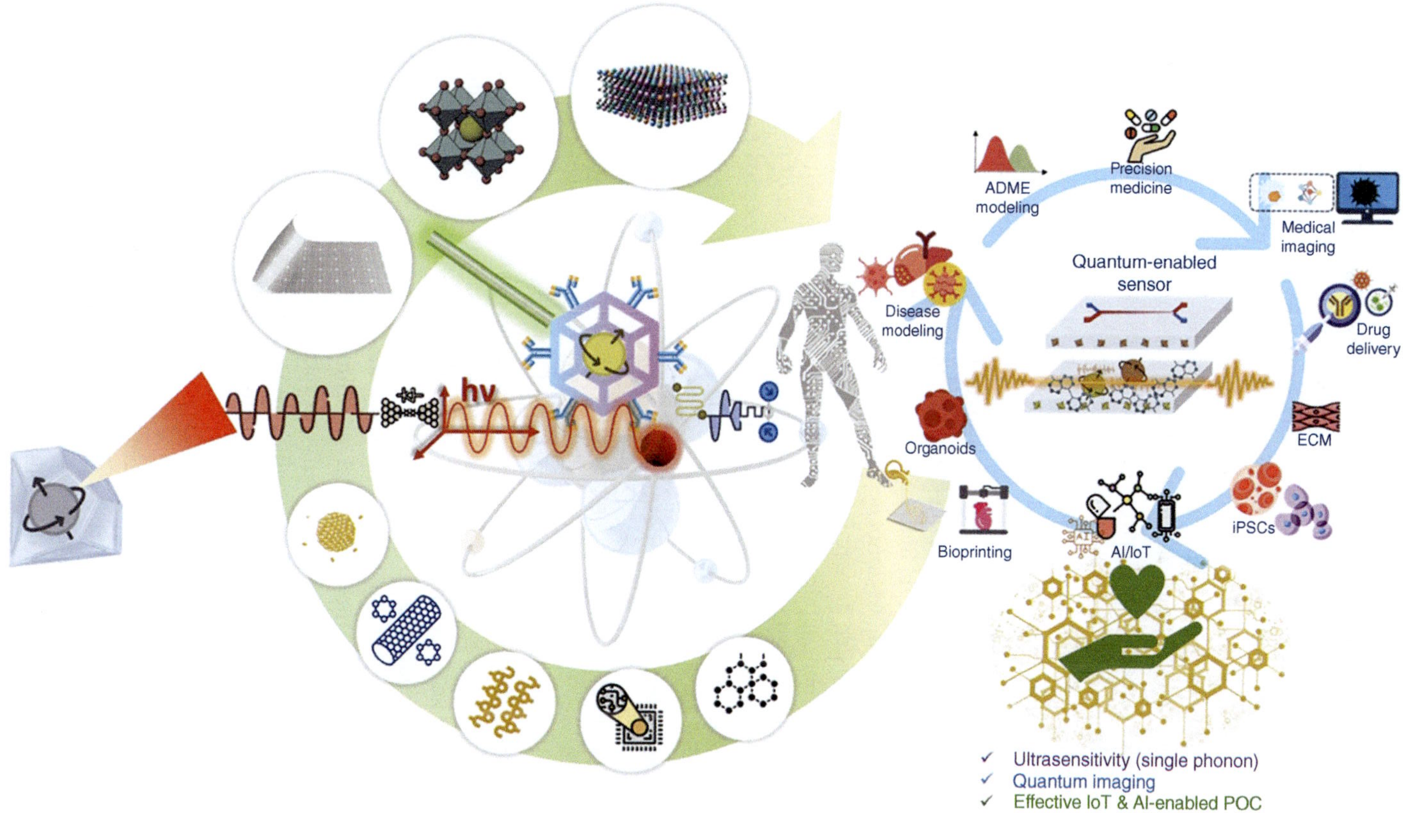

Figure 10.10 Futuristic application of quantum biosensor. Suparna et al. [14] / IOP Publishing / CC BY 3.0.

particularly powerful in high-throughput drug screening and biomarker detection. The integration of AI/ML into quantum sensor data has made it easier to translate raw sensor data into conclusions that could be useful for drug discovery/preclinical trials, thereby improving the potential value of quantum biosensor complexes used for identification of biological phenomena. New methodologies in nanomanufacturing, including roll-to-roll printing and scalable synthesis, are being created for mass production applications [66]. There are also collaborative opportunities being developed between academia and industry to standardize the manufacturing protocols to best improve reproducibility and minimize costs. There are multiple pharmaceutical and biotech companies that are beginning to implement quantum biosensing technologies as pilot projects, confirming the advancement and importance of quantum tools. Recently, a major pharmaceutical firm has engaged the use of quantum dot-based biosensors to enable multiplexed biomarker screening during the oncology drug development process, thus increasing detection sensitivity and throughput. Other start-ups are using portable quantum plasmonic sensors to enable rapid pathogen detection in infectious disease drug trials to ensure timely pathogen detection, which is key in this audience and for their specific monitoring. Certainly, there are collaborations between industry and research institutions exploring the use of quantum biosensors with AI-annotated analytics in order to assess pharmacokinetic characterization and detailed drug profiling in preclinical contexts [67]. These real-world applications highlight both opportunities for quantum biosensing as well as the challenges and obstacles to optimizing the implementation and the application of it. Overall, these case studies and experiences are informative in terms of what is needed to facilitate the broader use of quantum tools across the drug discovery and development pipeline.

10.4.3 Regulatory Pathway Considerations

The regulatory approval landscape for quantum biosensor-based diagnostic tools and drug-development devices is still in one of the early stages, with just a few examples available to guide evaluation. The introduction of new quantum mechanisms, nanomaterials, and other attributes only creates further complexity (i.e. safety, efficacy, and standardization), and the regulatory agencies need to assess these aspects in detail. It is important to prioritize early engagement with agencies, including the FDA and EMA, to address the issues outlined above. Collective efforts are being undertaken to create guidelines, establish harmonization and safety standards, and develop validation studies that specifically account for the attributes of quantum biosensors. This holistic approach, which incorporates feedback about standards for regulatory processes and development paths for quantum biosensors early on, will help to facilitate the approval pathway by ensuring that safety and performance requirements are met before broad implementation in a clinical setting.

10.5 Future Outlook and Commercial Potential

Upcoming advances in quantum biosensor designs are likely to have a real impact on the sensitivity, specificity, and multiplexing abilities of biosensing technologies. Developments in hybrid quantum and classical sensor platforms, integrated quantum light sources, and improvements in nanofabrication will likely allow us to advance substantially past the technical barriers that persist and develop stronger and scalable devices in the future. These devices should be able to usher in new paradigms of personalized medicine, allowing quantum biosensors a direct pathway to offer accurate and timely tracking of each patient's biomarkers, enabling us to tailor treatment or intervene earlier in the disease process [68]. The pathway to clinical translation includes multiple phases, including rigorous validation, integration with the existing health care structures, and engagement with the regulatory authorities to ensure a true demonstration of safety and efficacy. While private capital investment has not fully returned following the pandemic, increased interest from venture capitalists and existing pharmaceutical companies creates an exciting investment and commercialization horizon for quantum biosensors. As quantum biosensing evolves, we can expect the technology to disrupt the drug discovery and diagnostics landscape by supporting earlier detection and delivering faster drug screening processes and more precise patient monitoring and management that will support the development of safer and more efficacious therapies.

10.5.1 Next-Generation Quantum Biosensor Designs

New generations of quantum biosensor designs are all being developed to improve sensitivity, miniaturization, and multiplexing through more development and integration of quantum materials and quantum-type optics. The next generation is encompassing hybrid platforms that leverage the combination of quantum dots, nitrogen-vacancy (NV) centers, plasmolytic nanostructures, and photonic circuits to realize ultrasensitive single-molecule detection. New integrated quantum photonic chips, lab-on-a-chip quantum platforms, and entangled photon (twin-photon)-based detection systems are all being established that limit noise and enhance signal fidelity. All of these earlier designs with advanced nanofabrication are opening new opportunities to develop robust and reproducible sensor arrays with scalable architectures for POC or *in vivo* applications. New biosensor designs not only extend the physical limits of biosensing but also expand new frontiers for real-time diagnostics, HTS, and personalized therapies.

10.5.2 Emerging Applications in Personalized Medicine

Quantum biosensors have great potential to reshape personalized medicine by providing sensitive, immediate, and individualized diagnostics. Their unique large-area sensitivities at the single-molecule level are often well suited to diagnosing disease at the earliest stages, as well as enabling the monitoring of therapeutic response on an individual level. For example, multiplexed biosensors utilizing quantum dots have the ability to determine many disease markers at once in a single biological sample (which could provide more comprehensive molecular fingerprints for patients), and non-invasive biosensors using NV centers can provide additional monitoring of neural and cardiovascular activity (which can lead to personalized neurology or cardiology). Quantum plasmonic biosensors can provide real-time monitoring of molecular interactions, which can do things like allow drug responses to be monitored in terms of a person's individual biochemical profile and dosages adjusted accordingly. In this way, abilities like those mentioned above are moving us away from empirical medicine towards precision healthcare, where clinicians have significant patient-specific data that enhances diagnostic accuracy and treatment planning, leading to improved clinical outcomes.

10.5.3 Roadmap for Clinical Translation

The clinical translation of quantum biosensors requires a multi-step roadmap to navigate the pathway from high-throughput research to use in the hospital setting. The first step is preclinical validation. The sensitivity, specificity, and biocompatibility of the sensor need to be validated rigorously *in vitro* and in animal model studies. The second step is prototype development. The development of a quantum biosensor prototype should emphasize sensor scalability, robustness, and integration into relevant diagnostic pathways. The next step is the clinical trial phase, in which the quantum biosensor demonstrates efficacy and safety in human cohorts, which may vary in disease states. An important parallel activity is regulatory engagement with premarketing regulatory agencies such as the FDA and EMA to confirm that the sensor quality, safety, and performance comply with regulations. While research teams are completing preclinical and clinical trial activities, it is equally important to develop supporting infrastructures with appropriate data infrastructures (e.g. AI-powered data analysis) to distill the complex outputs generated by quantum sensors to provide clinical utility. The last step is commercial deployment, potentially with partnerships with private companies on manufacturing, distribution, or user training. Collaboration across stakeholders, whether they be academic, industry, or regulatory, will be needed throughout the clinical translation journey, not least to navigate technical, ethical, and operational issues. Finally, the scaling up of quantum biosensors into clinical practice is important for sustainability in future diagnostics.

10.5.4 Investment and Commercialization Landscape

Currently, the quantum biosensor market is in a period of strong growth fueled by new investments and a wave of commercialization. This growth is in part due to the promise that quantum sensor technology has to transform drug discovery, diagnostics, and personalized medicine. Worldwide, the quantum sensor market is expected to grow from USD 506 million in 2024 to USD 839 million by 2031, which represents a compound annual growth rate (CAGR) of 7.6% during the forecast period [69]. The demand for accurate measurement devices, especially in health care, is one of the driving forces behind this trend. This will be aided by investments from venture capital and corporate investments, which are leading the way. For instance, Novo Holdings (Novo Nordisk) has recently stated it intends to invest roughly USD 200 million in quantum computing startups, specifically those in life sciences companies from Nordic countries. This type of investment gives confidence in quantum technologies to disrupt the pharmaceutical and health care segments. A few companies are leading efforts to commercialize quantum biosensing technologies. For example, XtalPi is a biotech firm based in China that uses quantum algorithms and AI to decrease drug discovery time. Their work provides an example of taking quantum technology to the next level with real-world applications in the pharmaceutical sector. Furthermore, the integration of quantum sensing, AI, and machine learning is improving the ability of healthcare professionals to use real-time data and predictive modeling to increase the likelihood of adopting quantum biosensors in healthcare. Although the industry is extremely promising, there are still challenges to overcome (scalability, manufacturing challenges, and regulatory/external challenges). Fortunately, companies, academics, and regulatory bodies are working to overcome these challenges for the overall benefit of the field and target audiences for the stronger clinical integration of quantum biosensors.

10.6 Conclusions

Quantum Biosensors (QBs) usher in a new era of drug discovery and development to achieve drug developers' unprecedented accuracy, speed, and efficiency in identifying and optimizing therapeutic compounds. Quantum biosensors utilize principles of quantum mechanics, which enable very sensitive detection of molecular-binding interactions, monitoring of drug-target binding in real-time, and HTS of possible drug candidates. QBs can be applied to the identification of lead compounds, pharmacokinetic profiling, toxicity determination, personalized medicine, and other applications to facilitate the speed and efficiency of the drug development process. Although there is substantial promise, there are still many challenges that must be navigated to integrate QBs into pharmaceutical research. Some of the major challenges involve scalability, maintaining quantum coherence, and accessing the whole array of affordable partitions across multiple instruments, requiring transdisciplinary collaboration involving physicists, chemists, bioengineers, and pharmacologists. Some other challenges involve ensuring high specificity in complex biological systems and using robust AI data analytics to apply quantum technology to drug screening in the clinical setting.

In the future, quantum biosensors (QBs) hold great potential for improving drug discovery. In short, drug candidates will be identified faster and more accurately using quantum molecular profiling. Not only do quantum biosensors offer a unique advantage over classical biosensors with real-time measurements of drug efficacy and side effects, which would help enhance the safety assessments and optimization of therapeutics in regulatory-starting preclinical trials to eventual clinical uses, but they also facilitate the transition to personalized medicine by detecting biomarkers rapidly and developing ideal individualized treatment regimens. By avoiding the typical trial-and-error approaches to explore potential therapies, which are time-consuming and have further cost implications, quantum biosensors lead to greater streamlining of the drug discovery process and generating life-saving therapeutics. In time, as individual research progress develops, the merge of quantum biosensing with AI and nanotechnology will no doubt push drug development-related processes to the next generation of precision therapeutics. For quantum biosensor technology to develop successfully, global collaboration will be needed to address important factors related to replication, scalability, stability of energy signals, and sustainability/affordability. If properly executed, not just drug discovery, but also new paths of revolutionary innovations in pharmacology will materially benefit the safety, efficacy, and overall patient pharmacology around the world. The future of drug discovery will take place in the intersection of quantum technology and biomedicine, which means QBs are set to be a basis for intelligent, rapid, and more efficient therapeutic breakthroughs. Quantum biology sensors have the potential to revolutionize medicine development by offering ongoing innovation while considering the ethical and regulatory parameters for use, to focus on patient-centered health care solutions that are targeted and patient-specific.

References

1 Das, S., Mazumdar, H., Khondakar, K.R. et al. (2024). Quantum biosensors: principles and applications in medical diagnostics. *ECS Sensors Plus* 3: 025001.

2 Aydin, E.B., Aydin, M., and Sezginturk, M.K. (2019). Biosensors in drug discovery and drug analysis. *Current Analytical Chemistry* 15: 467–484.

3 Nandi, S., Adimule, V., Sharma, V. et al. (2024). Theoretical characteristics of biosensing technology. In: *Biosensing Technology for Human Health: Eco-friendly Materials and Real-world Applications*, 488. https://doi.org/10.1039/9781837676323-00051.

4 Yang, C., Xu, C., Wang, X., and Hu, X. (2012). Quantum-dot-based biosensor for simultaneous detection of biomarker and therapeutic drug: first steps toward an assay for quantitative pharmacology. *The Analysts Journal* 137: 1205–1209.

5 Swamy, N.K., Mohana, K.N.S., Madhusudana, A.M. et al. (2024). Graphene nanoribbon derived from multi-walled carbon nanotube: an efficient viral gene hosting and biosensing molecular platform for the electroanalysis of HIV-1 gene. *Results in Chemistry* 7: 101393.

6 Kendrekar, P., Adimule, V., and Hurst, T. (ed.) (2023). *Drug and Therapy Development for Triple Negative Breast Cancer*. Wiley.

7 Bandekar, S.S., Kerur, S.S., Kore, S.K., and Hegde, P.G. (2020). Synthesis and antimicrobial activity of plant based silver nanoparticles - a review. *Bioscience Biotechnology Research Communications* 13: 122–130.

8 Ding, R., Chen, Y., Wang, Q. et al. (2022). Recent advances in quantum dots-based biosensors for antibiotics detection. *Journal of Pharmaceutical Analysis* 12: 355–364.

9 Adimule, V., Sharma, K., and Lakshminarayana, P. (2024). Biomedical applications of surfactant modified graphene based electrochemical sensing devices. In: *Advances in Surfactant Biosensor and Sensor Technologies* (ed. J.G. Manjunatha), 149–167. Cham: Springer Nature Switzerland.

10 Pushpanjali, P.A., Manjunatha, J.G., and Hareesha, N. (2021). An overview of recent developments of carbon-based sensors for the analysis of drug molecules. *Journal of Electrochemical Science and Engineering* 11: 161–177.

11 Adimule, V., Batakurki, S., and Keri, R. (2023). Preparation of nanomaterials using coumarin and their various applications. In: *Secondary Metabolites Based Green Synthesis of Nanomaterials and Their Applications* (ed. A. Husen), 159–171. Singapore: Springer Nature Singapore.

12 Ma, F., Li, C.C., and Zhang, C.Y. (2018). Development of quantum dot-based biosensors: principles and applications. *Journal of Materials Chemistry B* 6: 6173–6190.

13 Adimule, V., Keri, R., Nandi, S., and Bagihalli, G. (2024). New approaches for glucose biosensors. In: *Real-Time Applications of Advanced Electrochemical Sensing Devices* (ed. J.G. Manjunatha), 14–11. Bristol: IOP Publishing.

14 Suparna, D., Hirak, M., Kamil, R.K. et al. (2024). Review—Quantum biosensors: principles and applications in medical diagnostics. *ECS Sensors Plus* 3: 025001.

15 Tigari, G., Manjunatha, J.G., Nagarajappa, H., and Prinith, N.S. (2022). Research developments in carbon materials-based sensors for determination of hormones. *Journal of Electrochemical Science and Engineering* 12: 3–23.

16 Kelly, T. and Yang, X. (2024). Application of fluorescence-and bioluminescence-based biosensors in cancer drug discovery. *Biosensors* 14: 570.

17 Kanthappa, B., Manjunatha, J.G., Aldossari, S.A., and Raril, C. (2025). Electrochemical determination of uric acid in the presence of dopamine and riboflavin using a poly(resorcinol)-modified carbon nanotube sensor. *Scientific Reports* 15: 5822.

18 Manhas, N., Kumar, L.S., and Adimule, V. (2023). Early-stage diagnosis of breast cancer: amelioration in approaches. In: *Drug and Therapy Development for Triple Negative Breast Cancer* (ed. P. Kendrekar, V. Adimule, and T. Hurst), 1–34. Wiley.

19 Mercy, D.J., Girigoswami, K., and Girigoswami, A. (2024). A mini review on biosensor advancements-emphasis on quantum dots. *Results in Chemistry* 7: 101271.

20 Manjunatha, J.G., Osman, S.M., and Ataollahi, N. (2024). Electrochemical polymerized DL-phenylalanine modified carbon nanotube sensor for the selective and sensitive determination of caffeic acid with riboflavin. *Scientific Reports* 14: 30950.

21 Chamkouri, H., Si, J., Chen, P. et al. (2024). A first step to develop quantum medicine: radiometers, detectors, and biosensors. *Sensing and Bio-Sensing Research* 44: 100658.

22 Nandi, S., Kerur, S.S., and Dhanalakshmi, S. (2024). Electrical and dielectric properties of polymer-metal hybrid nanocomposites-a short review. *DFMA* 35: 1–13.

23 Adimule, V., Sharma, K., Lakshminarayana, P. et al. (2024). Fabrication of carbon nanospheres functionalized Sm_2O_3/Co_3O_4 novel hybrid nanocomposites for selective and sensitive amperometric detection of hydrogen peroxide. *Journal of Applied Electrochemistry* 54: 323–340.

24 Hareesha, N. and Manjunatha, J.G. (2020). Surfactant and polymer layered carbon composite electrochemical sensor for the analysis of estriol with ciprofloxacin. *Materials Research Innovations* 24: 349–362.

25 Gorshkov, K., Susumu, K., Wolak, M., and Oh, E. (2022). Fluorescent quantum dots enable SARS-CoV-2 antiviral drug discovery and development. *Expert Opinion on Drug Discovery* 17: 225–230.

26 Nandi, S., Adimule, V., Kerur, S.S. et al. (2024). Recent advances in metal-organic framework derived nanocomposites in gas and chemical sensors: a review. *Sensing Technology* 2: 2317385.

27 Prinith, N.S. and Manjunatha, J.G. (2022). Electrochemical analysis of L-tryptophan at highly sensitive poly(glycine) modified carbon nanotube paste sensor. *Materials Research Innovations* 26: 134–143.

28 Lelit, M., Białecki, A., Gabler, T. et al. (2023). Selected advances of quantum biophotonics–a short review. *International Journal of Electronics and Telecommunications* 69: 399–405.

29 Charithra, M.M. and Manjunatha, J.G. (2021). Electroanalytical determination of acetaminophen using polymerized carbon nanocomposite-based sensor. *Chemical Data Collections* 33: 100718.

30 Adimule, V., Yallur, B.C., Bhowmik, D., and Gowda, A.H.J. (2021). Morphology, structural and photoluminescence properties of shaping triple semiconductor YxCoO: ZrO_2 nanostructures. *Journal of Materials Science: Materials in Electronics* 32: 12164–12181.

31 Singh, R.D., Shandilya, R., Bhargava, A. et al. (2018). Quantum dot based nano-biosensors for detection of circulating cell free miRNAs in lung carcinogenesis: from biology to clinical translation. *Frontiers in Genetics* 9: 616.

32 Pushpanjali, P.A., Manjunatha, J.G., Amrutha, B.M., and Hareesha, N. (2021). Development of carbon nanotube-based polymer-modified electrochemical sensor for the voltammetric study of curcumin. *Materials Research Innovations* 25: 12–420.

33 Yoosefian, M. and Ebrahimpour, Y. (2024). Edge-functionalized graphene quantum dot as a nano-biosensor for simultaneous detection of glucose, galactose and lactose. *Journal of Molecular Liquids* 413: 126044.

34 Vinayak, A. (2024). Developments in inorganic and organic based nanostructured materials for electrochemical biosensing applications. In: *Novel Nanostructured Materials for Electrochemical Bio-Sensing Applications* (ed. J.G. Manjunatha), 37–56. Elsevier.

35 Raril, C. and Manjunatha, J.G. (2020). Fabrication of novel polymer-modified graphene-based electrochemical sensor for the determination of mercury and lead ions in water and biological samples. *Journal of Analytical Science and Technology* 11: 1–10.

36 Kerur, S.S., Bandekar, S., Hanagadakar, M.S. et al. (2021). Removal of hexavalent chromium-industry treated water and wastewater: a review. *Materials Today: Proceedings* 42: 1112–1121.

37 Thorlabs (2024). Surface-enhanced Raman spectroscopy (SERS). https://www.thorlabs.com/newgrouppage9.cfm?objectgroup_id=16610

38 Pilot, R., Signorini, R., Durante, C. et al. (2019). A review on surface-enhanced Raman scattering. *Biosensors* 9 (2): 57.

39 Dey, D. and Goswami, T. (2011). Optical biosensors: a revolution towards quantum nanoscale electronics device fabrication. *Journal of Biomedicine & Biotechnology* 2011: 348218.

40 Hiroshi, Y., Kazuhide, S., and Yoshinobu, B. (2023). Theranostics applications of quantum dots in regenerative medicine, cancer medicine, and infectious diseases. *Advanced Drug Delivery Reviews* 200: 114863.

41 Li, S., Su, W., Wu, H. et al. (2020). Targeted tumour theranostics in mice via carbon quantum dots structurally mimicking large amino acids. *Nature Biomedical Engineering* 4: 704–716.

42 Rosddi, N.N.M., Fen, Y.W., Omar, N.A.S. et al. (2021). Glucose detection by gold modified carboxyl-functionalized graphene quantum dots-based surface plasmon resonance. *Optik* 239: 166779.

43 Kim, N., Bae, M., Cho, E. et al. (2025). Plasmonic biosensors in cancer-associated miRNA detection. *Biosensors* 15 (3): 165.

44 Riitta, H. and Riitta, S. (2012). Magnetoencephalography: from SQUIDs to neuroscience: neuroimage 20th anniversary special edition. *NeuroImage* 61: 386–396.

45 Venkatraman, H., Raveendra, B., Vandna, S. et al. (2025). Structural characterization of photoluminescent, light-driven fluorescence and phosphorescence decay properties of new azobenzene dyes containing alkoxy side chain and their DFT studies. *Journal of Molecular Structure* 1336: 142092.

46 Sanmartín-Matalobos, J., Bermejo-Barrera, P., Aboal-Somoza, M. et al. (2022). Semiconductor quantum dots as target analytes: properties, surface chemistry and detection. *Nanomaterials* 12 (14): 2501.

47 Zhu, C., Chen, Z., Gao, S. et al. (2019). Recent advances in non-toxic quantum dots and their biomedical applications. *Progress in Natural Science: Materials International* 29: 628–640.

48 Sarkar, S., Le, P., Geng, J. et al. (2020). Short-wave infrared quantum dots with compact sizes as molecular probes for fluorescence microscopy. *Journal of the American Chemical Society* 142: 3449.

49 Kirill, G., Kimihiro, S., Jiji, C. et al. (2020). Quantum dot-conjugated SARS-CoV-2 spike pseudo-virions enable tracking of angiotensin converting enzyme 2 binding and endocytosis. *ACS Nano* 14: 12234–12247.

50 Nandi, S.S., Adimule, V., Kadapure, S.A., and Kerur, S.S. (2022). Rare earth-based nanocomposite materials for prominent performance supercapacitor: a review. *AMM* 908: 3–18.

51 Nandi, S., Adimule, V., Kerur, S.S. et al. (2023). Electrochemical high-performance hybrid supercapacitors of carbon nanosphere doped 3D Zr (II) Linked 4-{[(1E)-1-Hydroxy-3-Oxoprop-1-En-2-Yl]Sulfanyl} benzoic acid metal organic frameworks. *Advances in Materials Research* 1177: 87–100.

52 Chakram, D.S., Nandi, S., Godavarti, U. et al. (2025). Recent progress in gas sensing characteristics of lithium doped sodium potassium niobate ($K_{0.5}$, $Na_{0.5-x}$ Li_x NbO_3) ceramics: microstructure, distinctive electrical applications. *Materials Today Communications* 44: 112072.

53 Adimule, V., Nandi, S., Kumar, P. et al. (2025). Enhancing photoluminescence dynamics, visible light driven photo catalytic dye degradation and distinctive electrical properties of cerium doped pyrimidine based covalent organic frameworks: sustainable synthesis and structural characterization. *Journal of Physics and Chemistry of Solids* 199: 112548.

54 Gordon, M.S., Christoph, B., Christian, W. et al. (2014). Synthesis of radioactively labelled CdSe/CdS/ZnS quantum dots for in vivo experiments. *Beilstein Journal of Nanotechnology* 5: 2383–2387.

55 Oetiker, N., Munoz-Villagr, C., Vasquez, C.C. et al. (2020). Bacterial phototoxicity of biomimetic CdTe-GSH quantum dots. *Journal of Applied Microbiology* 1633: 1–14.

56 Pacheco-Liñán, P.J., Bravo, I., Nueda, M.L. et al. (2020). Functionalized CdSe/ZnS quantum dots for intracellular pH measurements by fluorescence lifetime imaging microscopy. *ACS Sensors* 5 (7): 2106–2117.

57 Yuchen, T., Anggrek, C.N., Thamir, H. et al. (2022). Functionalized fluorescent nanodiamonds for simultaneous drug delivery and quantum sensing in HeLa cells. *ACS Applied Materials & Interfaces* 14 (34): 39265–39273.

58 Danila, D., Pardo, P.S., Misra, R.D.K., and Boriek, A.M. (2025). Liposomes as imaging agents of inflammation and oxidative stress in bone implants. *Current Issues in Molecular Biology* 47 (5): 295.

59 Pratiksha, S.H., Radheshyam, T.L., Aaditi, G.W. et al. (2025). Quantum dot nanotechnology: advancing target drug delivery in oncology. *Journal of Nanoscience and Nanotechnology* 7: 100172.

60 Mona, J., Paulina, C., Zuzanna, O. et al. (2023). Sensing of magnetic-field gradients with nanodiamonds on optical glass-fiber facets. *ACS Applied Nano Materials* 13 (6): 11077–11084.

61 Lemon, C.M., Karnas, E., Han, X. et al. (2015). Micelle-encapsulated quantum dot-porphyrin assemblies as in vivo two-photon oxygen sensors. *Journal of the American Chemical Society* 137 (31): 9832–9842.

62 Karnayana, P.M., Manjunatha, J.G., Edwin, S.D., and Ravishank, D.K. (2025). Batteries carbon waste used surface modified sensing platform for the electrochemical analysis of the antioxidant rutin. *Chemistry Africa* 8: 1–12.

63 Moulya, K.P., Manjunatha, J.G., and Aldossari, S.A. (2025). Advanced graphene composite paste electrode for amaranth dye detection with tartrazine. *Monatshefte fuer Chemie* 156: 595–605.

64 Wu, Y., Alam, M.N.A., Balasubramanian, P. et al. (2021). Nanodiamond theranostic for light-controlled intracellular heating and nanoscale temperature sensing. *Nano Letters* 21: 3780.

65 Vinayak, A., Rangappa, K., Vandna, S. et al. (2025). Investigation of non-linear optical and photophysical properties of fullerene decorated photochromic liquid crystals. *Journal of Solid State Chemistry* 343: 125151.

66 Manjunatha, J.G. and Hussain, C.M. (2022). Functionalized nanomaterial-based electrochemical sensors: principles, fabrication methods, and applications. In: *Fabrication Methods, and Applications*, 1e (ed. C.M. Hussain and J.G. Manjunatha), 309–335. Elsevier.

67 Adimule, V., Jagadeesha Gowda, A.H., Nandi, S.S., and Bowmik, D. (2022). Antimalarial activity of novel class of 1,3-Benzoxaborole derivatives containing 1,3,4-Oxadiazole moiety. In: *Drug Development for Malaria: Novel Approaches for Prevention and Treatment* (ed. P. Kendrekar), 285–302. Wiley.

68 Tyler, J., Choi, S.W., and Tewari, M. (2020). Real-time, personalized medicine through wearable sensors and dynamic predictive modeling: a new paradigm for clinical medicine. *Current Opinion in Systems Biology* 20: 17–25.

69 Dinesh, B., Sohini, P., Tania, A., and Shrimanata, S.R. (2024). Biosensors and their widespread impact on human health. *Sensors International* 5: 100257.

11

Quantum Biosensing for Environmental Monitoring and Public Health

Shridevi Salagare[1], Siddaramanna Ashoka[2], and Prashanth S. Adarakatti[3]

[1] *Department of Chemistry and Research Centre, NMKRV College for Women, Bengaluru, Karnataka, India*
[2] *Department of Chemistry, School of Applied Sciences, REVA University, Bengaluru, Karnataka, India*
[3] *Department of Chemistry, SVM Arts, Science and Commerce College, ILKAL – 587125 Affiliated to Bagalkot University, Jamkhandi, Karnataka, India*

11.1 Introduction

The innovative fusion of biological sensing and quantum physics, known as quantum biosensing, enables unprecedented advancements in environmental and public health monitoring [1–4]. Although traditional biosensing systems can be efficient, they frequently encounter limitations in sensitivity, selectivity, and their ability to operate across diverse environmental conditions [5]. By leveraging entanglement, quantum tunneling, and superposition, quantum biosensing offers a transformative approach to overcoming these challenges. It employs quantum systems such as superconducting qubits, nitrogen-vacancy centers in diamonds, and quantum dots to detect biological and chemical changes at the molecular level [6]. These quantum-based platforms can identify even trace amounts of contaminants, viruses, or biomolecules with remarkable sensitivity and precision [7]. Applying quantum mechanical principles to biosensing has enabled major innovations in real-time monitoring of environmental pollutants, disease biomarkers, and other critical analytes. A key advantage of this approach is its ability to function effectively in dynamic and complex environments. For example, quantum sensors can maintain high performance under conditions of extreme pressure, electromagnetic interference, and temperature fluctuations [8]. This makes them ideal for environmental monitoring applications where unpredictable conditions are common.

Quantum biosensing has the potential to transform personalized treatment and disease diagnostics within the field of public health [1]. Quantum-enhanced imaging and spectroscopy can detect diseases at very early stages with exceptional accuracy, enabling timely interventions [9]. In addition, the ability to monitor environmental factors such as air and water quality provides early warnings of harmful exposures, directly contributing to improved public health outcomes [10]. The advancement of quantum biosensing is also fostering interdisciplinary collaboration. Engineers, biologists, physicists, and chemists are working together to develop and refine quantum-based systems that address urgent global challenges [11]. These collaborative efforts are paving the way for innovations

Quantum Biosensing in Medical Diagnostics, First Edition. Edited by J. G. Manjunatha.

that extend beyond laboratory research, influencing sustainable development, healthcare infrastructure, and policymaking [12].

In summary, the rapidly developing field of quantum biosensing has the capacity to redefine how environmental and public health challenges are monitored and addressed. By harnessing the unique properties of quantum systems (Figure 11.1), this transformative technology promises unprecedented accuracy and reliability in tackling critical global issues [12].

11.2 Principles of Quantum Biosensing

Utilizing the special qualities of quantum physics, quantum biosensing detects biological and chemical changes with previously unheard-of sensitivity and accuracy. The fundamental ideas and important processes that make these discoveries possible are examined in this section.

11.2.1 Quantum Dots and Their Applications

Quantum dots are nanoscale semiconductor particles whose size-dependent optical and electrical properties arise from quantum confinement [13, 14]. Upon excitation, they emit light at specific wavelengths, making them highly suitable for biosensing applications [15, 16]. By tagging target analytes such as viruses, biomolecules, and trace contaminants, quantum dots enable highly sensitive detection. Their excellent photostability and tunable emission spectra also support multiplexing, allowing the simultaneous detection of multiple analytes. Consequently, quantum dots serve as a versatile platform for disease diagnosis and environmental monitoring [17, 18].

11.2.1.1 Nitrogen-Vacancy Centers in Diamonds

Defects in the diamond lattice that have a nitrogen atom next to a vacant lattice site are known as nitrogen-vacancy centers [19–21]. Long coherence times and sensitivity to electric and magnetic fields are just two examples of these centers' exceptional quantum characteristics. In biosensing, NV centers are used to identify subtle variations in environmental parameters like pH, temperature, and the presence of heavy metals [22]. They are perfect for real-time monitoring in complex environments because of their biocompatibility and adaptability to various environmental conditions (Figure 11.2).

11.2.1.2 Quantum Coherence and Superposition

The operation of quantum biosensors is based on the fundamental concepts of quantum coherence and superposition [1, 24]. Coherence refers to the preservation of a quantum system's phase relationships, which enables highly accurate measurements [25]. Superposition greatly increases the sensitivity of detection by enabling quantum systems to exist in numerous states at once. These characteristics are used to get high-resolution measurements of biological and environmental phenomena in sophisticated biosensing methods like quantum interferometry and spectroscopy. Quantum biosensors can detect environmental changes and tiny molecule interactions with remarkable accuracy by utilizing these concepts [26].

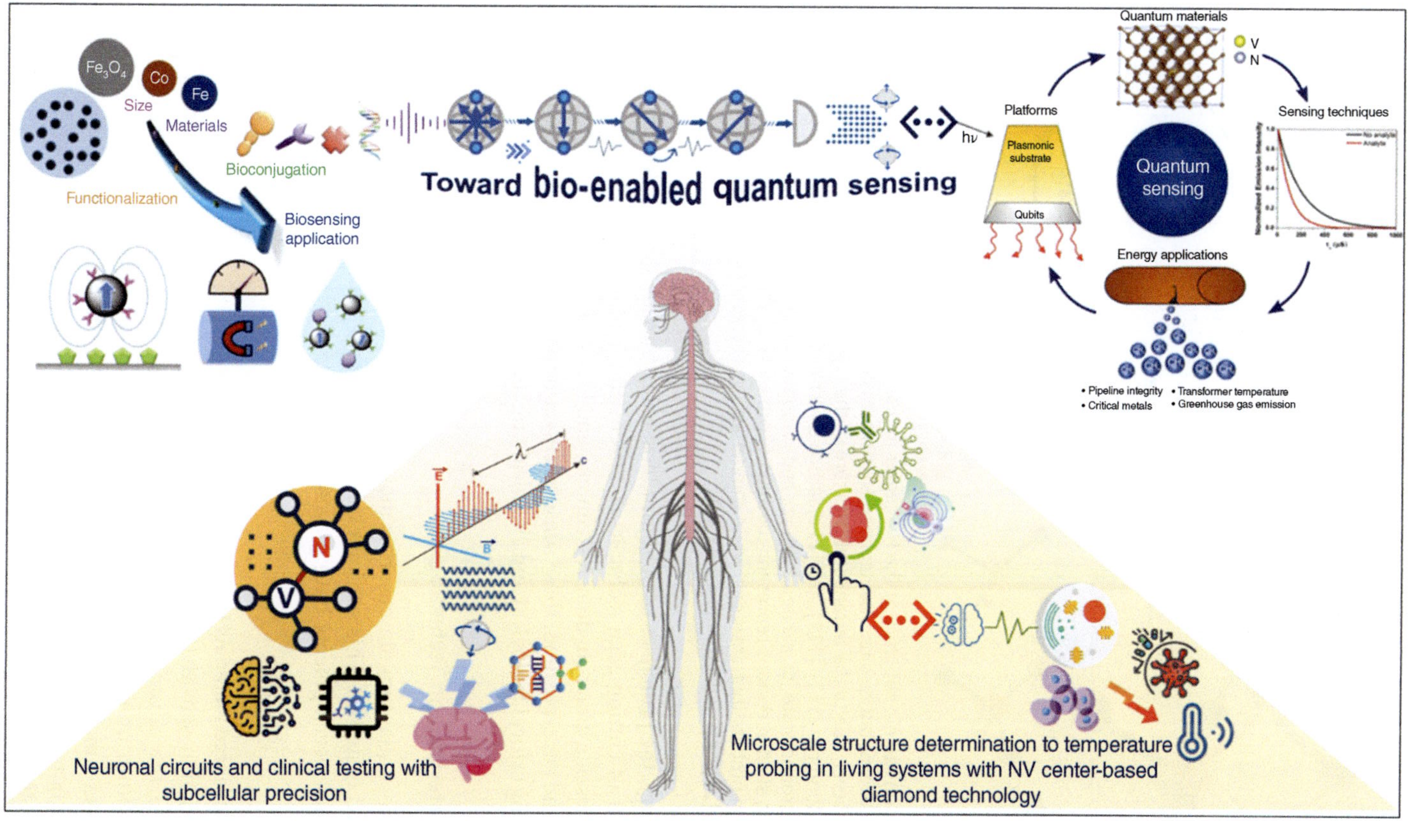

Figure 11.1 Visual overview of quantum biosensing. Das et al. [1] / IOP Publishing / CC BY 3.0.

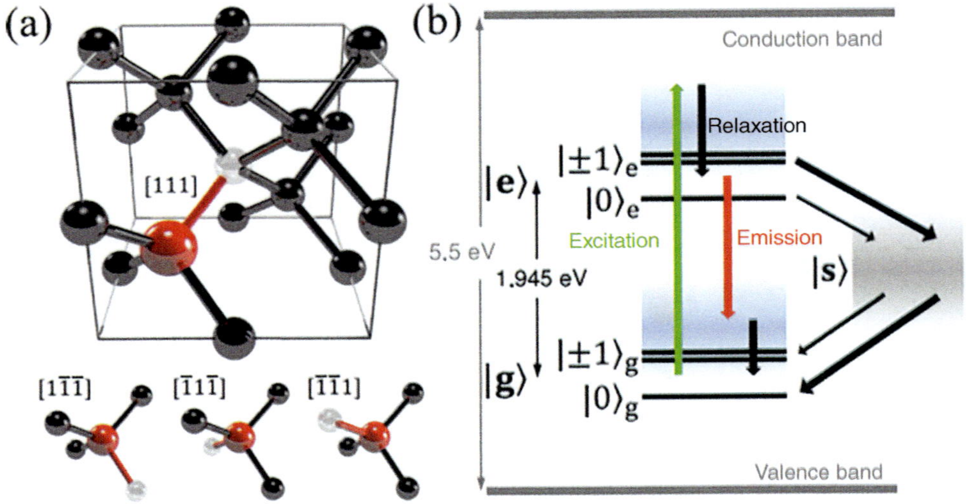

Figure 11.2 (a) Atomic structure of the NV center in diamond with a substitutional nitrogen atom adjacent to a vacant lattice site. The depicted NV center is oriented along the [111] lattice direction; however, due to the tetrahedral structure of the crystal lattice, the NV center axis can also be oriented along the [111], [111], and [111] lattice directions. (b) Energy levels and transitions of the NV, with ground state jgi, excited state jei, and intermediate singlet states jsi. The corresponding shaded regions represent continua of orbital and vibrational states. Thicker arrows represent high probability transitions, while thinner arrows represent low probability transitions. Sewani et al. [23] / AIP Publishing / CC BY 4.0.

In summary, the principles of quantum biosensing provide a robust foundation for the development of cutting-edge technologies in environmental and public health monitoring. By harnessing quantum dots, NV centers, and fundamental quantum phenomena like coherence and superposition, these sensors deliver unparalleled performance in detecting and analyzing critical biological and chemical targets.

11.2.2 Materials and Technologies

The creation of quantum biosensors is mostly dependent on cutting-edge technology and sophisticated materials. This section explores how novel materials, nanotechnology, and computational developments have played crucial roles in the development of quantum biosensing.

11.2.2.1 Advanced Materials for Quantum Biosensors

The exceptional sensitivity and specificity of quantum biosensors are primarily attributed to the advanced materials that constitute their core [27]. Owing to their unique quantum properties—including high electrical conductivity and tunable optical characteristics—materials such as graphene, transition metal dichalcogenides (TMDs), and hybrid nanocomposites are widely utilized. These substances improve detection capabilities by facilitating communication between biological targets and quantum sensors. For example, graphene is a great platform for immobilizing proteins in biosensors

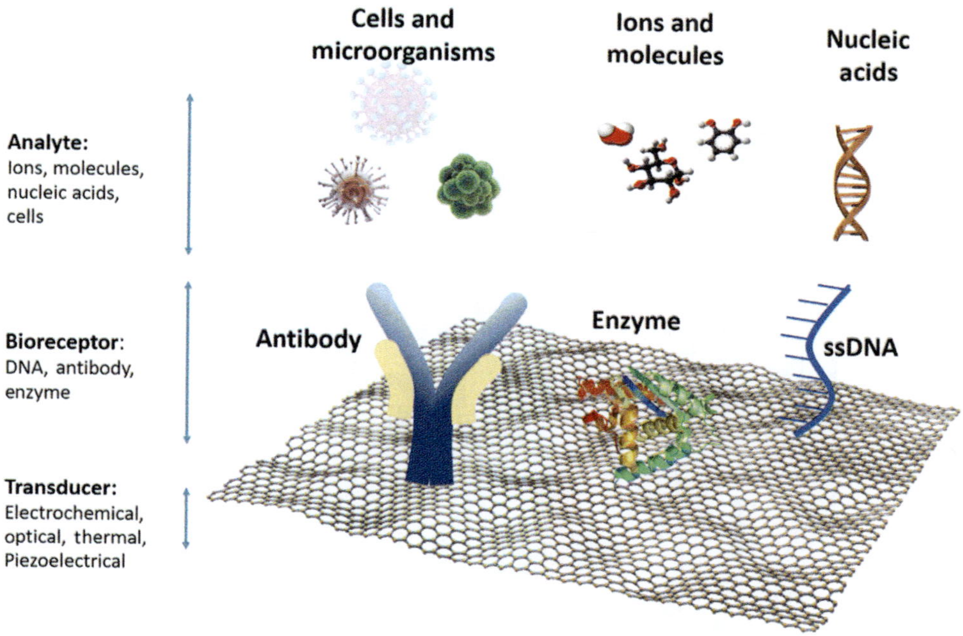

Figure 11.3 Examples of biosensors and components on a graphene platform. Peña-Bahamonde et al. [28] / Springer Nature / CC0 1.0.

due to its remarkable electrical characteristics and two-dimensional structure and is represented in Figure 11.3.

11.2.2.2 Integration with Nanotechnology

In order to make quantum biosensors portable and usable in practical settings, nanotechnology is essential to their miniaturization. Sensor components, including quantum dots, nanowires, and nanodiamonds, may be precisely controlled during production thanks to nanoscale engineering (Figure 11.4) [30]. By increasing the biosensors' surface area and signal responsiveness, this integration enhances their sensitivity for detecting minute amounts of contaminants, pathogens, or biomarkers [31]. Additionally, nanotechnology enables the creation of wearable and flexible quantum biosensors, which are essential for ongoing health and environmental monitoring.

11.2.2.3 Role of Artificial Intelligence and Machine Learning

In quantum biosensing, machine learning (ML) and artificial intelligence (AI) have become essential technologies [32]. The analysis of intricate datasets produced by quantum sensors is made possible by these computational tools, which make it easier to spot minute patterns and connections. By using quantum signals to forecast biological or environmental changes, machine learning algorithms can maximize sensor performance [4]. Furthermore, real-time data processing and interpretation are made possible by AI-driven systems, opening the door for automated decision-making in applications related to public health and environmental monitoring.

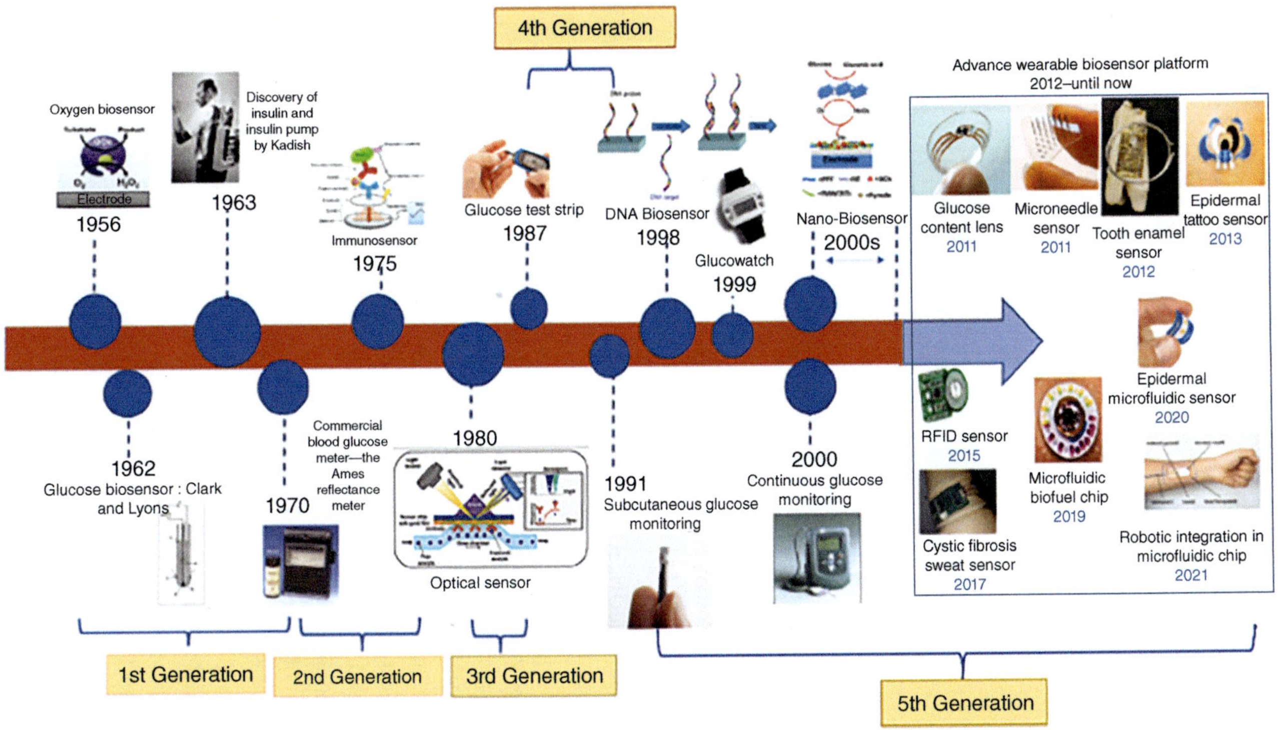

Figure 11.4 Biosensor development timeline. Ramesh et al. [29] / MDPI / CC BY 4.0.

In summary, quantum biosensing has been transformed by the combination of nanotechnology, sophisticated materials, and AI/ML technologies. These developments lay the groundwork for the creation of extremely sophisticated, portable, and sensitive sensors that tackle pressing issues in environmental and public health monitoring.

11.2.3 Technological Advances in Quantum Biosensors

Modern quantum biosensors are based on quantum materials, such as topological insulators, quantum dots, and NV centers in diamonds. NV centers are widely used in biosensing applications, particularly for the detection of environmental contaminants and biomarkers, because of their remarkable sensitivity to magnetic and electric fields [3]. The capacity of semiconductor nanoparticles known as quantum dots, which have optical characteristics that can be adjusted, to accurately identify particular analytes by fluorescence signals has made them well-known. In order to improve surface area, conductivity, and signal transduction efficiency, quantum sensors are also being integrated with emerging materials such as 2D layered materials (e.g. graphene). These developments have significantly enhanced quantum biosensors' performance and adaptability in a range of applications.

The combination of quantum biosensing and nanotechnology has revolutionized the field. Higher sensitivity and specificity can be achieved by improving the interface between analytes and quantum sensors using nanotechnology [1]. For example, the selective detection of particular infections or toxins is made possible by the functionalization of nanoparticles with biomolecules. Additionally, quantum biosensing devices are integrating AI algorithms for pattern identification and data analysis. Quantum biosensors can decipher complex data, spot trends, and increase the accuracy of decisions in real time by utilizing machine learning algorithms [33]. Additionally, biosensors can improve their detection skills over time thanks to AI's facilitation of adaptive learning.

The mobility and downsizing of quantum biosensors have been major areas of current research. The creation of small, portable devices with real-time monitoring capabilities has been made possible by developments in microfabrication and integration. For example, NV-diamond sensors are utilized in portable diagnostic devices for medical applications, while portable quantum dot-based fluorescence sensors are used for environmental quality evaluations [34]. These gadgets are perfect for on-site investigation in remote or resource-constrained environments since they provide real-time input. Furthermore, advancements in wireless communication technology have made it easier to send sensor data to centralized systems, allowing for remote analysis and monitoring.

All of these technical developments point to how quantum biosensors, which offer accurate, dependable, and real-time sensing solutions, have the potential to completely transform industries including healthcare, environmental monitoring, and food safety.

11.2.4 Applications in Environmental Monitoring

11.2.4.1 Pollutant Identification (Heavy Metals, Pesticides, etc.)

Because of their unmatched sensitivity and specificity, quantum biosensors have completely changed the way environmental contaminants are detected. Traces of heavy metals, pesticides, and other harmful compounds can be found with these sensors, which make use of quantum processes including spin coherence and quantum tunneling. For example, diamonds' NV centers are very good at identifying heavy metals in water sources, including lead (Pb^{2+}), cadmium (Cd^{2+}), and mercury (Hg^{2+}) [35–38]. Similarly, the fluorescent properties of quantum dot–based biosensors enable accurate pesticide detection, even at parts-per-billion (ppb) concentrations [39–42]. These capabilities are essential for ensuring regulatory compliance and environmental safety.

11.2.4.2 Tracking the Quality of the Air, Water, and Soil

Additionally, real-time soil, water, and air quality monitoring is made possible by quantum biosensors. Quantum sensors can identify pollutants like nitrogen oxides (NO_x), particulate matter, and volatile organic compounds (VOCs) in air quality monitoring, helping to provide early warnings for air pollution events [43]. Quantum-based fluorescence and electrochemical sensors offer precise measurements of contaminants such as phosphates, pathogens, and nitrates in water. Quantum sensors have made it possible to detect pollutants, pH, and nutrient levels that may have an impact on agricultural output, which has also improved soil monitoring. This thorough monitoring guarantees sustainable procedures and reduces environmental concerns.

11.2.4.3 Examples of Cases

The influence of quantum biosensing is illustrated by a number of practical applications. NV-diamond-based sensors were used to monitor heavy metal contamination in rivers in an industrial case study, allowing for prompt action to avert public health emergencies [44]. The application of quantum dot-based sensors in agricultural regions to identify pesticide runoff in water bodies, guaranteeing legal compliance and safeguarding aquatic ecosystems, is another noteworthy example. Furthermore, integrated quantum sensor networks have been used to track air pollution levels in metropolitan settings, offering useful information for urban planning and policymaking [45].

These uses highlight how revolutionary quantum biosensing may be in tackling difficult environmental issues, advancing sustainability, and protecting public health. Quantum biosensors have the potential to become essential instruments in environmental science with further development.

11.2.5 Applications in Public Health

11.2.5.1 Early Pathogen and Biomarker Identification

In public health, quantum biosensing technologies are becoming revolutionary instruments for early pathogen and biomarker detection. These sensors have remarkable sensitivity in detecting even the smallest amounts of bacteria, viruses, and disease-specific biomarkers by taking advantage of quantum phenomena like superposition and entanglement. For example, single-molecule detection of bacterial antigens or viral RNA sequences is possible with quantum dot-based biosensors, allowing for quick and precise diagnosis [46]. This is particularly beneficial in the management of infectious diseases, where early diagnosis is critical for preventing widespread outbreaks. Quantum-enhanced fluorescence imaging also aids in identifying biomarkers for long-term conditions such as diabetes, cardiovascular disease, and cancer.

11.2.5.2 Using Quantum Sensors to Diagnose Illnesses

Through the integration of sophisticated detecting methods with portable and user-friendly technologies, quantum sensors are revolutionizing the diagnosis of disease. The capacity of NV diamond sensors to identify trace magnetic fields linked to biomolecular activity has been shown, enabling noninvasive diagnosis of cardiac and neurological disorders [47]. In a similar vein, real-time examination of blood, saliva, or tissue samples using quantum-enhanced Raman spectroscopy ensures highly accurate point-of-care diagnostics. Antibiotic resistance detection is another noteworthy application, where quantum sensors can quickly detect resistant germs and direct efficient treatment plans [48].

11.2.5.3 Consequences for International Health Emergencies

Managing global health emergencies like pandemics or antibiotic resistance is one of the applications of quantum biosensing. Quantum-enhanced detection platforms were investigated for quick and accurate virus detection in field and clinical settings during the COVID-19 pandemic. In order to be ready for any emergencies in the future, these sensors may also be used to track new dangers like biohazards or zoonotic illnesses. Beyond diagnostics, quantum biosensors can measure immune responses at the molecular level to evaluate vaccine efficacy, improving public health tactics [1].

Quantum biosensing has the potential to transform early illness detection, enhance patient outcomes, and lessen the effects of health emergencies when it is included in public health infrastructure. This technology can be crucial to attaining global health fairness by tackling issues like cost and scalability.

11.3 Challenges and Limitations

11.3.1 Material and Technological Difficulties

Quantum biosensing has enormous material and technological obstacles in spite of its revolutionary potential. Complex materials like NV centers in diamonds, quantum dots, and superconducting circuits—all of which require elaborate production processes—are frequently used in the construction of quantum sensors. It is still very difficult to guarantee these quantum materials' stability and reproducibility in a variety of environmental settings. For example, it is very challenging to sustain quantum coherence, which is necessary for high sensitivity, in biological surroundings that are noisy. Additionally, the incorporation of quantum systems into traditional biosensing platforms necessitates advancements in readout mechanisms and signal processing, both of which are still in their infancy.

11.3.2 Both Cost-Effectiveness and Scalability

Another significant barrier to the development of quantum biosensing technology is scalability. The majority of quantum sensors are created in labs using expensive technology, which makes the switch to mass production difficult. High-quality quantum materials, including monodisperse quantum dots or diamonds with defects, are costly and resource-intensive to produce. Furthermore, some quantum systems' need for cryogenic cooling raises operating costs and restricts their use in environments with limited resources. Widespread adoption depends on finding affordable solutions without sacrificing performance, especially in developing nations with little funding for public health and environmental monitoring infrastructure.

11.3.3 Regulatory and Ethical Considerations

There are also moral and legal concerns with the use of quantum biosensing. In order to guarantee data privacy, accuracy, and dependability, the use of such cutting-edge technologies in environmental and public health monitoring calls for precise regulations. For instance, strict medical rules must be followed when using quantum sensors in disease diagnostics to avoid misdiagnosis or data exploitation. The fair distribution of these technologies also raises ethical questions because of the potential for access gaps between rich and developing nations due to their high cost. In order to solve these problems and promote innovation while guaranteeing responsible use, regulatory frameworks must change.

In conclusion, even though quantum biosensing has a lot of potential, overcoming these obstacles will call for cooperation from the scientific, business, and regulatory sectors. In order to successfully incorporate quantum biosensors into real-world applications and guarantee that their advantages are available to everyone, these constraints must be addressed.

11.3.4 Future Perspectives

Quantum biosensing has shown great promise in revolutionizing environmental health monitoring and public health issues [5]. The future development of the discipline is especially promising due to emerging technologies, multidisciplinary opportunities, and advances in global monitoring networks.

To improve sensitivity and selectivity, the next generation of quantum biosensors is probably going to incorporate cutting-edge materials like nanostructures and two-dimensional (2D) materials [49]. Real-time detection of low-concentration analytes, including heavy metals, diseases, and trace contaminants, is anticipated to be made possible by advancements in quantum dots, nanodiamonds, and NV centers in diamonds. Furthermore, unmatched precision in biological and environmental monitoring is promised by quantum-enhanced imaging and spectroscopic techniques, such as entanglement-based approaches and quantum coherence. When these technologies are combined with ML and AI algorithms, it will be easier to quickly comprehend complicated datasets and produce actionable insights for prompt response [50].

As urbanization accelerates, smart cities will increasingly rely on quantum sensing technologies to monitor environmental conditions and ensure public safety. Quantum biosensors can provide real-time, high-resolution data when integrated into smart grids, water and air quality monitoring systems, and medical facilities [6]. When connected through Internet of Things (IoT) networks, these sensors can form intelligent systems capable of detecting and responding to environmental threats and disease outbreaks. At a global scale, quantum-enhanced satellite networks could track pollutant dispersion, monitor climate change, and forecast pandemics by analyzing environmental biosignatures. These capabilities align with the Sustainable Development Goals (SDGs) and support the growing need for resilient infrastructure.

It will take interdisciplinary cooperation amongst physicists, chemists, biologists, engineers, and data scientists to advance quantum biosensing. New biosensing platforms that integrate quantum principles with molecular biology and material science can be developed through interdisciplinary efforts. For example, creating bio-hybrid sensors with unmatched detecting capabilities may be made possible by the fusion of synthetic biology and quantum mechanics. Furthermore, converting quantum biosensing research into useful applications such as public health and environmental policy initiatives will require collaborations between government, business, and academia.

In summary, quantum biosensing has a promising future thanks to new technologies, smart city applications, and interdisciplinary partnerships that are opening the door to game-changing solutions in public health and environmental monitoring. The field has the potential to address some of the most important global issues of our day as it develops.

11.4 Conclusions

At the cutting edge of scientific advancement, quantum biosensing has revolutionary promise for public health and environmental monitoring. Quantum physics and biosensing technologies have come together to create previously unheard-of possibilities for precisely identifying, tracking, and reducing health and environmental threats. The range of environmental diagnostics and public health interventions is being redefined by the development of real-time, high-resolution data acquisition made possible by quantum-enhanced materials and detection methods.

Quantum biosensors combined with digital technologies like AI, machine learning, and the IoT are predicted to transform data processing and decision-making in the future. In addition to improving the sensitivity and dependability of biosensing systems, this collaboration will make scalable applications possible in a variety of fields, including global health networks and smart cities. Quantum sensing's importance in promoting sustainable development and global resilience is further highlighted by its application to pollution prevention, pandemic preparedness, and climate change.

To fully utilize quantum biosensing, the future will need to be paved with investments in infrastructure, training, and interdisciplinary research and development. To close the gap between laboratory inventions and practical applications, academic institutions, businesses, and policymakers must work together in collaborative frameworks. As these technologies advance, they have the potential to change how society responds to environmental and public health issues by becoming more impactful, accurate, and proactive.

In summary, quantum biosensing is poised to redefine the landscape of environmental monitoring and public health, offering a beacon of hope in addressing some of the most pressing challenges of the twenty-first century. By embracing innovation, fostering collaboration, and aligning efforts with global sustainability goals, the field can ensure a safer, healthier, and more resilient future for all.

Acknowledgments

The author gratefully dedicates this chapter to his beloved son, Pradyum, on the joyous occasion of his fourth birthday. Heartfelt thanks are also extended to his wife, Praneeta, for her unwavering support, encouragement, and inspiration throughout the course of this work. The author further acknowledges Mr. Satish Basapur for his valuable discussions and insightful contributions that enriched the development of this chapter.

References

1 Das, S., Mazumdar, H., Khondakar, K.R. et al. (2024). Quantum biosensors: principles and applications in medical diagnostics. *ECS Sensors Plus* 3: 025001.

2 Martynenko, I., Litvin, A., Purcell-Milton, F. et al. (2017). Application of semiconductor quantum dots in bioimaging and biosensing. *Journal of Materials Chemistry B* 5: 6701–6727.

3 Petrini, G., Moreva, E., Bernardi, E. et al. (2020). Is a quantum biosensing revolution approaching? Perspectives in NV-assisted current and thermal biosensing in living cells. *Advanced Quantum Technologies* 3: 2000066.

4 Degen, C.L., Reinhard, F., and Cappellaro, P. (2017). Quantum sensing. *Reviews of Modern Physics* 89: 035002.

5 Das, S.K., Nayak, K.K., and Krishnaswamy, P. (2023). Progression in quantum sensing/bio-sensing technologies for healthcare. *Quantum* 7: 9.

6 Khondakar, K.R. and Kaushik, A.K. (2024). *Next-Generation Smart Biosensing: Nano-Platforms, Nano-Microfluidics Interfaces, and Emerging Applications of Quantum Sensing*. Academic Press.

7 Mpofu, K.T. and Mthunzi-Kufa, P. (2024). Recent advances in quantum biosensing technologies. In: *Current Developments in Biosensor Applications and Smart Strategies*. https://doi.org/10.5772/intechopen.1009461.

8 Crawford, S.E., Shugayev, R.A., Paudel, H.P. et al. (2021). Quantum sensing for energy applications: review and perspective. *Advanced Quantum Technologies* 4: 2100049.

9 Shams, M., Choudhari, J., Reyes, K. et al. (2023). The quantum-medical nexus: understanding the impact of quantum technologies on healthcare. *Cureus* 15: e48077.

10 Ratre, P., Thareja, S., and Mishra, P.K. (2024). Quantum imaging-based nanobiosensors: pioneering point-of-care approach for early diagnosis of environmental-linked breast cancer. *Journal of Reproductive Healthcare and Medicine* 5: 1.

11 Casotti, M.C., Meira, D.D., Zetum, A.S.S. et al. (2024). Integrating frontiers: a holistic, quantum and evolutionary approach to conquering cancer through systems biology and multidisciplinary synergy. *Frontiers in Oncology* 14: 1419599.

12 Simone, G. (2024). Trends of biosensing: plasmonics through miniaturization and quantum sensing. *Critical Reviews in Analytical Chemistry* 54: 2183–2208.

13 Lim, S.Y., Shen, W., and Gao, Z. (2015). Carbon quantum dots and their applications. *Chemical Society Reviews* 44: 362–381.

14 Bacon, M., Bradley, S.J., and Nann, T. (2014). Graphene quantum dots. *Particle & Particle Systems Characterization* 31: 415–428.

15 Bera, D., Qian, L., Tseng, T.-K., and Holloway, P.H. (2010). Quantum dots and their multimodal applications: a review. *Materials* 3: 2260–2345.

16 Jamieson, T., Bakhshi, R., Petrova, D. et al. (2007). Biological applications of quantum dots. *Biomaterials* 28: 4717–4732.

17 Bailey, R.E., Smith, A.M., and Nie, S. (2004). Quantum dots in biology and medicine. *Physica E: Low-dimensional Systems and Nanostructures* 25: 1–12.

18 Wagner, A.M., Knipe, J.M., Orive, G., and Peppas, N.A. (2019). Quantum dots in biomedical applications. *Acta Biomaterialia* 94: 44–63.

19 Gali, Á. (2019). Ab initio theory of the nitrogen-vacancy center in diamond. *Nanophotonics* 8: 1907–1943.

20 Green, B.L., Collins, A.T., and Breeding, C.M. (2022). Diamond spectroscopy, defect centers, color, and treatments. *Reviews in Mineralogy and Geochemistry* 88: 637–688.

21 Nebel, C.E. (2020). Nitrogen-vacancy doped CVD diamond for quantum applications: a review. *Semiconductors and Semimetals* 103: 73–136.

22 Wu, Y., Jelezko, F., Plenio, M.B., and Weil, T. (2016). Diamond quantum devices in biology. *Angewandte Chemie International Edition* 55: 6586–6598.

23 Sewani, V.K., Vallabhapurapu, H.H., Yang, Y. et al. (2020). Coherent control of NV−centers in diamond in a quantum teaching lab. *American Journal of Physics* 88: 1156–1169.

24 Chaurasia, R. and Dhabekar, B. (2024). An overview of quantum biology. In: *Handbook on Radiation Environment, Volume 1: Sources, Applications and Policies* (ed. D.K. Aswal), 667–686. Springer Nature.

25 Shapiro, J.H. and Shepard, S.R. (1991). Quantum phase measurement: a system-theory perspective. *Physical Review A* 43: 3795.

26 Anandampillai, S., Zhang, X., Sharma, P. et al. (2008). Quantum dot–DNA interaction: computational issues and preliminary insights on use of quantum dots as biosensors. *Computer Methods in Applied Mechanics and Engineering* 197: 3378–3385.

27 Ma, F., Li, C.-C., and Zhang, C.-Y. (2018). Development of quantum dot-based biosensors: principles and applications. *Journal of Materials Chemistry B* 6: 6173–6190.

28 Peña-Bahamonde, J., Nguyen, H.N., Fanourakis, S.K., and Rodrigues, D.F. (2018). Recent advances in graphene-based biosensor technology with applications in life sciences. *Journal of Nanobiotechnology* 16: 75.

29 Ramesh, M., Janani, R., Deepa, C., and Rajeshkumar, L. (2022). Nanotechnology-enabled biosensors: a review of fundamentals, design principles, materials, and applications. *Biosensors* 13: 40.

30 Kumar, S., Nehra, M., Kedia, D. et al. (2019). Nanodiamonds: emerging face of future nanotechnology. *Carbon* 143: 678–699.

31 Qureshi, A. and Niazi, J.H. (2020). Biosensors for detecting viral and bacterial infections using host biomarkers: a review. *Analyst* 145: 7825–7848.

32 Mpofu, K.T. and Mthunzi-Kufa, P. (2025). Recent advances in artificial intelligence and machine learning based biosensing technologies. In: *Current Developments in Biosensor Applications and Smart Strategies*. IntechOpen. *Selcan Karakuş*. 332–357.

33 Cui, F., Yue, Y., Zhang, Y. et al. (2020). Advancing biosensors with machine learning. *ACS Sensors* 5: 3346–3364.

34 Anders, J., Babaie, M., Bardin, J.C. et al. (2023). CMOS integrated circuits for the quantum information sciences. *IEEE Transactions on Quantum Engineering* 4: 1–30.

35 Yin, H., Truskewycz, A., and Cole, I.S. (2020). Quantum dot (QD)-based probes for multiplexed determination of heavy metal ions. *Microchimica Acta* 187: 1–25.

36 Hareesha, N. and Manjunatha, J. (2021). Electro-oxidation of formoterol fumarate on the surface of novel poly(thiazole yellow-G) layered multi-walled carbon nanotube paste electrode. *Scientific Reports* 11: 12797.

37 Pushpanjali, P., Manjunatha, J., Hareesha, N. et al. (2021). Voltammetric analysis of antihistamine drug cetirizine and paracetamol at poly(L-leucine) layered carbon nanotube paste electrode. *Surfaces and Interfaces* 24: 101154.

38 Monnappa, A.B., Manjunatha, J.G., and Bhatt, A.S. (2020). Design of a sensitive and selective voltammetric sensor based on a cationic surfactant-modified carbon paste electrode for the determination of alloxan. *ACS Omega* 5: 23481–23490.

39 Nsibande, S. and Forbes, P. (2016). Fluorescence detection of pesticides using quantum dot materials–a review. *Analytica Chimica Acta* 945: 9–22.

40 Raril, C., Manjunatha, J.G., Ravishankar, D.K. et al. (2020). Validated electrochemical method for simultaneous resolution of tyrosine, uric acid, and ascorbic acid at polymer modified nano-composite paste electrode. *Surface Engineering and Applied Electrochemistry* 56: 415–426.

41 Pushpanjali, P.A., Manjunatha, J.G., and Hareesha, N. (2021). An overview of recent developments of carbon-based sensors for the analysis of drug molecules. *Journal of Electrochemical Science and Engineering* 11: 161–177.

42 Tigari, G. and Manjunatha, J. (2020). Poly(glutamine) film-coated carbon nanotube paste electrode for the determination of curcumin with vanillin: an electroanalytical approach. *Monatshefte für Chemie-Chemical Monthly* 151: 1681–1688.

43 Galstyan, V. (2021). "Quantum dots: perspectives in next-generation chemical gas sensors" – a review. *Analytica Chimica Acta* 1152: 238192.

44 Das, D., Raj, R., Jana, J. et al. (2022). Diamond—the ultimate material for exploring physics of spin-defects for quantum technologies and diamondtronics. *Journal of Physics D: Applied Physics* 55: 333002.

45 Ahmad, N. and Jas, S. (2025). Quantum-inspired neural networks for time-series air pollution prediction and control of the most polluted region in the world. *Quantum Machine Intelligence* 7: 1–13.

46 Zhang, Y. and Wang, T.-H. (2012). Quantum dot enabled molecular sensing and diagnostics. *Theranostics* 2: 631.

47 Zhang, T., Pramanik, G., Zhang, K. et al. (2021). Toward quantitative bio-sensing with nitrogen–vacancy center in diamond. *ACS Sensors* 6: 2077–2107.

48 Sudewi, S., Sai Sashank, P.V., Kamaraj, R. et al. (2024). Understanding antibiotic detection with fluorescence quantum dots: a review. *Journal of Fluorescence*. 35(5): 1–25.

49 Verma, A. and Yadav, B.C. (2024). Comprehensive review on two dimensional nanomaterials for optical biosensors: present progress and outlook. *Sustainable Materials and Technologies* 40: e00900.

50 Rane, N.L., Paramesha, M., Choudhary, S.P., and Rane, J. (2024). Artificial intelligence, machine learning, and deep learning for advanced business strategies: a review. *Partners Universal International Innovation Journal* 2: 147–171.

12

Nanoparticles in Quantum Biosensing

Nikiwe Mhlanga

Department of Science Technology and Innovation(DSTI)/Mintek Nanotechnology Innovation Centre (NIC), Advanced Materials Division, Mintek, Randburg, South Africa

12.1 Quantum Mechanics Theory

The basis of quantum biosensing is quantum mechanics (QM) theory. The QM theory is largely founded on the many-bodied Schrodinger equation, compressed to $H\varphi - E\varphi$. The H-Hamiltonian operator represents the kinetic and potential energy stemming from electrons and ions interaction, and E is the system's energy. The QM theory describes the interaction of ions and electrons in a system, and different assumptions are considered to simplify the equation. Born-Oppenheimer simplified the quantum state of molecules with the assumption of ions evolving slowly in space while electrons respond instantaneously to the ionic motion. This assumption results in a wave function dependent on electronic degrees of freedom alone. The Born-Oppenheimer approximation neglects the movement of the atomic nuclei [1, 2]. Subsequently, Hartree separates the H variables and assumes eigenenergy from one electron. Fock modifies Hartree's assumption by replacing the one-electron eigenfunction from antisymmetrized wavefunctions and assumes involvement of all electrons in the creation of an electric field and yields the Hartree-Fock theory [3]. The QM is the basis of all materials, including nanoparticles and their application in biosensing.

In nanoparticles, quantum confinement is observed with particle size reduction to nanoscale. The reduction restricts or confines the electrons into a smaller space, and their energy levels are quantized as opposed to continuous in bulk materials. Several merits are associated with quantum confinement, such as enhanced optical and electrical properties and increased bandgap, which paves the way for applications in quantum computing, solar energy, health, and biosensing [4, 5]. The type of quantum confinement is determined by the characteristic dimensions of the particle, i.e. zero-, one-, and two-dimensional quantum confinement. For example, quantum dots are zero-dimensional materials with electrons confined in all three dimensions while quantum wires are confined in two dimension and allow free movement in one, and lastly, quantum wells allow movement in two dimensions and are confined in one [4].

Quantum Biosensing in Medical Diagnostics, First Edition. Edited by J. G. Manjunatha.

12.2 Quantum Biosensing

Quantum biosensing (QB) entails the application of the QM theory in the sensing and detection of biomolecules. This results in the fabrication of biosensors capitalizing on the QM phenomenon merited with high sensitivity, precision, and single-atom detection [6]. The architecture of the QM-based biosensors includes materials such as quantum dots and nanoparticles that are engineered with biological entities. The targeting biological entities include antibodies, DNA strands, peptides, and aptamers and locate and bind the targeted biomolecules [6]. The interaction of the capture and captured entity evokes changes in optical properties such as fluorescence, and this is indicative of their interaction. This indicator confirms a successful detection of the captured biomolecule, and the wavelength/intensity of the emitted light qualifies and quantifies the biomolecule. A myriad fields use the QB phenomenon, such as biosecurity, environmental monitors, and disease diagnosis [6]. The interferences and entanglement in the vicinity of the molecules are the foundation of QB [6].

12.3 Nanoparticles in Quantum Biosensing

The application of nanoparticles in QB is driven by their properties. Nanoparticles have a high-to-surface volume ratio and tunable physicochemical properties. These attributes merit surface functionality with targeting moieties that enhances interaction with biological entities. Also, fluorescence amplification and its potential for multiplexed detection of biomolecules allures to the QB application [7–12].

The detection or diagnosis of disease causatives involves the use of biosensors or testing devices. The World Health Organisation (WHO) standard for these devices is **A**ffordable, **S**ensitive, **S**pecific, **U**ser-friendly, **R**apid, **E**quipment-free, and **D**elivered at bedside (ASSURED). Diagnosis is preliminary and important in the management of sicknesses, and it informs subsequent treatment and management steps. Predominately, effective treatment and disease management hinge on accurate prognosis. Point-of-care testing (POCT) emulates the WHO ASSURED standard. POCT entails field-based testing or availability of tests in closer proximity to the patients. The POCT attributes of low cost, user friendliness, and robustness align with the ASSURED standard and are popular in low-resourced WHO regions [13]. Nanomaterials, especially quantum nanoparticles, are some of the drivers of the POCT. For example, lateral flow assays (LFAs) use gold nanoparticles (Au NPs) for visualization of the test and control lines. The LFAs, shown in Figure 12.1, are the standard of the POCT and are based on a technique of eluting an immunoassay along a membrane driven by capillary force. The membrane has set positions for positive and control test lines, and they are visualized by the use of colorimetric materials [15, 16]. Typically, the test and control lines are immobilized with relevant biological molecules that will react with the immunoassay and give a line or negative result in the absence of the line. Various colorimetric materials have been used, such as enzymes, fluorophores, quantum dots, and nanoparticles (carbon, gold, silver) [15]. Au NPs, merited by easy synthesis, vibrant color, stability, biocompatibility, high optical properties and biofunctionalization, are mainly used in LFAs [15]. Recently, the Au NPs are coupled with other nanoparticles for

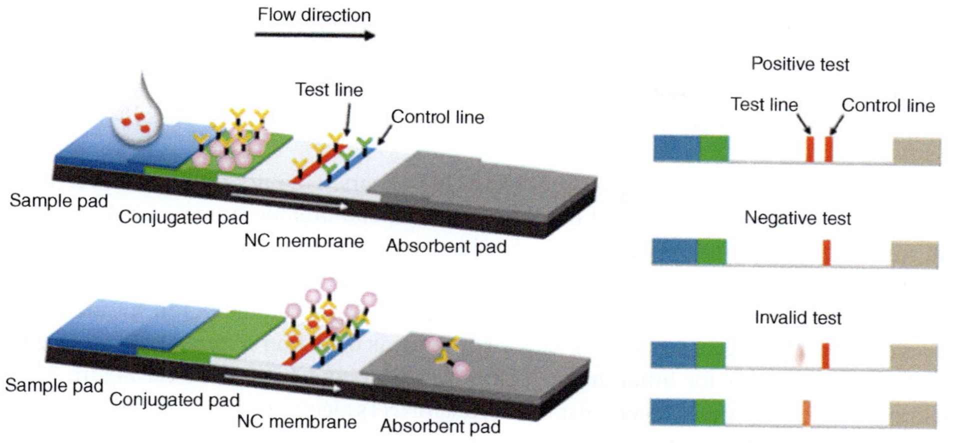

Figure 12.1 LFA schematic representation, open access journal. Kim et al. [14] / MDPI / CC BY 4.0.

Table 12.1 LFAs of Au NPs coupled with various nanoparticles.

Au NPs-based colorimetry	Test line bioreceptor	Detection limit	References
Gold@platinum nanorods	Hepatitis B virus DNA	8.5 pM	[17]
Au/Ti_3C_2TX	HIV DNA	0.01 nM	[18]
SiO_2@Au	Ricin Staphylococcal enterotoxin B, botulinum neurotoxin type A	0.1 0.05 0.1 ng/mL	[19]
Au@silica	Cancer cells microRNA-21	1.0 pM	[20]
SiO_2@Au/QDs	SARS-CoV-2 virus	1 and 0.033 ng/mL	[21]
Au@Fe_3O_4	SARS-Cov-2 spike	1.22 pg/mL	[22]
Au@Cu	Clenbuterol		[23]

even better sensitivity and multimodal biosensing, i.e. coupling with other biosensing modalities such as surface-enhanced Raman spectroscopy (SERS). Table 12.1 lists a few studies of Au NPs coupled with other nanoparticles for LFAs.

The application of Au NPs is not limited to LFAs; it extends to other biosensors, i.e. electrochemical, optical, and surface plasmon resonance. The popularity of Au NPs in the mentioned biosensors is driven by their high optical properties, surface plasmon resonance, fluorescence quenching ability, and robust electron transfer affinity [24]. The biosensor is characteristically made of a bioreceptor, a transducer, and the analyte. The bioreceptor, antibodies, enzymes, proteins, cells, nucleic acids, are immobilized on the surface of the Au NPs and will recognize and interact with the analyte, producing biological signals. The transducer converts the biological signals to electrical signals that are displayed on a screen [24]. The biosensors are merited with *in situ* real-time sensing and multiplexity and are gaining popularity in the global market driven by the healthcare needs and customized medicines. The versatility of Au NPs is of interest in the biosensing field. Au

NPs are synthesized using various techniques, yielding different sizes and shapes with a plethora of applications. This versatility in size and shape proffers different sensing mechanisms in biosensing. For example, smaller-sized Au NPs offer high sensitivity and a quenching effect, while their larger counterparts have better amplification properties [24] for use in SERS.

The Au NPs are explored in electrochemical biosensors. Huang et al. [25] developed an electrochemical biosensor based on a conductive metal-organic framework decorated with Au nanoparticles for early detection of cancer. Real-time *in situ* testing was conducted for detection of H_2O_2 released by human colon cancer cells and proved its capability in biosensing with sensitivity. Park et al. [26] developed a Au-decorated boron nitride nanotubes (Au@BNNTs) electrochemical biosensor towards detection of metalloproteinase-9 (MMP-9), a biomarker for inflammatory, cardiovascular, and lung diseases. Figure 12.2 illustrates the sensor and its mechanism. The Au@BNNTs electrode was immobilized with the MMP-9 aptamer to capture the MMP-9, and it displayed sensitive performance via electrochemical impedance spectroscopy and cyclic voltammetry. Also, a linear detection of MMP-9 was achieved with a detection limit of 0.0218 ng/mL.

Optical biosensors benefit from plasmonic nanoparticles. Nanoparticles such as Cu, Au, and Ag interestingly improve the signal of traditional Raman spectroscopy via their roughened surface, and this phenomenon is SERS. In biosensing, the SERS immunoassay/probe inherits the format of an enzyme-linked immunosorbent assay (ELISA). SERS probes can be labeled with a SERS tag or nonlabeled; the former confirms the biological molecule

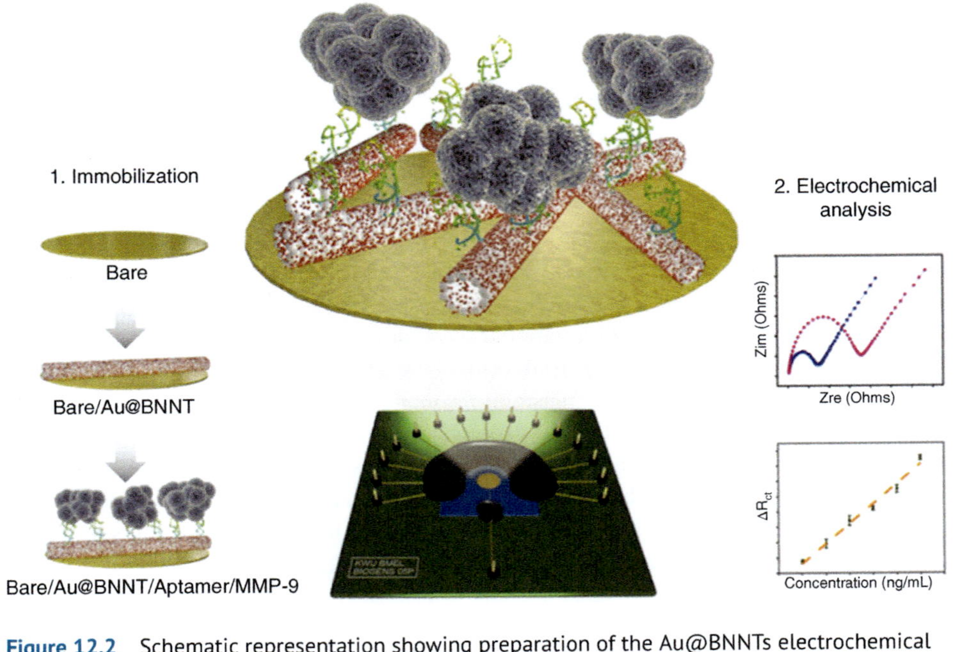

Figure 12.2 Schematic representation showing preparation of the Au@BNNTs electrochemical biosensor for MMP9, sensing mechanism and snippets of results. Park et al. [26] / with permission of Elsevier.

via the presence of the SERS tag, while the latter gives the Raman characterize peaks of the direct target/biomolecule [27]. The author and coworkers developed a bimetallic SERS substrate [27] used to immobilize *Plasmodium falciparum* sandwich for quantitative and qualitative detection of malaria [28].

Carbon materials such as quantum dots (QDs), carbon nanotubes, and graphene are used in biosensing owing to their properties [29]. QDs are nanosized, zero-dimensional particles endowed with unique optoelectronic properties attractive for QB and bioimaging. The QDs are merited by a broad absorption band with a high extinction coefficient, narrow symmetric emission bands, high photoluminescence, high fluorescence quantum yield, and enhanced photostability [30–32]. A size reduction to nanoscale in QDs results in quantum confinement and versatile optical properties, i.e. an increase in confinement results in an increased band gap and a blue shift. This enables alteration of the photoluminescence properties by modification of the size [31]. Ultraviolet excitation of QDs excites some electrons from the valence band, and they migrate to the conduction band, where they move freely and conduct electricity. When the electrons lose the excitation energy, they revert to their positions in the valence band, and this phenomenon yields photons. The emitted photons exhibit colors determined by the QDs' energy band gap [32]. The smaller-sized QDs with a high energy band gap emit blue photons while larger QDs have low energy gap and are characterized by red emissions [32]. QDs exist in different compositions, such as CdSe, CdTe, CQDs, PbQDs, and AgQDs, which determines their spectral range emission [31] and confirms their vast biosensing potential.

QDs are applied in POCT, environmental, and medical imaging. In QB, they are used as transducers and proffer the following merits: high sensitivity resulting in lower detection limits when compared with other QB nanoparticles such as Au NPs. Also, based on their myriad emission wavelength, they have multiplexing potential. However, issues such as toxicity and instability negate their use [32]. Advancement in surface chemistry offers mitigation of the issues and yields different architectures of QDs with enhanced stability and toxicity.

QB utilizes the fluorescent capability of QDs in sensing. The fluorescent-based biosensors, when compared to fluorophore dyes, offer more stability and extended fluorescence durations with an enhanced yield. An interaction between the QDs and biomolecules results in a loss or decreased fluorescence. The interaction of the two entities gives four mechanisms: static—interactions yield a nonfluorescing complex; dynamic—rapid energy is transferred between the quencher and QDs; inner filter effect—absorption of emission by molecule and the Forster resonance energy transfer—decreased donor emission; and increased acceptor emission [32].

QDs have potential in both *in vitro* and *in vivo* biosensing, although the latter is hindered by their high toxicity. However, traditional *in vivo* biosensing is revolutionized with wearable biosensors. Wearable biosensors seek to detect biomarkers with zero/minimal pain, leveraging the abundant body excretion, sweat. Chang et al. [33] explored the ability of QDs in wearable sensors. Nitrogen-doped graphene (N-GQDs) coupled with metal-organic framework-derived NiCo-based double-layered hydroxides were precursors in the fabrication of the wearable. The wearable monitors hypoxia from sweat during anaerobic activities. Inclusion of the NGQDs improved charge transfer and enhanced the sensitivity of the sensor. Other carbon materials are also widely studied in biosensing; carbon paste sensor electrodes are applied in the detection of riboflavin [34], drugs such as cetirizine and paracetamol [35], and alloxan, a toxic glucose analogue [36].

Magnetic nanoparticles (MNPs) are of particular interest in the biosensing field, driven by their key properties of large surface area, fluorescing ability, and superparamagnetism. MNPs in biosensing utilize the already mentioned architecture/mechanism of biosensors, i.e. optical, electrochemical, magnetic, fluorescence, and colorimetric [37]. With the urgency to promote POCT for such times, MNPs are excellent POCT candidates, and their application is even observed in lab-on-chip, LFAs, and smaller and larger field gadgets. The interest in MNPs is sparked by their high biomolecule affinity, selectivity, large surface area, low toxicity, biocompatibility, affordability, magnetic enrichment, and separation [37]. These strengths have resulted in the fabrication of immunoassays not only limited to biosensing and bioimaging but also extending to drug targeting. MNPs in biosensing of biological pathogen proffer rapidity, sensitivity, selectivity, and user-friendliness. MNPs are also useful in sample preparation, which informs the quality of sensing. Typically, biomolecules exist in complex natural media; MNPs merited with magnetic fields are utilized for separation and enrichment [37].

12.3.1 Miniaturized and Smart Biosensors

Optical and electrochemical biosensors mainly employ lab-based spectroscopy or microscopy, which negates their field application. These biosensors, unlike LFAs, are

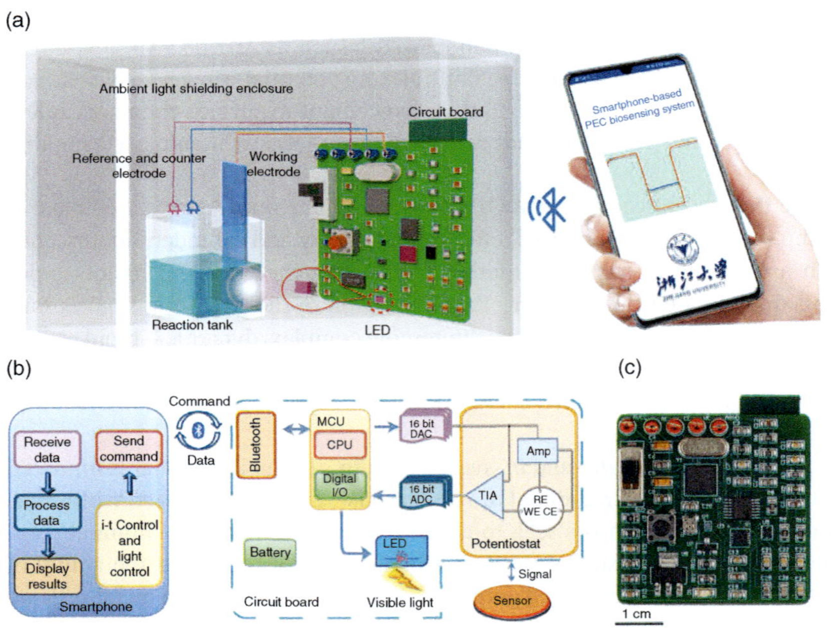

Figure 12.3 Depiction of PEC smartphone-integrated sensor. (a) PEC system with smartphone, enclosure cabin, circuit board with chip LED, 3-electrode system, and reaction tank. (b) The working principle of the PEC system. (c) The designed circuit board. Reproduced with permission from Zhang et al. [38] / Elsevier.

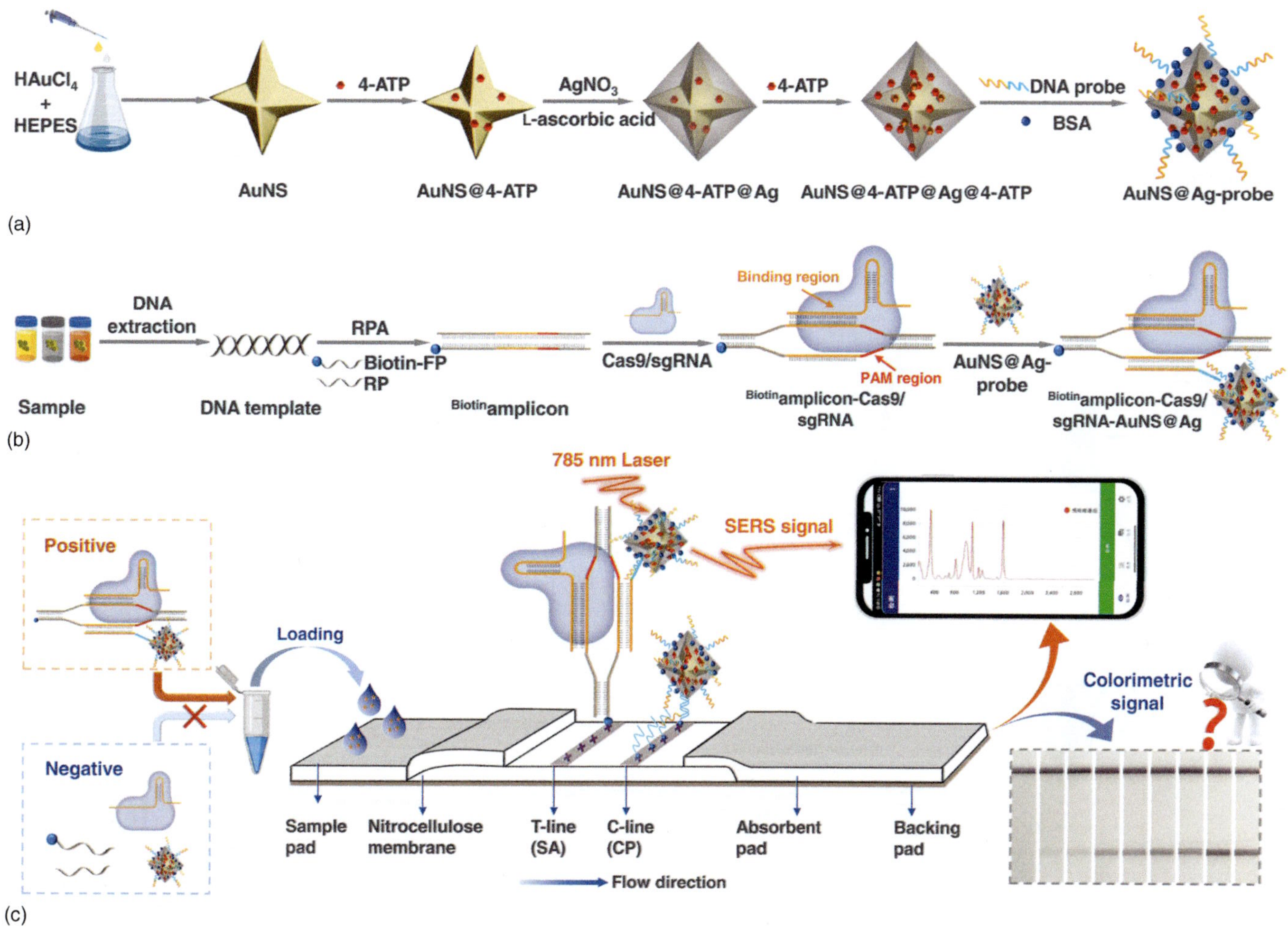

Figure 12.4 Depiction of the dual LFA/SERS platform for detecting *Staphylococcus aureus*. (a) Schematic of the Ag-decorated Au and their bioconjugation to the DNA probe. (b) DNA extraction and synthesis depiction of the formation of a positive test immunoassay sandwich. (c) LFA/SERS integrated system test flow [40]. Reproduced with permission from Wang et al. [40] / Elsevier.

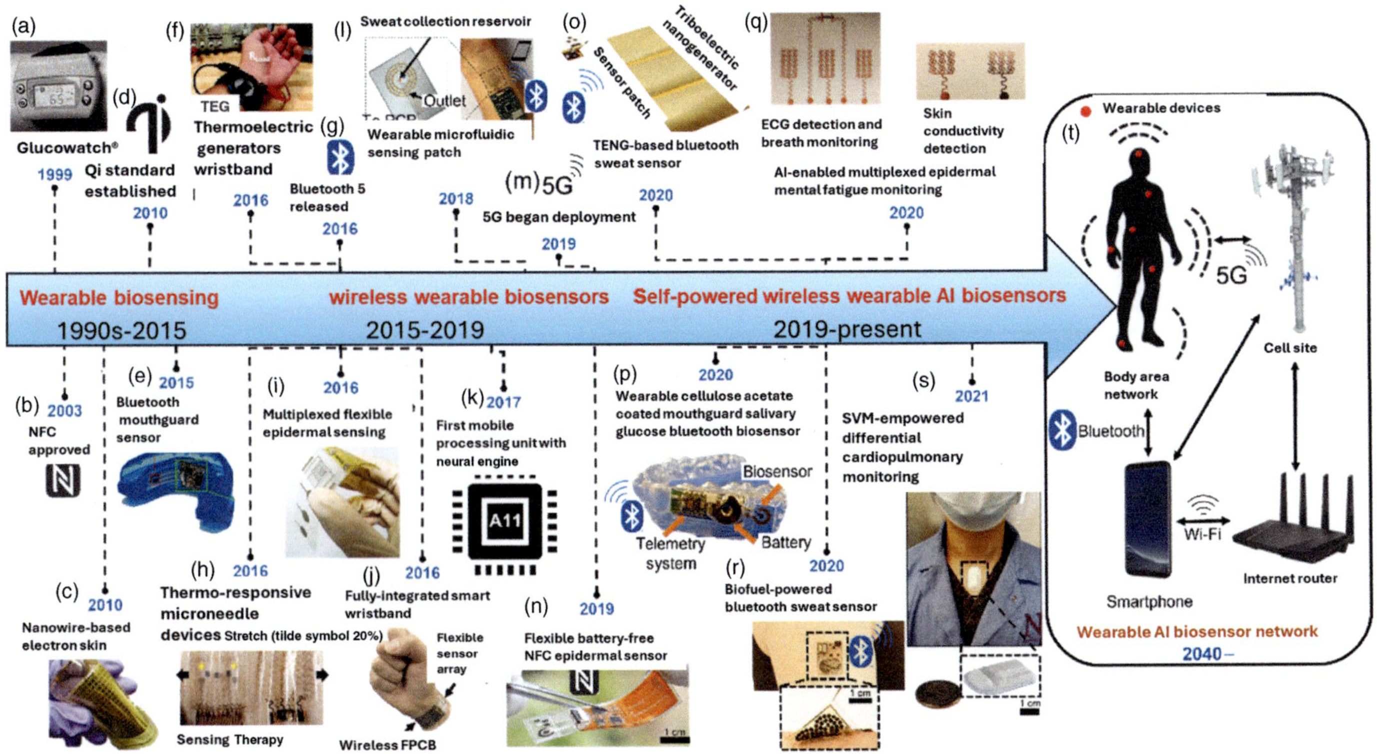

Figure 12.5 A timeline of the development of wearable biosensors from 1990s to 2015 wearable biosensing e.g. GlucoWatch; 2015–2019, wireless wearable biosensors e.g. wearable microfluidic sensing path; 2019-present, self-powered wearable AI biosensors. Zhang et al. [42] / Elsevier / CC BY 4.0.

endowed with quantification capabilities that could benefit field application. Smart biosensors and miniaturization of larger microscopies are envisioned to facilitate this transition. For example, a portable and versatile smartphone-based photoelectrochemical (PEC) biosensor for the detection of matrix mettaloproteinase-2 (MMP-2) proves the introduction of lab-based tests into the field. The sensor has a miniaturized circuit board and is integrated into a smartphone. Experimentally it is made of graphitic carbon nitride and Au NPs loaded with indium tin oxide for photoactive activity signal amplification. The Au NPs, because of their superior surface biofunctionalization, are conjugated with MMP-2-specific peptide modified with bovine serum albumin (BSA) on the terminal. On exposure to the target, part of the peptide hydrolyzes and is cleaved off together with BSA, and this yields a steric hindrance that triggers an increase in PEC current. The system was tested on buffer and artificial serum and had enhanced sensitivity with a detection limit of 2.48 and 0.55 pg/mL, respectively [38]. Figure 12.3 schematics show the smartphone-integrated system.

A novel one-pot SERS immunoassay for detecting SARS-CoV-2 spike protein was developed by Kaladharan et al. [39]. The system, via the use of miniaturized Raman and the elimination of washing steps, qualifies as a POCT. Solid support, digital versatile disks, moulded intopillars were coated with SERS-labeled Ag nanoparticles for double clamping. SERS activities were tested on spike protein and virus-like protein (VLP) in buffer and had a detection limit of 5 pg/mL, while untreated saliva spiked with VLP had a detection limit of 400 pg/mL. Wang et al. [40] report on a colorimetric LFA coupled with SERS for ultrasensitive detection of *Staphylococcus aureus*. The test is designed for POCT via integration of Raman into a smartphone as shown in Figure 12.4.

Artificial intelligence (AI) in biosensing is an avenue of interest in affording accurate, precise, and sensitive diagnostics. AI algorithms are used in biosensing for data interpretation and manipulation. Cheng et al. [41] coupled SERS biosensing with deep learning and DNN visualization algorithms to develop a digital biomarker of hepatocellular carcinoma for visualization and spectral characterization. An extension of AI to wearable biosensors is set to revolutionize traditional biosensing. Figure 12.5, reproduced from ref [42], shows a timeline of the growth of wearable biosensors to the future.

12.4 Nanoparticles-Enabled Bioimaging

Bioimaging encompasses the noninvasive real-time recording of biological processes and image capture of biological tissues enabled by the use of different modalities via contrast agents. The imaging modalities—magnetic resonance, optical, ultrasound, photoacoustic, X-ray computed, and nuclear imaging—use unique materials and contrast agents to produce images or videos of the biological sites [43, 44]. Bioimaging fundamentally yields early, accurate detection of disease and informs the treatment plan and monitors the response of the patient to the treatment [44]. Various quantum nanoparticles are eligible imaging contrast agents merited by biocompatibility, low toxicity, longer blood circulation half-life, high quantum yield, and surface functionalization [44, 45]. QDs are known for their high photoluminescence quantum yield, optoelectrical properties, a wide range of photon emission wavelengths determined by a tunable bandgap, high photostability, bright fluorescence with a long half-life, and strong light absorbance [46]. These alluring attributes are suitable for bioimaging. As such, reports on the application of both *in vitro* and *in vivo* QDs in bioimaging are well documented. However, *in vivo* demand use of biocompatible and nontoxic QDs is inverse to *in vitro*, and their toxicity hinges on size, charge, concentration, capping agents, functional groups, and photolytic and mechanical stability [46]. These attributes are mainly controlled during the synthesis and surface functionalization stages.

Surface functionalization with appropriate agents yields biocompatible QDs. Carbon QDs, a class of QDs, proffer high biocompatibility and low toxicity coupled with low cost and upscalable production. Surface functionalization of the CQDs with β and γ cyclodextrines improved their toxicity and maintained their luminescence quantum yield and absorption molar coefficient [47]. The green synthesis approach of nanoparticles is attractive for imaging purposes, although its upscaling is still a bottleneck. An eco-friendly synthesis approach of one-step hydrothermal conversion of sustainable *Coccinia grandis* into CQDs was adopted by Karuppasamy et al. [48]. Banana biomass via heat treatment also resulted in high fluorescing biocompatible carbon dots [49]. Chen et al. [50], using the green synthesis vehicle, successfully synthesis red-emitting carbon dots (rCD) merited with low toxicity, impressive biocompatibility, and water solubility. They used a one-step hydrothermal method using neutral red and L-arginine precursors to prepare the rCD with an average diameter of 2.34 nm. The rCD was tested for bioimaging and antibacterial properties. For bioimaging, zebrafish and Swiss mice were both exposed to rCD: zebrafish were fed with 1 mg/mL at ambient temperature for 30 minutes; Swiss mice were intraperitoneally injected with varying concentrations (5, 10, and 15 mg/kg) for 48 hours. Imaging was done with a fluorescence microscope and confirmed the fluorescing ability of the rCD as shown in Figure 12.6. Similarly, their cytotoxicity, tested on A549 cells, proffered insignificant cell changes with exposure to a concentration of 0–1.0 mg/mL rCD.

Plasmonic nanoparticles, Au, Ag, and Cu, exhibit fluorescence properties enabling application in fluorescent imaging in addition to optical imaging. Fluorescence imaging is a highly sensitive modal that uses near-infrared wavelength (700–750 nm) excitation and an emission within the range of 750–822 nm. It is negated by failure to penetrate to deep tissues, which is possible with an NIR-IIb range of 1500–1700 nm [45]. The rich attributes of the nanoparticles afford multimodal imaging, i.e. the amalgamation of the different

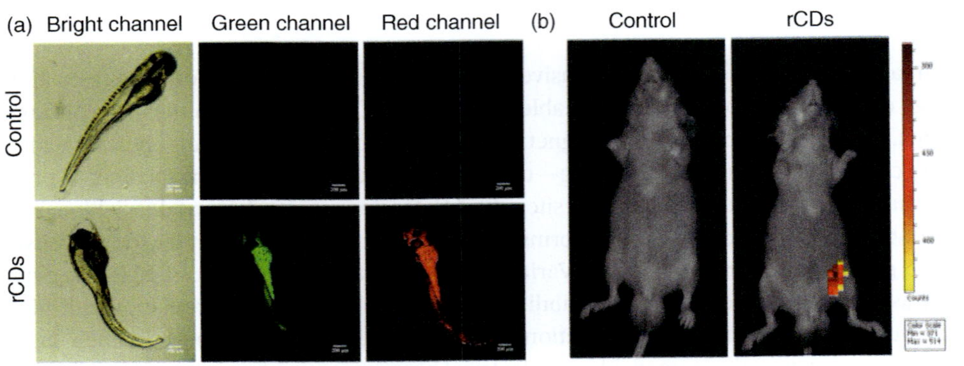

Figure 12.6 Imaging of (a) rCD-treated zebrafish and (b) Swiss mice. Reproduced with permission from Chen et al. [50] / Elsevier.

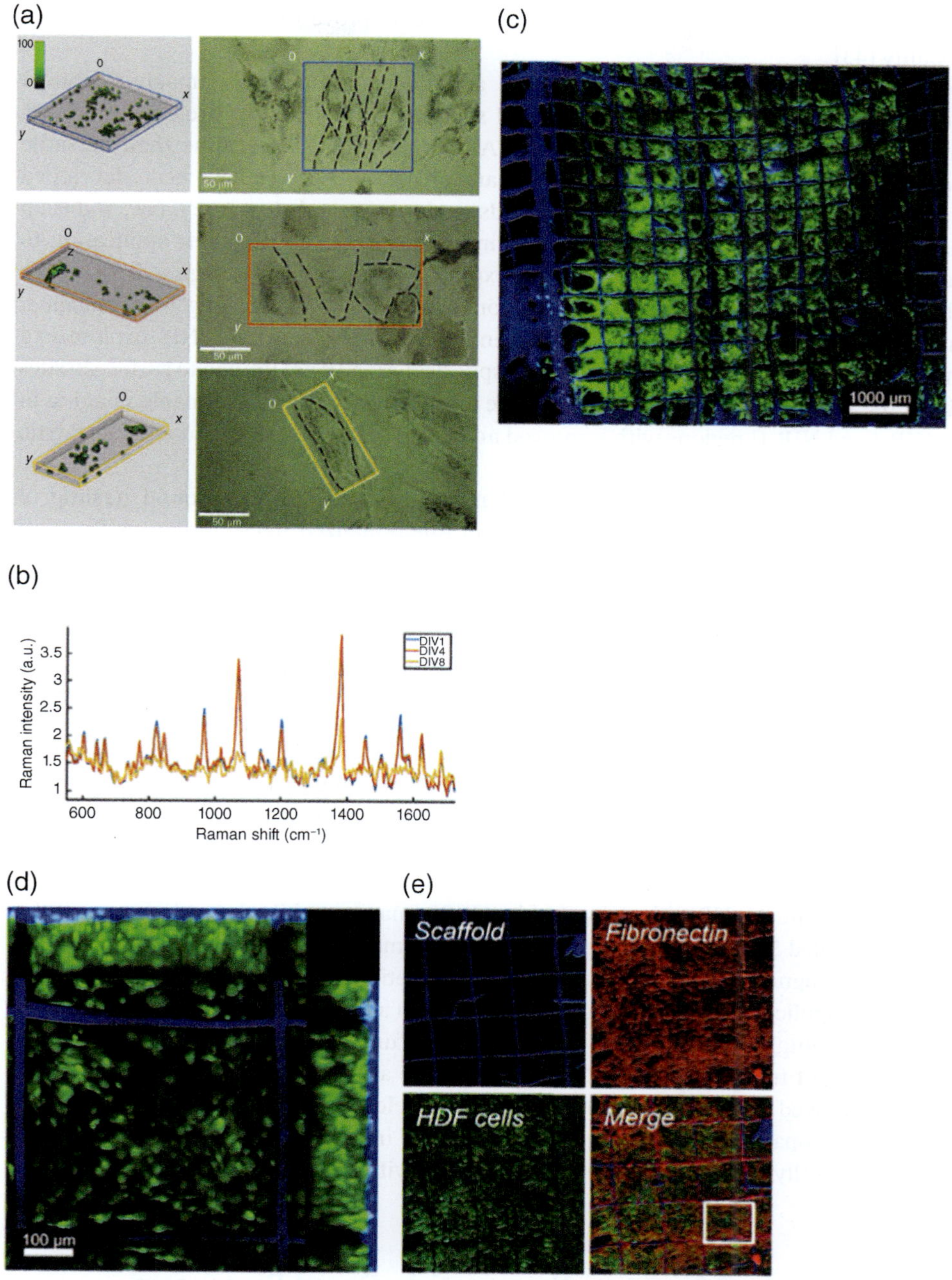

Figure 12.7 The dual 3D SERS-fluorescence imaging. (a) Left-volume reconstructions for SERS confocal imaging; right-bright field optical images. (b) Average SERS spectra calculated from imaging volume. Cell networks formed in 3D jet-printed scaffolds after. (c) Entire scaffold tile image filled with cells. (d) Maximum intensity projection (MIP) in the scaffold. (e) MIPs of Z-stacks, individual components, and merged image. Lenzi et al. [54] / American Chemical Society / CC BY-NC-ND 4.0.

imaging modalities to yield enhanced imaging and improve on challenges of a single modality [43].

Of the plasmonic nanoparticles, Au NPs are the most studied bioimaging contrast agents owing to their outstanding chemical stability, biocompatibility, easier synthesis, and versatile optical properties. The many Au configurations determine their versatile properties, and a plethora of different sizes and shapes of Au NPs have been developed to date. Their shapes range from spheres, rods, and stars to rods to name a few, and each shape at a given size yields different appealing properties. As such, their application to many diverse fields is unlimited and worth exploring. Au NPs in bioimaging drive on the following modalities: localized surface plasmon resonance (LSPR) imaging, SERS imaging, and luminescence imaging [51]. The different shapes and sizes of Au NPs explore and improve challenges in imaging; for example, spherical Au NPs are limited to surface tissues within the near-infrared region. However, the anisotropic counterparts enable imaging in the NIR-I and NIR-II regions with improved and better localized electron density, catalytic effect, and electromagnetic field [51].

Guo et al. [52] report on a multi-metal nanotheranostic toward wound healing of diabetic wounds. The nanotheranostic, aptamer functionalized Ag@cupriferous Prussian blue (ACPA) imaging system mobilizes the ACPA for programmable cooperative wound healing, while near-infrared mild hyperthermia triggers the release of the ions that are used for antipathogen, cell migration, and revascularization. The wound healing process is monitored via SERS imaging.

SERS imaging is also coupled with fluorescence imaging, and traditionally lanthanide-doped fluoride nanoparticles fluorescent nanoprobes are functionalized with Raman reporters via aryl diazonium salt chemistry. The dual-modal SERS-coded lanthanide-doped fluoride nanoprobes showed SERS activity for cell bioimaging [53].

Advancements in imaging are progressing to 3D imaging. Lenzi et al. [54] proposed a 3D imaging platform that inherently monitors real-time proliferation of cells using 3D scaffolds of poly(lactic-*co*-glycolic acid) immobilized with fluorescent labels and SERS-active Au NPs (Figure 12.7). The Au NPs primarily enabled SERS imaging, while the architecture of the scaffolds enabled 3D growth of the cells for confocal imaging.

Superparamagnetic nanoparticles have rich magnetic and optical properties attractive for imaging applications. In imaging they have shown a long blood half-life that promotes extended imaging duration, *in vivo* stability, and biofunctionalization to enhance uptake and targeting of imaging tissues [43]. The MNPs are also coupled with other nanoparticles for improved application. Gupta et al. [55] developed multifunctional Fe_3O_4@ZnO core-shell nanoparticles toward the treatment and imaging of cancer cells. The core shells successfully induced reduction on HeLa cells with potential for fluorescent cellular imaging.

12.5 Outlook of Quantum Nanoparticle-Based Biosensors

POCT expedites diagnosis of disease causatives and adheres to the WHO ASSURED standard. Commercialized POCT and LFAs drive qualitative detection of diseases, while expensive lab-based assays such as polymerase chain reactions and spectro-/microscope prowess both qualitative and quantitative testing. The LFAs mobilizes quantum nanoparticles for colorimetric purposes. Quantum nanoparticles, due to their quantum confinement phenomenon, exhibit properties harnessed for biosensing and bioimaging. Nanoparticles are drivers of the future of biosensing owing to their versatile properties merited for myriad biosensing and bioimaging architectures and mechanisms. Their different shapes and sizes find application in the synthesis of POCT and lab-based assays, and the latter are extending to POCT. Smart and miniaturized biosensors are the future of diagnostics.

The quantum nanoparticle-based biosensors stand to revolutionize biosensing. Fundamentally, POCT tests should be highly sensitive, accurate, and rapid in addition to other key measures. The biosensors driven by quantum nanoparticles can attain ultrasensitivity, rapidity, and accuracy through integration of artificial intelligence (AI). Current studies are exploring the integration of AI algorithms in optical biosensors to enhance data manipulation and processing. Another avenue in biosensing that is going to benefit from quantum nanoparticles is wearables. Wearables offer rapid, precise, real-time monitoring, and the optoelectrical properties of the nanoparticles benefit their fabrication. In conclusion, quantum nanoparticles merited with excellent versatile optoelectrical properties are the future for precise, accurate, ultrasensitive field-based biosensors. Although challenges such as toxicity, up scalability, and many more still need attention, their promising application should not be overlooked but explored in parallel with challenge mitigations.

Acknowledgments

The author acknowledges Mintek, Department of Science and Technology and Innovation, South Africa.

References

1 Mhlanga, N. and Ntho, T. (2021). Computational theories used in the study of quantum dots. *Quantum Dots: Properties and Applications* 96: 113–144.

2 Sutcliffe, B.T. and Woolley, R.G. (2012). On the quantum theory of molecules. *The Journal of Chemical Physics* 137 (22): 22A544.

3 Onishi, T. (2022). Hartree–Fock Equation. In: *Ferroelectric Perovskites for High-Speed Memory: A Mechanism Revealed by Quantum Bonding Motion* (ed. T. Onishi), 105–123. Springer.

4 Ramalingam, G., Kathirgamanathan, P., Ravi, G. et al. (2020). Quantum confinement effect of 2D nanomaterials. In: *Quantum Dots-Fundamental and Applications* (ed. F. Divsar), 1–12. IntechOpen.

5 Bera, D., Qian, L., Tseng, T.-K., and Holloway, P.H. (2010). Quantum dots and their multimodal applications: a review. *Materials* 3 (4): 2260–2345.

6 Das, S., Mazumdar, H., Khondakar, K.R. et al. (2024). Quantum biosensors: principles and applications in medical diagnostics. *ECS Sensors Plus* 3 (2): 025001.

7 Acharya, K., Salvi, B.V., and Shende, P. (2025). Strategic aspect for improving the sensitivity of nanoparticles in biomedical sciences. *International Journal of Polymeric Materials and Polymeric Biomaterials* 74 (1): 54–66.

8 Chen, F., Zhang, R., Xing, H. et al. (2023). Highly sensitive metal-enhanced fluorescence platform based on high-quality CdSe/ZnS quantum dots and gold nanodendrites. *Chemical Engineering Journal* 477: 147240.

9 Shahazi, R., Saddam, A.I., Islam, M.R. et al. (2024). Recent progress in Nanomaterial based biosensors for the detection of cancer biomarkers in human fluids. *Nano Carbons* 2 (2): 1254.

10 Li, L., Wang, T., Zhong, Y. et al. (2024). A review of nanomaterials for biosensing applications. *Journal of Materials Chemistry B* 12 (5): 1168–1193.

11 Holzinger, M., Le Goff, A., and Cosnier, S. (2014). Nanomaterials for biosensing applications: a review. *Frontiers in Chemistry* 2: 63.

12 Naresh, V. and Lee, N. (2021). A review on biosensors and recent development of nanostructured materials-enabled biosensors. *Sensors* 21 (4): 1109.

13 Thwala, L.N., Ndlovu, S.C., Mpofu, K.T. et al. (2023). Nanotechnology-based diagnostics for diseases prevalent in developing countries: current advances in point-of-care tests. *Nanomaterials* 13 (7): 1247.

14 Kim, J., Shin, M.-S., Shin, J. et al. (2023). Recent trends in lateral flow immunoassays with optical nanoparticles. *International Journal of Molecular Sciences* 24 (11): 9600.

15 Guliy, O.I. and Dykman, L.A. (2024). Gold nanoparticle–based lateral-flow immunochromatographic biosensing assays for the diagnosis of infections. *Biosensors and Bioelectronics: X* 17: 100457.

16 Chen, X., Zhan, W., Duan, H. et al. (2024). Crescent-shaped Janus nanoassemblies of gold nanoparticles and AIEgens: enhancing plasmonic and fluorescent activities for colorimetric and ratiometric fluorescence lateral flow immunoassay. *Nano Today* 58: 102452.

17 Li, N., Xi, X., Zhu, J. et al. (2022). High sensitivity and rapid detection of hepatitis B virus DNA using lateral flow biosensors based on Au@Pt nanorods in the absence of hydrogen peroxide. *The Analyst* 147 (3): 423–429.

18 Cheng, S., Luo, L., Bao, M. et al. (2025). A dual-mode colorimetric/photothermal lateral flow biosensor based on Au/Ti_3C_2TX for HIV-DNA detection. *Analytica Chimica Acta* 1338: 343588.

19 Jia, X., Wang, K., Li, X. et al. (2022). Highly sensitive detection of three protein toxins via SERS-lateral flow immunoassay based on SiO_2@ Au nanoparticles. *Nanomedicine: Nanotechnology, Biology and Medicine* 41: 102522.

20 Dong, T., Yin, R., Yu, Q. et al. (2021). Sensitive detection of microRNA-21 in cancer cells and human serum with Au@ Si nanocomposite and lateral flow assay. *Analytica Chimica Acta* 1147: 56–63.

21 Han, H., Wang, C., Yang, X. et al. (2022). Rapid field determination of SARS-CoV-2 by a colorimetric and fluorescent dual-functional lateral flow immunoassay biosensor. *Sensors and Actuators B: Chemical* 351: 130897.

22 Guo, G., Zhao, T., Sun, R. et al. (2024). Au-Fe_3O_4 dumbbell-like nanoparticles based lateral flow immunoassay for colorimetric and photothermal dual-mode detection of SARS-CoV-2 spike protein. *Chinese Chemical Letters* 35 (6): 109198.

23 Hu, H., Tian, J., Shu, R. et al. (2024). A cheaper substitute for HRP: ultra-small Cu–Au bimetallic enzyme mimics with infinitesimal steric hindrance to promote catalytic lateral flow immunodetection of clenbuterol. *Lab on a Chip* 24 (8): 2272–2279.

24 Parkhe, V.S. and Tiwari, A.P. (2024). Gold nanoparticles-based biosensors: pioneering solutions for bacterial and viral pathogen detection—a comprehensive review. *World Journal of Microbiology and Biotechnology* 40 (9): 269.

25 Huang, W., Xu, Y., Wang, Z. et al. (2022). Dual nanozyme based on ultrathin 2D conductive MOF nanosheets intergraded with gold nanoparticles for electrochemical biosensing of H_2O_2 in cancer cells. *Talanta* 249: 123612.

26 Park, J.A., Park, E., Kwon, Y. et al. (2024). Synthesis of Au-decorated boron nitride nanotubes and its application to pretreatment-free electrochemical biosensor for matrix metalloproteinase9 in clinical sample. *Sensors and Actuators B: Chemical* 399: 134876.

27 Mhlanga, N., Domfe, T., and Skepu, A. (2019). Fabrication of surface enhanced Raman spectroscopy substrates on solid supports. *Applied Surface Science* 476: 1108–1117.

28 Mhlanga, N., Domfe, T., Skepu, A., and Ntho, T.A. (2020). Sandwich-based surface-enhanced Raman scattering probes for detection and quantification of malaria. *Journal of Raman Spectroscopy* 51 (12): 2416–2424.

29 Joshi, P., Mishra, R., and Narayan, R.J. (2021). Biosensing applications of carbon-based materials. *Current Opinion in Biomedical Engineering* 18: 100274.

30 Shen, Y., Jia, F., He, Y. et al. (2024). A CRISPR-Cas12a-powered, quantum dot-based and magnetic nanoparticle-assisted (QD-CRISPR-MNP) biosensor for the screening of *Salmonella*. *Microchemical Journal* 200: 110438.

31 Wegner, K.D. and Hildebrandt, N. (2024). Near infrared quantum dots for biosensing and bioimaging. *TrAC Trends in Analytical Chemistry* 180: 117922.

32 Quesada-González, D. and Merkoçi, A. (2025). Quantum dots for biosensing: classification and applications. *Biosensors and Bioelectronics* 273: 117180.

33 Chang, L.-Y., Rinawati, M., Guo, Y.-T. et al. (2024). Nitrogen-doped graphene quantum dots incorporated into MOF-derived NiCo layered double hydroxides for nonenzymatic lactate detection in noninvasive biosensors. *ACS Applied Nano Materials* 7 (12): 14431–14442.

34 Hareesha, N. and Manjunatha, J. (2020). A simple and low-cost poly(DL-phenylalanine) modified carbon sensor for the improved electrochemical analysis of Riboflavin. *Journal of Science: Advanced Materials and Devices* 5 (4): 502–511.

35 Pushpanjali, P., Manjunatha, J., Hareesha, N. et al. (2021). Voltammetric analysis of antihistamine drug cetirizine and paracetamol at poly(L-leucine) layered carbon nanotube paste electrode. *Surfaces and Interfaces* 24: 101154.

36 Monnappa, A.B., Manjunatha, J.G., and Bhatt, A.S. (2020). Design of a sensitive and selective voltammetric sensor based on a cationic surfactant-modified carbon paste electrode for the determination of alloxan. *ACS Omega* 5 (36): 23481–23490.

37 Wang, M., Jin, L., Hang-Mei Leung, P. et al. (2024). Advancements in magnetic nanoparticle-based biosensors for point-of-care testing. *Frontiers in Bioengineering and Biotechnology* 12: 1393789.

38 Zhang, Q., Chen, Z., Shi, Z. et al. (2021). Smartphone-based photoelectrochemical biosensing system with graphitic carbon nitride/gold nanoparticles modified electrodes for matrix metalloproteinase-2 detection. *Biosensors and Bioelectronics* 193: 113572.

39 Kaladharan, K., Chen, K.-H., Chen, P.-H. et al. (2023). Dual-clamped one-pot SERS-based biosensors for rapid and sensitive detection of SARS-CoV-2 using portable Raman spectrometer. *Sensors and Actuators B: Chemical* 393: 134172.

40 Wang, J., Jiang, H., Chen, Y. et al. (2024). CRISPR/Cas9-mediated SERS/colorimetric dual-mode lateral flow platform combined with smartphone for rapid and sensitive detection of *Staphylococcus aureus*. *Biosensors and Bioelectronics* 249: 116046.

41 Cheng, N., Lou, B., and Wang, H. (2023). Discovering the digital biomarker of hepatocellular carcinoma in serum with SERS-based biosensors and intelligence vision. *Colloids and Surfaces B: Biointerfaces* 226: 113315.

42 Zhang, Y., Hu, Y., Jiang, N., and Yetisen, A.K. (2023). Wearable artificial intelligence biosensor networks. *Biosensors and Bioelectronics* 219: 114825.

43 Mhlanga, N., Mphuthi, N., Van der Walt, H. et al. (2024). Nanostructures and nanoparticles as medical diagnostic imaging contrast agents: a review. *Materials Today Chemistry* 40: 102233.

44 Habeeb, M., Vengateswaran, H.T., Tripathi, A.K. et al. (2024). Enhancing biomedical imaging: the role of nanoparticle-based contrast agents. *Biomedical Microdevices* 26 (4): 1–18.

45 Mohammadi, T., Gheybalizadeh, H., Rahimpour, E. et al. (2024). Advanced photoluminescent nanomaterials for targeted bioimaging of cancer cells. *Heliyon*. 11: e41566.

46 Nabil, M. and Megahed, F. (2024). Quantum dot nanomaterials: preparation, characterization, advanced bio-imaging and therapeutic applications. *Journal of Fluorescence* 34 (6): 2467–2484.

47 Rodríguez-Varillas, S., Fontanil, T., Casado, J.E. et al. (2024). Surface modification of carbon dots with cyclodextrins as potential biocompatible photoluminescent delivery/bioimaging nanoplatform. *Analytica Chimica Acta* 1318: 342948.

48 Karuppasamy, B.D., Perumal, S., Atchudan, R. et al. (2024). Eco-friendly *Coccinia Grandis*-derived excitation-dependent fluorescence of carbon dots for in-vivo bioimaging and fluorescent ink applications. *Materials Science and Engineering B* 303: 117300.

49 Patel, D.K., Won, S.-Y., Jung, E. et al. (2024). Assessment of the biocompatibility and bioimaging potential of fluorescent carbon dots derived from waste biomass. *Materials Letters* 361: 136152.

50 Chen, J., Li, T., Lin, C. et al. (2025). Green synthesis of red-emitting carbon dots for bioimaging, sensing, and antibacterial applications. *Spectrochimica Acta Part A: Molecular and Biomolecular Spectroscopy* 328: 125458.

51 Si, P., Razmi, N., Nur, O. et al. (2021). Gold nanomaterials for optical biosensing and bioimaging. *Nanoscale Advances* 3 (10): 2679–2698.

52 Guo, Y., Li, Y., Fan, R. et al. (2023). Silver@ Prussian blue core–satellite nanostructures as multimetal ions switch for potent zero-background SERS bioimaging-guided chronic wound healing. *Nano Letters* 23 (18): 8761–8769.

53 Chen, H., Nizard, P., Decorse, P. et al. (2024). Dual-mode nanoprobes based on lanthanide doped fluoride nanoparticles functionalized by aryl diazonium salts for fluorescence and SERS bioimaging. *Small* 20 (10): 2305346.

54 Lenzi, E., Jimenez de Aberasturi, D., Henriksen-Lacey, M. et al. (2022). SERS and fluorescence-active multimodal tessellated scaffolds for three-dimensional bioimaging. *ACS Applied Materials & Interfaces* 14 (18): 20708–20719.

55 Gupta, J., Hassan, P., and Barick, K. (2021). Core-shell Fe_3O_4@ZnO nanoparticles for magnetic hyperthermia and bio-imaging applications. *AIP Advances* 11 (2): 025207-6.

13

Quantum Optical Sensors for Biomedical Applications

Cigdem Kanbes-Dindar[1,2], Nazife Aslan[1], and Bengi Uslu[2]

[1] *Chemistry Department, Polatlı Science and Arts Faculty, Ankara Hacı Bayram Veli University, Ankara, Turkey*
[2] *Department of Analytical Chemistry, Faculty of Pharmacy, Ankara University, Ankara, Turkey*

13.1 Introduction

Optical sensors are instruments that utilize the interplay between light and matter to detect alterations in the physical or chemical environment. These sensors may detect alterations in optical properties such as wavelength, phase, polarization, light intensity, and refractive index, subsequently analyzing the chemical or biological processes that ensue. These devices employ light for sensing, enabling rapid, sensitive, and non-contact measurements [1]. Optical sensors are utilized throughout various domains, including medicine, biotechnology, food safety, environmental monitoring, and military applications.

Also, the operation of optical sensors typically dictates their categorization. For studies involving DNA, proteins, or cells, fluorescence sensors are a common tool. They monitor the light that fluorophores give off when they are exposed to a given type of light. Surface plasmon resonance (SPR) sensors measure changes in refractive index caused by molecules that stick to a metal surface. They can do this without using any labels and in real time. Fiber optic sensors can detect changes or losses that happen when light moves through an optical fiber. These sensors are great for delicate jobs like measuring things inside the body because they are small, flexible, and safe. Photonic crystal sensors are also very picky because their periodic structures let them control the flow of light at certain wavelengths. Raman spectroscopy sensors, on the other hand, find things by looking at how light scatters when molecules vibrate in a certain way. One of the best ways to see how optical sensors have changed healthcare is in diagnostic systems. They provide quick, dependable, and less invasive treatments in a number of fields, including keeping an eye on medications, finding cancer biomarkers, measuring blood glucose levels, and identifying bacteria and viruses. These systems are also becoming more important because they can do important things like check for pollutants in the environment and watch food go bad. Optical sensors have a lot of great features. For example, they are very sensitive, not affected by electromagnetic interference, very small, able to analyze data in real time, and able to handle many signals at once. These traits make it possible to design biosensor systems that are cheap, convenient to carry, and simple to operate.

Quantum Biosensing in Medical Diagnostics, First Edition. Edited by J. G. Manjunatha.

Nanotechnology has significantly enhanced the performance of optical sensors. Nanotechnology enables the manipulation of material properties at the nanoscale, encompassing their optical, electrical, and chemical characteristics. Nanomaterials are employed in numerous sensor systems due to their significant advantages, including high sensitivity, specificity, quantum effects, and extensive surface area [2–4]. Specifically, they possess advantages including reduced detection limits, enhanced signals, and improved capture of target biomolecules [5, 6]. Consequently, optical sensors are now capable of detecting biomarkers at the nanoscale.

When examining the nanomaterials used in optical systems, gold and silver nanoparticles, carbon nanotubes, graphene oxide, quantum nanomaterials (QDs), metal-organic frameworks (MOFs), and silica nanoparticles are the most important nanomaterials used in optical systems. When examining the reasons for choosing these nanomaterials based on their intended use, gold and silver nanoparticles are used in optical sensors to facilitate the detection of target molecules because these nanoparticles provide high optical resonance on the surface. QDs, on the other hand, are quite popular in fluorescent sensors because they emit light in a narrow band and have high quantum efficiency. Furthermore, carbon quantum dots (CQDs) have high conductivity and biocompatibility, and their surfaces, which allow for modification of biorecognition elements, make these sensors more selective. The main reason for the interest in QDs is their narrow emission spectrum, which is necessary to meet the narrow emission range required for multiple imaging applications and to reduce spectral overlap [7]. Furthermore, the emission wavelength of QDs can be precisely modulated by varying their size, shape, or composition. For example, as the size of a QD increases, the band gap decreases due to decreased quantum confinement and greater orbital overlap, shifting the emitted light from the blue to the red region of the spectrum [8, 9]. Thus, this property, also known as the quantum size effect, allows QDs to emerge as nanomaterials capable of signaling across the entire visible spectrum and into the near-infrared (NIR) region, making them highly versatile for optical sensing and imaging applications. In addition, to overcome the optical stability problem, one of the reasons why nanomaterials are not suitable for optical applications, and to increase their photoluminescence efficiency, QDs are often synthesized with a core-shell architecture. In addition, to overcome the optical stability problem, which is one of the reasons why nanomaterials are not suitable for optical applications, and to increase their photoluminescence efficiency, QDs are often synthesized with a core-shell architecture. For example, CdSe/ZnS and InAs/GaAs, where a wider bandgap shell covers a narrow bandgap core, are common systems because their structural designs increase quantum efficiency and resistance to oxidation while suppressing surface trap states [10].

In another important application example, QDs can be tailored to the application by adjusting parameters such as core diameters, shell thicknesses, and core-shell band offsets in the biologically relevant NIR window (800–840 nm), where tissue absorption and autofluorescence are minimal [11]. However, this limits their direct applicability in biological environments because, despite their many superior optical properties, QDs are generally hydrophobic and cytotoxic in nature [12]. Therefore, to overcome this problem, surface modification strategies are applied to increase aqueous solubility, colloidal stability, and biocompatibility [13]. Among these strategies, encapsulation with amphiphilic polymers and ligand exchange with hydrophilic molecules are commonly preferred. Furthermore, these modified QDs enable selective labeling of specific cellular structures, molecular biomarkers, or target analytes. This selectivity facilitates both high-sensitivity imaging and quantitative analysis. Because these unique properties are not found in many other nanomaterials, QDs stand out as promising nanoplatforms for biological imaging and diagnostics, particularly for applications requiring long-term monitoring and multiple target detection.

This chapter elucidated the utilization of quantum optical sensors in biomedical applications. First, some important information about quantum materials, such as their characteristics and synthesis methods, was explained. The rationale for selecting quantum nanoparticles for biosensing applications was discussed. Furthermore, current studies illustrating quantum-based sensor applications in fluorescence spectroscopy, Raman spectroscopy, and imaging approaches were examined.

13.2 Properties, Synthesis, and Importance of Quantum Nanoparticle

The conventional definition of nanoparticles indicates that nanomaterials possess dimensions between 1 and 100 nm, with this size range significantly influenced by their elevated surface area-to-volume ratio and the existence of surface atoms [14]. For a nanomaterial to demonstrate quantum effects, it must be within the 1–10 nm range [15]. QDs are notable semiconductor nanostructures typically measuring between 2 and 10 nm in diameter [16]. The electronic and optical properties are significantly influenced by the size and shape of QDs [16–18]. Quantum nanoparticles, specifically QDs, exhibit unique size-dependent properties that render them highly valuable across various fields, particularly in biomedical applications (Figure 13.1). QDs demonstrate tunable optical properties characterized by size-dependent absorption and emission spectra, elevated photoluminescence quantum yields, and remarkable photostability [19, 20]. These characteristics render them suitable for applications that demand high sensitivity and specificity, including imaging and diagnostics [21, 22]. A nanomaterial needs to be low toxicity in order to be used in biosensors. QDs can be used in *in vivo* applications by altering their surface to make them more biocompatible and less toxic [22–24]. Furthermore, subsequent conjugation of QDs with biomolecules such as antibodies, peptides, proteins, or nucleic acids provides molecular specificity, facilitating the targeted detection of cellular markers or analytes. Therefore, QDs are biocompatible materials that facilitate real-time imaging of tissues and cells and provide high-resolution images for diagnostic purposes [25]. They can also be designed for targeted drug delivery, thereby increasing treatment efficacy while reducing side effects [26, 27]. Their small size and customizable surfaces enable precise delivery and controlled release of therapeutic agents. These distinctive properties enable QDs to be used in therapeutic applications, integrating therapeutic and diagnostic functions, enabling them to simultaneously treat and monitor diseases [28].

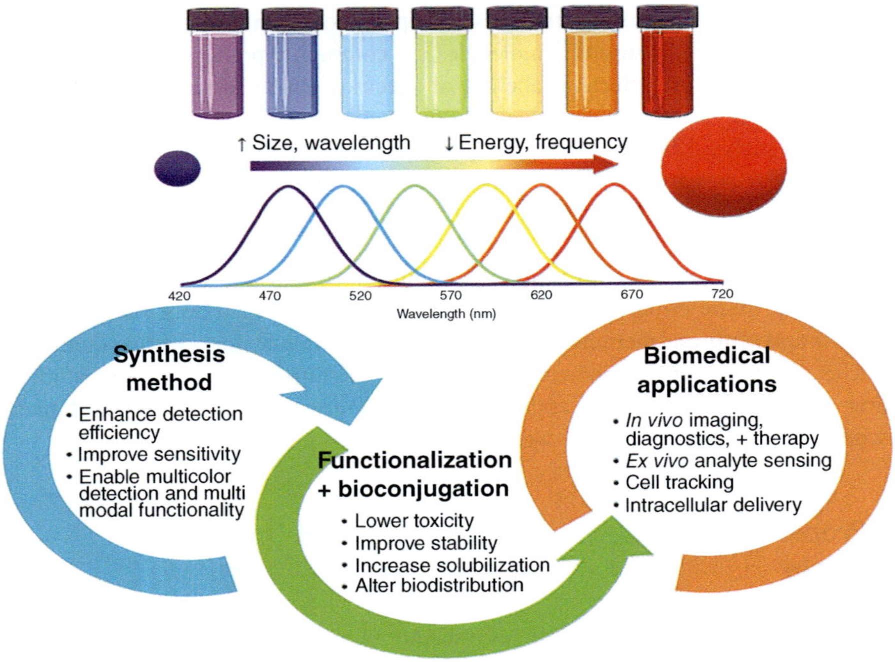

Figure 13.1 Sustainable design strategies for QDs for biosensing applications. Wagner et al. [12] / with permission of Elsevier.

There are several ways to make synthesis QDs, and each has its own pros and cons. Some methods may be better for certain uses than others. One reason high-temperature colloidal synthesis is so popular is that it can make QDs with the same structure and size distribution every time [29]. However, colloidal synthesis has more practical and cost-effective applications in solar cell, laser, and LED technologies [30]. According to Jangid and Kothari [31], the sol-gel technique enables precise regulation of particle size, porosity, and content via the two stages of hydrolysis and polycondensation. However, precise regulation of solvent, pH, and temperature is necessary for the easy and cheap production of quantum dots by co-precipitation [31]. The ability to precisely regulate nucleation and growth rates makes the hot injection approach critical for the fabrication of perovskite colloidal quantum dots. In contrast, laser ablation is a physical, top-down approach that produces high-purity quantum dots through extensive material degradation [32]. The solvothermal method synthesizes quantum dots with distinct optical properties by applying high-pressure and high-temperature conditions in organic solvents [33]. In the vapor deposition method, QDs are synthesized by depositing gaseous precursors onto substrates via organometallic nucleation centers where the cluster-seed mechanism is active, and then by chemical deposition, which can vary their doping and photophysical properties [34]. Mechanical fabrication and plasma synthesis provide fine-grained control over the size and optical properties of quantum dots [35]. Injection-free technologies are used in industrial applications such as photovoltaics and solid-state lighting because they allow stirred tank reactors to increase output more easily [36].

13.3 Applications of Quantum Optical Sensors for Biomedical Sensing

13.3.1 Utilizing Raman Spectroscopy for Quantum Optical Sensors for Biomedical Sensing

Raman spectroscopy is a good alternative for quantum optical biomedical sensors because it uses inelastic photon scattering to study the chemical structure and behavior of living organisms [37]. The Raman effect modifies the energy and wavelength of photons, causing inelastic scattering and energy transfer of molecular vibrational modes [38]. Each molecule has a unique fingerprint that indicates its structure, chemical environment, and vibrational frequencies based on the Raman shift (in wavenumbers) and the intensity of the scattered light [39]. Raman spectroscopy's sensitivity to molecular vibrational modes enables it to detect a wide range of chemical species as well as small structural changes in complex biological samples [40]. Biomedical applications benefit from the label-free nature of the technique, which eliminates the need for exogenous labels or contrast agents and allows for the study of dynamic processes in real time [41]. The low probability of Raman scattering events means that the signal strengths used in Raman spectroscopy are moderate [42]. Thanks to technological innovations in data processing, signal amplification, and light sources, the biomedical applications of Raman spectroscopy have increased significantly [43]. For example, it is possible to amplify weak Raman signals using signal enhancement techniques such as resonance Raman spectroscopy and surface-enhanced Raman scattering. Thanks to these signal enhancements, Raman spectroscopy has become applicable in a variety of applications, including the detection and measurement of disease biomarkers, monitoring drug efficacy and drug distribution in the human body, elucidating the structures of various tissues, and monitoring metabolic processes. In this context, amplifying the Raman signal and modifying the substrate surface with QDs have become popular options in recent years because the addition of quantum dots increases the sensitivity of the method. This combines quantum technology with Raman spectroscopy to provide highly selective and sensitive diagnostic tools for the detection of biological agents [44].

For instance, Lu and his team designed a rapid, sensitive, and noninvasive method for the detection of Helicobacter pylori using a hybrid nanostructure composed of gold nanorods (AuNRs), nanomica platelets (NMPs), and ZnO QDs that enhance Raman signals by both electromagnetic and chemical means [13]. Figure 13.2 illustrates the simple and effective steps of using surface-enhanced Raman spectroscopy (SERS) for the detection of *Helicobacter pylori*. To emphasize the importance of sampling and to obtain more realistic results, this study used a special sampling swab to puncture the stool at three different locations. The sample was thoroughly mixed, and then the AuNRs/NMPs/ZnO QDs SERS substrate was added. The mixture was shaken vigorously for 20 cycles and then left to stand for 30 minutes to allow the analyte and substrate to mix and interact evenly. A drop of the sample-substrate mixture is subsequently applied to an aluminum plate and allowed to air-dry (Figure 13.2a). Following the drying process, SERS analysis is conducted, facilitating rapid, noninvasive, and highly sensitive detection of *Helicobacter pylori* biomarkers while ensuring minimal patient discomfort (Figure 13.2b). Adenine was detected with great sensitivity using these nanohybrids (AuNRs/NMPs), which create a well-organized 3D hotspot

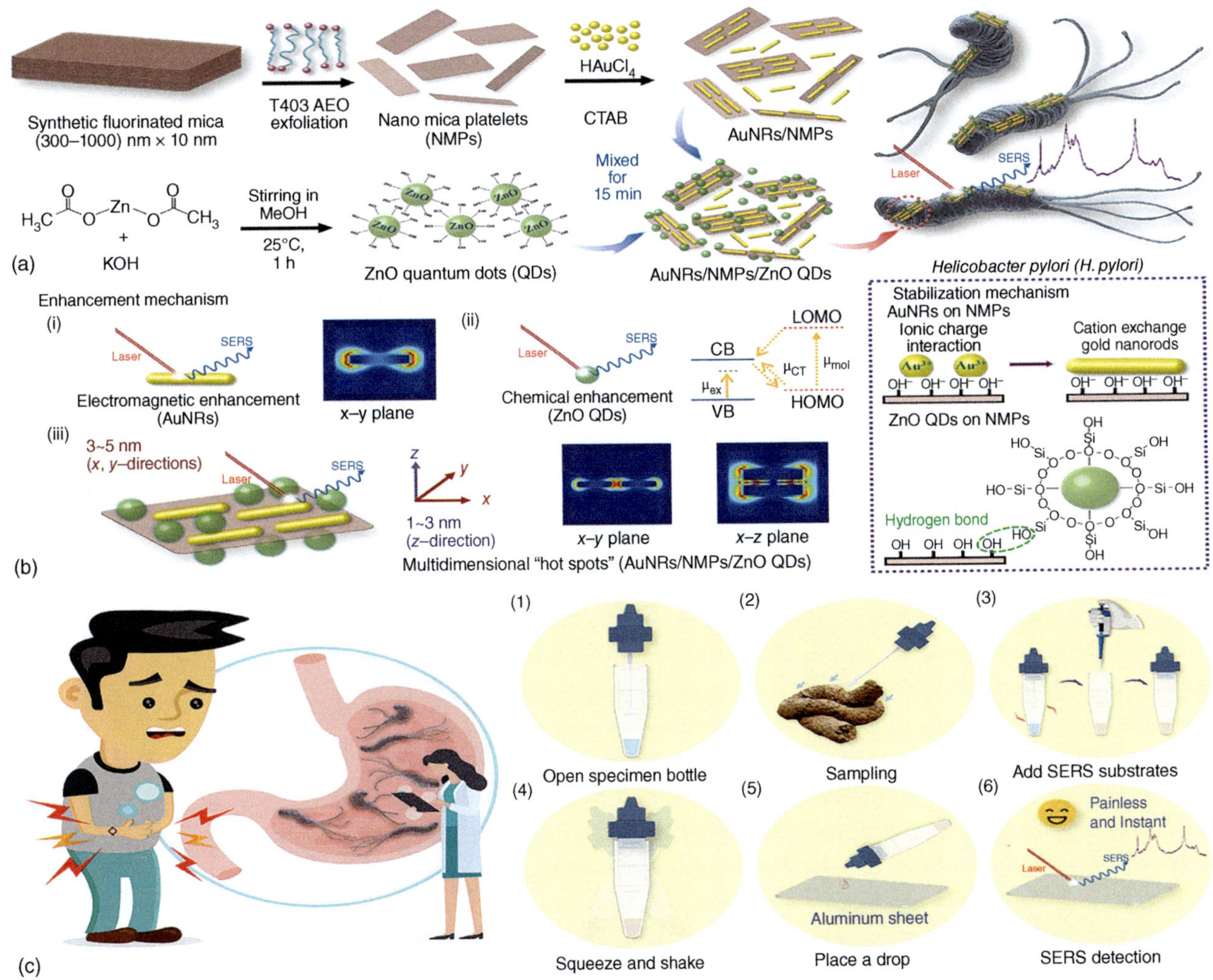

Figure 13.2 Illustration of the synthesis strategy and SERS enhancement principles involving AuNRs/NMPs/ZnO QDs nanohybrids, along with their application in improving fecal antigen detection protocols. Lu et al. [13] / John Wiley & Sons / CC BY 4.0.

structure with an enhanced Raman signal intensity (LOD = 10^{-9} M, enhancement factor (EF) = 2.0×10^8, reproducibility of platform (RSD) = 8.82%). The NMPs not only stabilize AuNR arrangement but also aid hotspot formation, while the integration of ZnO QDs introduces chemical enhancement via charge transfer mechanisms, boosting sensitivity even further (LOD = 10^{-10} M, EF = 1.6×10^9, RSD = 7.66%). A potential alternative to current diagnostic approaches utilizing electromagnetic and chemical amplification, this synergistic platform enables the dependable detection of *H. pylori* in stool samples up to 90 CFU/mL using SERS (Figure 13.2c).

Keshavarz et al. examined a unique organic, nitrogen-doped quantum-scale graphite nanostructure created in a nitrogen-rich environment using femtosecond laser-induced ionization, applicable in biosensing and the development of an optical *Raman* sensor [45]. The resultant self-assembled structure operates as an SERS-active biosensor with multifunctional properties. It attains an enhancement factor of 3.2×10^7 for crystal violet dye at a detection limit of 100 nM, while yielding about 6-fold and 12-fold enhancements for intracellular biomolecules, including DNA/RNA and proteins, respectively. This device facilitates live-cell biomolecular fingerprinting and real-time observation of transitory biochemical fluctuations. It further facilitates tissue-like development in fibroblast cells and specifically produces apoptotic-like, non-proliferative behavior in cervical cancer cells. These findings illustrate the potential of the N-G template for integrated cancer diagnostics and targeted therapeutic applications, underscoring a viable avenue for metal-free, biocompatible biosensor technology.

Staphylococcus aureus is a highly prevalent foodborne pathogen linked to numerous illness outbreaks. Therefore, its swift and precise identification is essential for protecting public health, particularly in the realms of food safety and clinical diagnostics. Lu et al. developed a novel nanohybrid biosensing substrate consisting of silver nanocubes (AgNCs), NMPs, and zinc ZnO QDs, aimed at ultrasensitive bacterial detection through SERS [46]. In a recently published study by Lu and coworkers, a novel nanohybrid biosensor substrate composed of surface-enhanced SERS silver nanocubes (AgNCs), NMPs, and zinc ZnO QDs was developed for the detection of *Staphylococcus aureus* bacteria at trace levels [46]. The hybrid nanomaterial selection was due to the three-dimensional structure of AgNC/NMP/ZnO QDs synthesized via a one-step polyol-mediated method, which is necessary for the activation of synergistic electromagnetic and chemical enhancement mechanisms, exhibiting a lightning rod effect and enabling efficient charge transfer. Subsequent optimizations resulted in significantly improved detection limits: 10^{-10} M for adenine and 10 CFU/mL for *Staphylococcus aureus*. Considering the potential impact of these results on clinical, environmental, and food safety sectors, it offers a scalable and multifunctional platform for real-time biosensor applications.

Aflatoxin B_1 (AFB_1), a potent carcinogenic mycotoxin produced by Aspergillus species, poses significant risks to food safety and public health, and highly sensitive detection strategies are required. Therefore, a dual-signal hybrid aptasensor-based biosensor combining SERS and fluorescence mechanisms was developed for the quantitative and qualitative determination of AFB_1 [47]. The sensor measurement mechanism utilized satellite nanostructures formed by hybridizing CdTe quantum dot-decorated silica nanoparticles with DTNB-labeled gold nanoparticles to maximize the use of localized surface plasmon resonance (LSPR) and Förster resonance energy transfer (FRET) effects. Thus, the developed

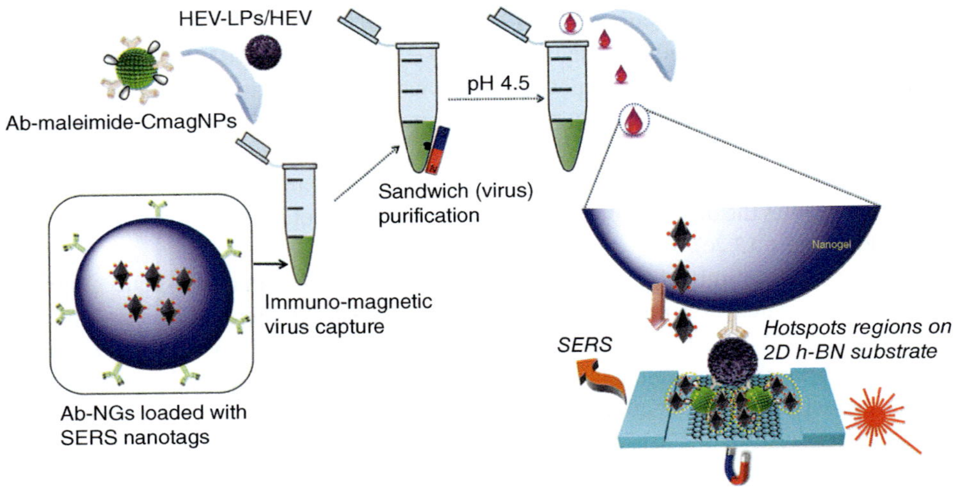

Figure 13.3 Nanogel-driven encapsulation and pH-controlled release of SERS nanotags for enhanced virus immunoassays via hotspot generation on two-dimensional h-BN substrates. Achadu et al. [48] / with permission of American Chemical Society.

measurement strategy enabled binding to the target, recovery of fluorescence intensity, and a reduction in SERS signals, achieving sensitive detection in a linear range from 10^{-4} to 10^{3} ng/mL. Furthermore, this approach resulted in a sensor sensitivity with a very low detection limit of 0.094 pg/mL. Therefore, the method proposed in this study stands out as a logical platform for AFB_1 screening, demonstrating its potential to exhibit significant sensitivity, reproducibility, and applicability to real samples.

Achadu et al. designed a novel biosensor utilizing SERS technology, which incorporates molybdenum trioxide quantum dot (MoO_3-QD)-encapsulated nanogels (NGs) for the highly sensitive detection of hepatitis E virus (HEV) and norovirus (NoV) [48]. Given the essential requirement for pathogen-specific sensors and the limitations of traditional SERS nanotags in complex matrices, MoO_3-QDs were chosen for their strong plasmonic and charge-transfer characteristics. The pH-responsive nanogels functioned as protective carriers, facilitating the controlled release of SERS-active nanotags and the subsequent formation of hotspots mediated by maleimide-functionalized magnetic nanoparticles (Figure 13.3). The dual substrate system that integrates MoO_3-QDs with two-dimensional hexagonal boron nitride nanosheets demonstrated a significant enhancement factor of approximately 10^{8}, enabling detection limits as low as 6.5 and 8.2 fg/mL for HEV and NoV virus-like particles, respectively. The platform exhibited effective detection in clinical samples, characterized by high selectivity and reproducibility, underscoring its potential for dependable viral diagnostics.

13.3.2 Utilizing Fluorescence Spectroscopy for Quantum Optical Sensor Sensing

Optical imaging techniques have become indispensable in biomedical research and diagnostics due to their ability to noninvasively visualize molecular processes *in vivo*.

These approaches provide critical insights into disease progression, pharmacodynamics, and biomarker identification. Optical signals arise primarily from fluorescence or bioluminescence; fluorescence involves photon emission at longer wavelengths than excitation, whereas bioluminescence results from enzymatic reactions [49]. For example, a classic example of QDs in cancer detection was demonstrated by Gao et al. [11]. In their study, QDs were conjugated with antibodies specific to prostate-specific membrane antigen to label human prostate cancer cells. This study highlighted the potential of QD-based probes for selective and efficient cancer cell detection [11].

Neurotransmitters are vital biochemical messengers that regulate physiological processes and behavioral responses in both the central and peripheral nervous systems [50]. They enable communication between and within cells. Unlike electrical signals, detecting chemical signals like neurotransmitters requires high spatial and temporal precision. In neuroscience, QDs are particularly valuable due to their excellent photostability, which makes them well-suited for long-term neuronal labeling and monitoring of synaptic activity. For instance, in 2015, Zhou et al. developed a fluorescent sensor for dopamine detection using core/shell hybrids of polypyrrole and graphene quantum dots, which provided high sensitivity and selectivity. These hybrid materials exhibited fluorescence intensity up to three times higher than that of pure graphene quantum dots. The sensor detected dopamine by monitoring fluorescence quenching with increasing dopamine concentration, showing a linear response in the 5–8000 nM range and achieving a detection limit as low as 10 pM (S/N = 3). Furthermore, this method was successfully applied to real samples, demonstrating its potential for diagnostic applications [51].

Moreover, in a study reported by Kruss and colleagues in 2014, researchers developed a NIR fluorescent nanosensor composed of single-walled carbon nanotubes and a DNA aptamer that specifically binds serotonin to determine serotonin, a key neurotransmitter that is difficult to detect in real time and with high accuracy using traditional sensors. The sensor emits light at 995 nm, and its brightness increases up to 1.8 times when serotonin is present. It can reversibly detect serotonin in the range of 100 nM to 1 μM, with a detection sensitivity of about 301 nM. The sensor successfully monitored the release of serotonin from individual blood platelets and identified the areas where serotonin was released with a time delay of about 21–30 seconds from stimulation. Overall, this sensor shows promise for visualizing how cells communicate using neurotransmitters [52].

13.3.3 Optical Sensor for Biomedical Imaging

The unique tunable and stable luminescence of QDs makes them highly versatile probes for biological imaging, allowing precise visualization of cellular and molecular processes. This is especially valuable for early disease diagnosis, where highly sensitive and specific imaging tools are needed [8, 41].

Overexpressed tumor-specific markers and tumor-associated markers are ideal targets for imaging and treatment, and they have been widely used in both *in vitro* and *in vivo* applications for various cancers. QDs conjugated with targeting agents, such as antibodies or peptides, are commonly employed to specifically bind to these markers, enabling targeted imaging and sustained fluorescence visualization of cancer cells [8].

Manjubaashini et al. synthesized CQDs from agro-waste biomass, specifically pista and groundnut shells, using a hydrothermal method [53]. These CQDs serve as innovative probes for bioanalysis and *in vivo* bioimaging, fulfilling the demand for biocompatible, cost-effective, and sustainable materials designed for biologically relevant analytes. The research underscores the importance of CQDs in fluorescence-based sensing, attributed to their adjustable optical characteristics. These include notable fluorescence quantum yields (9.5% for pista shell extraction [PSE] and 9.0% for groundnut shell extraction [GNSE]), pronounced excitation-dependent emission profiles, and enhanced cell viability (90%) alongside hemocompatibility (0.3% for PSE and 0.1% for GNSE). The structural characterization validated the amorphous nature (XRD), nanoscale particle sizes (~5 nm for PSE and ~12 nm for GNSE), and surfaces rich in functional groups, thereby enhancing interactions with biological targets. The improved antibacterial activity and ability to stimulate growth factor expression (IGF-1, Col-1) highlight their therapeutic potential. In Figure 13.4, zebrafish were incubated with CQDs at doses of 10, 50, 75, and 100 μL. CQDs fluoresced intensely blue (365 nm) in the zebrafish yolk sac, but the control group did not. In addition, increasing CQDs concentration caused strong fluorescence. At all doses, PSE CQDs (Figure 13.4a) absorbed more in the yolk sac and ocular region than GNSE CQDs. At greater concentrations, the yolk sac and ocular area absorbed PSE CQD (Figure 13.4b). The uptake research examined CQDs bioaccumulation in zebrafish larvae. In this study, CQDs can penetrate the cell membrane and stain the cytoplasm, proving their potential in biological and medical research by proving that synthesized CQDs can be used as a probe for *in vivo* biological imaging, especially in zebrafish larvae. Therefore, the application

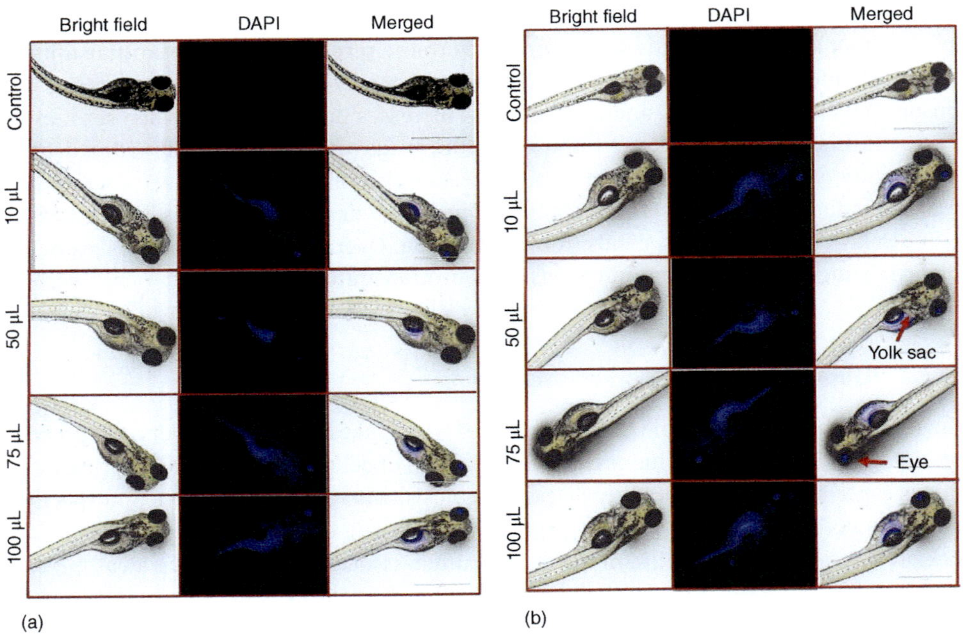

Figure 13.4 *in vivo* fluorescence imaging of zebrafish demonstrating tissue-specific distribution of CQDs. Manjubaashini et al. [53] / with permission of Elsevier.

of zebrafish models revealed effective fluorescence imaging with no associated toxicity, confirming these CQDs as robust sensor platforms designed for live imaging scenarios where accurate visualization and biological responsiveness are essential.

Mishra et al. indicated biocompatible, fluorescent GOQDs synthesized via pyrolysis of *Glycyrrhiza glabra* root, aiming to develop a low-cost, eco-friendly biosensing probe tailored for imaging *Drosophila melanogaster*, a model organism with significant genomic similarity to humans [54]. The significance of analyte-specific detection and bioimaging in early disease diagnostics is underscored by the engineering of GOQDs, which possess a mean diameter of 2.6 nm and a graphitic interlayer spacing of 0.36 nm, demonstrating excitation-dependent photoluminescence across blue to red regions. Surface functionalities, including hydroxyl, carbonyl, and carboxylic acid groups, enhanced stable aqueous dispersion and biological compatibility. Cytotoxicity assessments, which included reactive oxygen species quantification, diamidino-2-phenylindole staining for genotoxicity, and larval crawling assays, confirmed the absence of DNA damage or developmental abnormalities, thereby validating the non-toxic profile of GOQDs at a concentration of 3.0 mg/mL. The findings underscore the potential of GOQDs as safe, sensitive, and analyte-specific bioimaging agents for diagnostic purposes.

In addition, in recent years, imaging studies have investigated the use of combined drugs for targeted drug delivery in cancer treatment. For example, Li et al. [55] investigated the use of mitochondria as carriers for CQDs *in vivo* imaging and the delivery of the anticancer drug doxorubicin, considering that mitochondria share an evolutionary bond with bacteria and function as semi-autonomous organelles. His work showed that mitochondria can effectively load CQDs without altering their optical properties. Furthermore, this system improved the biodistribution of CQDs between organs and prolonged their persistence after intravenous injection [55]. Similarly, Duan et al. [56] developed a pH-sensitive fluorescent drug delivery system using doxorubicin-laden carbon dots (CDs) to target human gastric cancer cells. They proved that CDs were biocompatible, finding that they maintained over 90% cell viability in both cancerous and normal stomach cells. The CDs-Doxorubicin formulation inhibited cancer cell growth approximately twice as effectively as in normal cells. This system has achieved a high drug loading efficiency of 75.3% and has enabled controlled drug release suitable for intravenous cancer treatment. Fluorescence imaging confirmed selective uptake by cancer cells with significantly lower uptake in normal cells, whereby it was decided that the time it took to monitor drug delivery was 48 hours [56].

13.4 Challenges of Quantum Optical Sensors for Biomedical Applications

On the other hand, quantum dot dyes should not deteriorate when exposed to light. Although researchers have shown that quantum dots are able to maintain their fluorescence intensity for a long time when excited relative to their fluorescence dyes. Quantum dot dyes, for example, have been proven to be 100 times more resistant to photobleaching than normal dyes such as Rhodamine 6G [57]. Despite these properties, QDs can sometimes be degraded by light or fail to maintain their activity for extended periods. Surface modification methods such as capping are being developed to address this [58].

Moreover, SERS is highly advantageous for biosensors due to its exceptional sensitivity, ability to identify specific compounds, and potential for label-free operation in complex biological environments. However, several significant issues hinder its widespread deployment in clinical and real-world biosensor contexts. A significant issue is the difficulty in reproducing SERS signals, exacerbated by the nanostructured substrates employed to enhance the Raman signal. The heterogeneity of hotspots, irregular nanoparticle production, and inter-batch variability render signal intensities unreliable, complicating quantitative analysis. Furthermore, achieving consistent enhancement throughout an extensive detection area remains challenging, complicating large-scale diagnostics. Some QDs may pose safety concerns for usage in living organisms, raising apprehensions over *in vivo* applications [12]. In clinical contexts, it is crucial to ensure that nanostructures do not elicit immunological responses or alter physiological functions; nevertheless, many existing studies fail to conduct comprehensive biocompatibility assessments. Biological matrices may also obstruct progress. Fluorescent backgrounds, protein adsorption, and signal attenuation in complex materials such as blood or serum can obscure the Raman signal, hence diminishing sensitivity and specificity. Despite advancements in surface changes and enhanced signal processing techniques, completely eliminating matrix effects remains challenging. The integration of microfluidics and lab-on-a-chip technologies remains nascent, facing challenges in reliable sample processing, signal standardization, and cost-effectiveness. From a translational perspective, the regulatory clearance pathways for SERS-based diagnostic devices are not clearly defined. This is mostly due to the novelty of nanomaterials and the absence of proven methodologies. Clinical validation requires extensive studies with statistically meaningful cohorts, which remain challenging to identify [59].

13.5 Conclusion

Quantum optical sensors represent a transformative advancement in biological imaging and diagnostics due to their exceptional sensitivity, selectivity, and capability for real-time data analysis. This chapter discussed the primary optical sensing techniques, including fluorescence and Raman spectroscopy. It demonstrated how QDs, such as quantum dots, carbon-based nanostructures, and metallic hybrids, enhance signal clarity, facilitate target identification, and improve compatibility with biological systems. The use of quantum dots into biosensing systems can alter the emission spectra, enhance photostability, and facilitate multifunctionality, such as imaging and diagnostics. Their application in Raman and fluorescence-based sensing has demonstrated significant potential for detecting infections, cancer biomarkers, neurotransmitters, and foodborne toxins at extremely low detection thresholds.

The integration of quantum nanotechnology and optical biosensing has facilitated novel methods for observing living organisms, targeting drug delivery, and personalizing medical treatments. SERS faces challenges with reproducibility, certain nanomaterials may pose cytotoxic risks, and clinical validation protocols lack standardization. Recent advancements in nanomaterial surface engineering, signal amplification, and microfluidic integration continue to address these persistent challenges.

The development of biocompatible, stable, and scalable quantum nanostructures, coupled with advanced optical detection systems, holds significant potential for next-generation biosensors applicable in clinical and point-of-care environments, demonstrating high efficacy in disease diagnosis. Future research should prioritize aligning legislation, conducting translational studies, and developing environmentally sustainable nanomaterials to fully utilize quantum optical sensors for diagnostic applications in biomedical sciences.

References

1 Hareesha, N. and Manjunatha, J.G. (2021). Electro-oxidation of formoterol fumarate on the surface of novel poly(thiazole yellow-G) layered multi-walled carbon nanotube paste electrode. *Scientific Reports* 11 (1): https://doi.org/10.1038/s41598-021-92099-x.

2 Raril, C., Manjunatha, J.G., Ravishankar, D.K. et al. (2020). Validated electrochemical method for simultaneous resolution of tyrosine, uric acid, and ascorbic acid at polymer modified nano-composite paste electrode. *Surface Engineering and Applied Electrochemistry* 56 (4): 415–426. https://doi.org/10.3103/S1068375520040134.

3 Pushpanjali, P.A., Manjunatha, J.G., Hareesha, N. et al. (2021). Voltammetric analysis of antihistamine drug cetirizine and paracetamol at poly(L-leucine) layered carbon nanotube paste electrode. *Surfaces and Interfaces* 24: 101154. https://doi.org/10.1016/J.SURFIN.2021.101154.

4 Tigari, G. and Manjunatha, J.G. (2020). Poly(glutamine) film-coated carbon nanotube paste electrode for the determination of curcumin with vanillin: an electroanalytical approach. *Monatshefte Fur Chemie* 151 (11): 1681–1688. https://doi.org/10.1007/s00706-020-02700-8.

5 Manjunatha, J.G., Pushpanjali, P.A., and Hareesha, N. (2021). An overview of recent developments of carbon-based sensors for the analysis of drug molecules. *Journal of Electrochemical Science and Engineering* https://doi.org/10.5599/jese.999.

6 Monnappa, A.B., Manjunatha, J.G., and Bhatt, A.S. (2020). Design of a sensitive and selective voltammetric sensor based on a cationic surfactant-modified carbon paste electrode for the determination of alloxan. *ACS Omega* 5 (36): 23481–23490. https://doi.org/10.1021/acsomega.0c03517.

7 Mazumder, S., Dey, R., Mitra, M.K. et al. (2009). Biofunctionalized quantum dots in biology and medicine. *Journal of Nanomaterials* 2009 (1): 815734.

8 Chan, W.C.W., Maxwell, D.J., Gao, X. et al. (2002). Luminescent quantum dots for multiplexed biological detection and imaging. *Current Opinion in Biotechnology* 13 (1): 40–46. https://doi.org/10.1016/S0958-1669(02)00282-3.

9 Mattoussi, H., Mauro, J.M., Goldman, E.R. et al. (2000). Self-assembly of CdSe–ZnS quantum dot bioconjugates using an engineered recombinant protein. *Journal of the American Chemical Society* 122 (49): 12142–12150.

10 Hu, D., Zhang, P., Gong, P. et al. (2011). A fast synthesis of near-infrared emitting CdTe/CdSe quantum dots with small hydrodynamic diameter for in vivo imaging probes. *Nanoscale* 3 (11): 4724. https://doi.org/10.1039/c1nr10933b.

11 Gao, X., Cui, Y., Levenson, R.M. et al. (2004). in vivo cancer targeting and imaging with semiconductor quantum dots. *Nature Biotechnology* 22 (8): 969–976. https://doi.org/10.1038/nbt994.

12 Wagner, A.M., Knipe, J.M., Orive, G., and Peppas, N.A. (2019). Quantum dots in biomedical applications. *Acta Biomaterialia* 94: 44–63. https://doi.org/10.1016/j.actbio.2019.05.022.

13 Lu, M.C., Yang, Y.C., Lee, C.J., and Chiu, C.W. (2025). Helicobacter pylori detection based on synergistic electromagnetic and chemical enhancement of surface-enhanced Raman scattering in 3D hotspot-activated gold nanorods/nano mica platelets/ZnO quantum dots. *Advanced Science* https://doi.org/10.1002/advs.202503562.

14 Alexander Kamyshny, S.M. (2010). Aqueous dispersions of metallic nanoparticles preparation, stabilization, and application. In: *Nanoscience*, 1e (ed. V.M. Starov), 32. CRC Press.

15 Lisiecki, I. (2005). Size, shape, and structural control of metallic nanocrystals. *The Journal of Physical Chemistry B* 109 (25): 12231–12244. https://doi.org/10.1021/jp058018p.

16 Reimann, S.M. and Manninen, M. (2002). Electronic structure of quantum dots. *Reviews of Modern Physics* 74 (4): 1283.

17 Bacon, M., Bradley, S.J., and Nann, T. (2014). Graphene quantum dots. In: *Particle and Particle Systems Characterization*, vol. 31, 415–428. Wiley-VCH Verlag https://doi.org/10.1002/ppsc.201300252.

18 Kong, J., Wei, Y., Zhou, F. et al. (2024). Carbon quantum dots: properties, preparation, and applications. *Molecules* 29 (9): 2002. https://doi.org/10.3390/molecules29092002.

19 Bera, D., Qian, L., Tseng, T.K., and Holloway, P.H. (2010). Quantum dots and their multimodal applications: a review. *Materials* 3 (4): 2260–2345. https://doi.org/10.3390/ma3042260.

20 Uslu, B. (2021). *Electroanalytical Applications of Quantum Dot-Based Biosensors.* Elsevier https://doi.org/10.1016/C2019-0-04090-3.

21 Bozal-Palabiyik, B., Kurbanoglu, S., Erkmen, C., and Uslu, B. (2021). Future prospects and concluding remarks for electroanalytical applications of quantum dots. In: *Electroanalytical Applications of Quantum Dot-Based Biosensors* (ed. B. Uslu), 427–450. Elsevier https://doi.org/10.1016/B978-0-12-821670-5.00008-7.

22 Rempel, A.A., Ovchinnikov, O.V., Weinstein, I.A. et al. (2024). Quantum dots: modern methods of synthesis and optical properties. *Russian Chemical Reviews* 93 (4): RCR5114. https://doi.org/10.59761/rcr5114.

23 Cepeda-Pérez, E.I., López-Luke, T., Pérez-Mayen, L. et al. (2015). Wet chemical synthesis of quantum dots for medical applications. *Clinical and Biomedical Spectroscopy and Imaging IV* 95371H. https://doi.org/10.1364/ECBO.2015.95371H.

24 Zhao, M.-X. and Zhu, B.-J. (2016). The research and applications of quantum dots as nano-carriers for targeted drug delivery and cancer therapy. *Nanoscale Research Letters* 11 (1): 207. https://doi.org/10.1186/s11671-016-1394-9.

25 Biju, V., Itoh, T., and Ishikawa, M. (2010). Delivering quantum dots to cells: bioconjugated quantum dots for targeted and nonspecific extracellular and intracellular imaging. *Chemical Society Reviews* 39 (8): 3031. https://doi.org/10.1039/b926512k.

26 Nair, A., Haponiuk, J.T., Thomas, S., and Gopi, S. (2020). Natural carbon-based quantum dots and their applications in drug delivery: a review. *Biomedicine and Pharmacotherapy* 132: 110834. https://doi.org/10.1016/j.biopha.2020.110834.

27 Pareek, A., Kumar, D., Pareek, A., and Gupta, M.M. (2025). Advancing cancer therapy with quantum dots and other nanostructures: a review of drug delivery innovations, applications, and challenges. *Cancers* 17 (5): https://doi.org/10.3390/cancers17050878.

28 Ahmad, J., Garg, A., Mustafa, G. et al. (2023). Hybrid quantum dot as promising tools for theranostic application in cancer. *Electronics* 12 (4): 972. https://doi.org/10.3390/electronics12040972.

29 Brichkin, S.B., Spirin, M.G., and Razumov, V.F. (2024). Colloidal quantum dots: 2. Methods for the synthesis of colloidal quantum dots. *High Energy Chemistry* 58 (S1): S10–S23. https://doi.org/10.1134/S0018143924700152.

30 Pandit, P. and Chandra, A. (2022). Quantum dot: a boon for biological and biomedical research. In: *Application of Quantum Dots in Biology and Medicine*, 209–228. Springer Nature Singapore https://doi.org/10.1007/978-981-19-3144-4_11.

31 Jangid, M. and Kothari, S. (2023). Coprecipitation synthesis of quantum dots. In: *Quantum Dots*, 77–91. Elsevier https://doi.org/10.1016/B978-0-12-824153-0.00005-7.

32 Shorgar, N., Bhati, I., and Jhalora, P. (2023). Laser ablation synthesis of quantum dots. In: *Quantum Dots*, 53–75. Elsevier https://doi.org/10.1016/B978-0-12-824153-0.00004-5.

33 Cheng, H., Wang, Y., Dai, H. et al. (2015). Nonlinear optical properties of PbS colloidal quantum dots fabricated via solvothermal method. *The Journal of Physical Chemistry C* 119 (6): 3288–3292. https://doi.org/10.1021/jp510214x.

34 Jawaid, A.M., Chattopadhyay, S., Wink, D.J. et al. (2013). Cluster-seeded synthesis of doped CdSe:Cu_4 quantum dots. *ACS Nano* 7 (4): 3190–3197. https://doi.org/10.1021/nn305697q.

35 Hunter, K.I., Held, J.T., Mkhoyan, K.A., and Kortshagen, U.R. (2017). Nonthermal plasma synthesis of core/shell quantum dots: strained Ge/Si nanocrystals. *ACS Applied Materials & Interfaces* 9 (9): 8263–8270. https://doi.org/10.1021/acsami.6b16170.

36 Hu, M.Z. and Zhu, T. (2015). Semiconductor nanocrystal quantum dot synthesis approaches towards large-scale industrial production for energy applications. *Nanoscale Research Letters* 10 (1): 469. https://doi.org/10.1186/s11671-015-1166-y.

37 Choo-Smith, L.-P., Edwards, H.G.M., Endtz, H.P. et al. (2002). Medical applications of Raman spectroscopy: from proof of principle to clinical implementation. *Biopolymers* 67 (1): 1–9. https://doi.org/10.1002/bip.10064.

38 Poddar, R., Andrews, J.T., Shukla, P., and Sen, P. (2008). Non-invasive glucose monitoring techniques: a review and current trends. http://arxiv.org/abs/0810.5755.

39 Liao, Z., Lizio, M.G., Corden, C. et al. (2020). Feasibility of integrated high-wavenumber Raman imaging and fingerprint Raman spectroscopy for fast margin assessment in breast cancer surgery. *Journal of Raman Spectroscopy* 51 (10): 1986–1995. https://doi.org/10.1002/jrs.5937.

40 Lauwerends, L.J., Abbasi, H., Bakker Schut, T.C. et al. (2022). The complementary value of intraoperative fluorescence imaging and Raman spectroscopy for cancer surgery: combining the incompatibles. *European Journal of Nuclear Medicine and Molecular Imaging* 49 (7): 2364–2376. https://doi.org/10.1007/s00259-022-05705-z.

41 Jokerst, J.V., Pohling, C., and Gambhir, S.S. (2013). Molecular imaging with surface-enhanced Raman spectroscopy nanoparticle reporters. *MRS Bulletin* 38 (8): 625–630. https://doi.org/10.1557/mrs.2013.157.

42 Lee, J.-H., Cho, H.-Y., Choi, H.K. et al. (2018). Application of gold nanoparticle to plasmonic biosensors. *International Journal of Molecular Sciences* 19 (7): 2021. https://doi.org/10.3390/ijms19072021.

43 Gupta, S., Huang, C.H., Singh, G.P. et al. (2020). Portable Raman leaf-clip sensor for rapid detection of plant stress. *Scientific Reports* 10 (1): 20206. https://doi.org/10.1038/s41598-020-76485-5.

44 Kouri, M.A., Spyratou, E., Karnachoriti, M. et al. (2022). Raman spectroscopy: a personalized decision-making tool on clinicians' hands for in situ cancer diagnosis and surgery guidance. *Cancers* 14 (5): 1144. https://doi.org/10.3390/cancers14051144.

45 Keshavarz, M., Chowdhury, A.K.M.R.H., Kassanos, P. et al. (2020). Self-assembled N-doped Q-dot carbon nanostructures as a SERS-active biosensor with selective therapeutic functionality. *Sensors and Actuators B: Chemical* 323: 128703. https://doi.org/10.1016/j.snb.2020.128703.

46 Lu, M.-C., Chen, C.-H., Yang, Y.-C., and Chiu, C.-W. (2025). Silver nanocubes/nano mica platelets flexible nanohybrid substrates modified by zinc oxide quantum dots with synergistic 3D lightning rod effect, and electromagnetic and chemical enhancements for highly sensitive SERS bacterial biosensor. *Materials Today Nano* 29: 100601. https://doi.org/10.1016/j.mtnano.2025.100601.

47 Wei, J., He, Y., Song, Z. et al. (2024). Satellite nanostructures composed of CdTe quantum dots and DTNB-labeled AuNPs used for SERS-fluorescence dual-signal detection of AFB1. *Food Control* 156: 110112. https://doi.org/10.1016/j.foodcont.2023.110112.

48 Achadu, O.J., Abe, F., Li, T.-C. et al. (2021). Molybdenum trioxide quantum dot-encapsulated nanogels for virus detection by surface-enhanced Raman scattering on a 2D substrate. *ACS Applied Materials & Interfaces* 13 (24): 27836–27844. https://doi.org/10.1021/acsami.1c04793.

49 Wu, X., Deng, Y., Xu, Y. et al. (2024). Activatable fluorescence and bio/chemiluminescence probes for aminopeptidases: from design to biomedical applications. *Advanced Materials* 36 (45): https://doi.org/10.1002/adma.202409893.

50 Kanbes-Dindar, C., Demirtaş, T.T., and Uslu, B. (2024). Nanostructured materials-modified electrochemical biosensing devices for determination of neurochemicals. In: *Novel Nanostructured Materials for Electrochemical Bio-Sensing Applications*, 331–365. Elsevier https://doi.org/10.1016/B978-0-443-15334-1.00012-2.

51 Zhao, M.-X. and Zeng, E.-Z. (2015). Application of functional quantum dot nanoparticles as fluorescence probes in cell labeling and tumor diagnostic imaging. *Nanoscale Research Letters* 10 (1): 171. https://doi.org/10.1186/s11671-015-0873-8.

52 Kruss, S., Landry, M.P., Vander Ende, E. et al. (2014). Neurotransmitter detection using corona phase molecular recognition on fluorescent single-walled carbon nanotube sensors. *Journal of the American Chemical Society* 136 (2): 713–724. https://doi.org/10.1021/ja410433b.

53 Manjubaashini, N., Bargavi, P., and Balakumar, S. (2024). Carbon quantum dots derived from agro waste biomass for pioneering bioanalysis and in vivo bioimaging. *Journal of Photochemistry and Photobiology A: Chemistry* 454: https://doi.org/10.1016/j.jphotochem.2024.115702.

54 Mishra, S.R., Mandal, T., Sahu, S. et al. (2024). Biocompatible fluorescent graphene oxide quantum dots for imaging of *Drosophila melanogaster*. *ACS Omega* 9 (37): 38916–38924. https://doi.org/10.1021/acsomega.4c05244.

55 Li, W.-Q., Wang, Z., Hao, S. et al. (2018). Mitochondria-based aircraft carrier enhances *in vivo* imaging of carbon quantum dots and delivery of anticancer drug. *Nanoscale* 10 (8): 3744–3752. https://doi.org/10.1039/C7NR08816G.

56 Duan, Q., Ma, Y., Che, M. et al. (2019). Fluorescent carbon dots as carriers for intracellular doxorubicin delivery and track. *Journal of Drug Delivery Science and Technology* 49: 527–533. https://doi.org/10.1016/j.jddst.2018.12.015.

57 Wegner, K.D. and Hildebrandt, N. (2015). Quantum dots: bright and versatile in vitro and in vivo fluorescence imaging biosensors. *Chemical Society Reviews* 44 (14): 4792–4834. https://doi.org/10.1039/C4CS00532E.

58 Abdellatif, A.A., Younis, M.A., Alsharidah, M. et al. (2022). Biomedical applications of quantum dots: overview, challenges, and clinical potential. *International Journal of Nanomedicine* 17: 1951–1970. https://doi.org/10.2147/IJN.S357980.

59 Lin, L.L., Alvarez-Puebla, R., Liz-Marzán, L.M. et al. (2025). Surface-enhanced Raman spectroscopy for biomedical applications: recent advances and future challenges. *ACS Applied Materials & Interfaces* 17 (11): 16287–16379. https://doi.org/10.1021/acsami.4c17502.

14

Quantum Sensing in Metabolomics

Gnanesh Rao[1], *Priya Tiwari*[4], *Raghu Ningegowda*[2], *Belakatte P. Nandeshwarappa*[3], *and Sandeep Chandrashekharappa*[4]

[1] *Department of Biochemistry, Bangalore University, Bangalore, Karnataka, India*
[2] *Department of Chemistry, Jyoti Nivas College Autonomous, Bangalore, Karnataka, India*
[3] *Department of PG Studies and Research in Chemistry, Shivagangothri, Davangere University, Davanagere, Karnataka, India*
[4] *Department of Medicinal Chemistry, National Institute of Pharmaceutical Education and Research Raebareli (NIPER-R), Lucknow, Uttar Pradesh, India*

14.1 Introduction

Quantum sensing represents a groundbreaking and swiftly advancing domain that exploits the distinctive characteristics of quantum systems, such as spin coherence, quantum entanglement, and superposition, to attain sensitivity and spatial resolution levels that significantly exceed those of conventional measurement methodologies. Within the realm of biomedical sciences, quantum sensing is set to transform our comprehension of intricate biological systems, with metabolomics, a field dedicated to the thorough analysis of small-molecule metabolites present in cells, tissues, and organisms, emerging as an exceptionally promising avenue for its implementation [1, 2]. The major constraints of the traditional metabolomic methodologies, including nuclear magnetic resonance (NMR) spectroscopy and mass spectrometry (MS), block real-time *in vivo* investigations and limit the ability to capture dynamic metabolic alterations at the single-cell or subcellular levels, the necessity for extensive sample preparation, and dependence on ensemble measurements that obscure the metabolic diversity of individual cells.

These limitations hinder real-time, *in vivo* examinations and constrain our capacity to observe dynamic metabolic changes at the resolution of single cells or subcellular domains [3]. Quantum sensing technologies offer transformative approaches to mitigate these challenges. Nitrogen-vacancy (NV) centers within diamond matrices, for example, have demonstrated the capability to perform nanoscale NMR and electron spin resonance (ESR) measurements on picoliter volumes of samples, thereby facilitating the detection and characterization of metabolites at the precision of individual cells and even isolated molecules [4].

Quantum Biosensing in Medical Diagnostics, First Edition. Edited by J. G. Manjunatha.

The optical readout capabilities of NV centers facilitate precise measurements of local parameters, such as temperature, magnetic fields, and chemical composition, rendering them exceptionally suitable for probing the intracellular milieu and functional heterogeneity within cellular populations. In addition to NV centers, various other quantum sensing platforms, including optically pumped atomic magnetometers (OPMs), molecular quantum nanosensors (MQNs), and quantum diamond microscopes, broaden the spectrum of biomedical applications. These advanced technologies enable the detection of weak magnetic signals with femto-level sensitivity, support sophisticated imaging modalities with sub-diffraction spatial resolution, and facilitate the examination of biological processes at the nanoscale. The amalgamation of quantum sensors with cutting-edge imaging techniques, such as confocal-atomic-force microscopy, has further augmented spatial resolution, permitting the localization of NV centers with nanometer precision and unveiling new opportunities for the investigation of cellular functions and disease mechanisms at atomic scales [5].

The significant sensitivity presented by biosensors serves as a salient characteristic, promoting the discovery of biologically important concentrations of metabolites and numerous analytes, which is essential for the preliminary diagnosis of diseases and the surveillance of subtle biochemical transformations [6–8].

The stability and biocompatibility inherent in nanodiamond-based sensors, in particular, render them well-suited for the longitudinal observation of biological processes and for *in vivo* applications, as evidenced by recent innovations in biocompatible integrated chips [9, 10]. These innovations have laid the groundwork for portable chemical assays and point-of-care diagnostics that leverage the enhanced sensitivity and reduced operational costs associated with quantum sensors, thereby improving their availability for both research endeavors and clinical settings.

14.2 Current Challenges in Metabolomics

Metabolomics faces several core challenges (Figure 14.1), such as variability in sample preparation and lack of standardized protocols leading to inconsistent results; no single analytical technique can detect all metabolites due to their diverse chemical nature; the field generates massive data that's hard to interpret and integrate with other biological layers; biological fluctuations from diet, age, and environment reduce reproducibility; and translating discoveries into clinical practice is slow due to regulatory and validation hurdles. Ethical concerns around data privacy also grow as metabolomics enters personalized medicine. Researchers are actively working to improve data analysis, standardization, and integration with other omics fields to overcome these hurdles [11–14]; however, it encounters significant analytical obstacles stemming from the intrinsic complexity of biological specimens and the limitations inherent in traditional detection methodologies.

14.2.1 Complexity of Biological Samples

Biological matrices, which include blood, tissues, or cellular extracts, encompass a diverse spectrum of metabolites, proteins, and lipids, thus complicating the accurate identification and quantification of low abundance biomolecules [15–17].

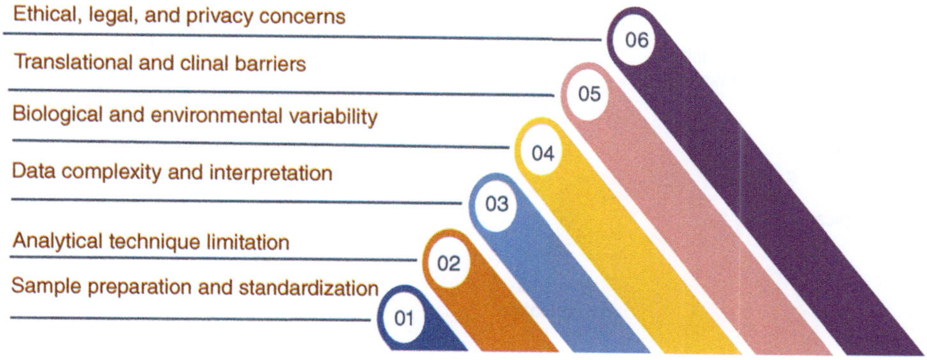

Figure 14.1 Major challenges in metabolomics.

14.2.2 Sensitivity and Selectivity of Detection Methods

Established platforms such as NMR spectroscopy and mass spectrometry (MS) frequently demonstrate insufficient sensitivity, particularly in the context of single-molecule or single-cell analysis, and necessitate labor-intensive sample preparation procedures that may introduce biases or lead to the volatilization of certain metabolites [18]. Furthermore, ensemble measurements obscure the heterogeneity present at the cellular level, and the majority of conventional methodologies lack the requisite spatial and temporal resolution for real-time or *in vivo* investigations, thereby constraining their applicability in dynamic biological frameworks [19, 20].

14.3 Quantum Sensing

Recent advances in quantum sensing technologies offer promising solutions to overcome key limitations in metabolomics, especially in the areas of detection sensitivity, precision, and selectivity. These technologies tap into fundamental quantum phenomena, including spin states, quantum entanglement, and quantum coherence, to measure extremely subtle changes in physical systems.

14.3.1 How Quantum Sensing Enhances Metabolomics

- ***Unmatched sensitivity:*** Quantum sensors, such as NV centers in diamond, can detect magnetic or electric fields associated with even trace-level metabolites. This allows for the identification of compounds previously undetectable with conventional techniques.
- ***Spin state manipulation:*** By controlling quantum spin states, sensors can isolate specific molecular interactions, improving accuracy in distinguishing between closely related metabolites.
- ***Quantum coherence and entanglement:*** These properties enable measurements with minimal noise and exceptional precision, especially useful in complex biological samples where metabolite signals often overlap or are masked.

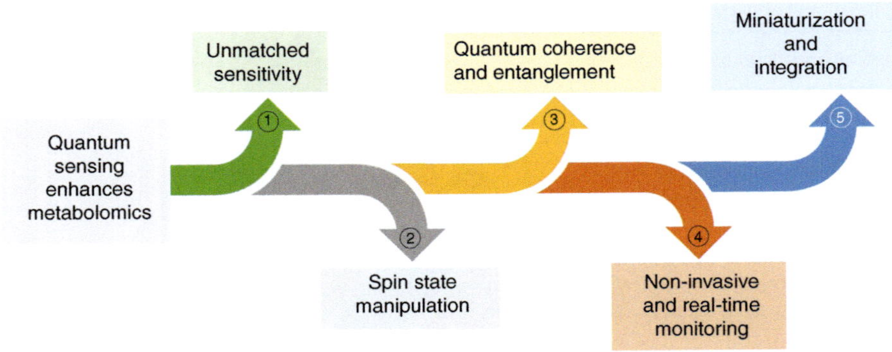

Figure 14.2 Quantum sensing enhances metabolomics.

- ***Non-invasive and real-time monitoring:*** Quantum sensors can operate at room temperature and in aqueous environments, making them suitable for live-cell or tissue studies without destructive sampling.
- ***Miniaturization and integration:*** Many quantum sensors are compatible with compact designs, paving the way for portable metabolomics devices and potential point-of-care diagnostics (Figure 14.2).

14.3.2 Types of Quantum Sensing Relevant to Metabolomics

Quantum sensing approaches, specifically those that leverage NV centers situated in diamond lattices, are markedly altering the landscape of metabolomics through the provision of nanoscale NMR and single-cell spectroscopy characterized by exceptional spatial resolution. Quantum sensing technologies are increasingly being explored to overcome analytical limitations in metabolomics. While not all quantum sensors are directly applied to metabolomics yet, several types (Figure 14.3) show strong potential for ultrasensitive detection of metabolites and molecular interactions:

1) **Quantum Magnetometers**
   - Detect extremely weak magnetic fields generated by molecular structures.
   - Useful for identifying spin-active metabolites or monitoring biochemical reactions.
   - NV centers in diamond are a leading platform for nanoscale magnetic sensing.
2) **Quantum Thermometers**
   - Measure minute temperature changes at the cellular or molecular level.
   - Can track metabolic heat production or enzyme activity in real time.
   - Based on entangled atoms or ions that respond to thermal fluctuations.
3) **Quantum Chemical Sensors**
   - Designed to detect specific chemical signatures with high selectivity.
   - Use quantum coherence to distinguish between similar molecular species.
   - Potentially applicable for identifying low-abundance metabolites in complex samples.
4) **Quantum Interferometers**
   - Measure molecular interactions and structural changes via interference patterns.
   - Used to study conformational shifts in biomolecules or metabolite binding events.

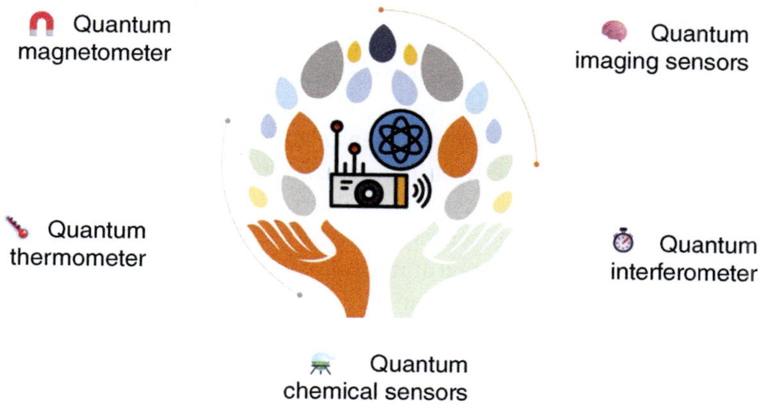

Figure 14.3 Quantum sensors in metabolomics.

5) **Quantum Imaging Sensors**
   - Provide high-resolution imaging of metabolic processes at the nanoscale.
   - Enhance visualization of metabolite distribution in tissues or cells.

Other recent developments in the field of quantum sensors being utilized in metabolomics are discussed in detail below:

14.3.3 Nitrogen-Vacancy (NV) Centers in Diamond

NV centers in diamond, because of its remarkable quantum characteristics, have emerged as an exceptionally promising platform tool for sensing applications. Atomic-scale defects, consisting of a nitrogen atom adjacent to a lattice vacancy, demonstrate spin-dependent photoluminescence that facilitates the optical manipulation and readout of their quantum states [21–23].

This distinctive attribute renders NV centers extraordinarily sensitive quantum sensors, allowing for the detection of electromagnetic fields with nanoscale spatial resolution [24]. In contrast to numerous quantum systems that necessitate cryogenic environments, NV centers function effectively at ambient temperatures, a critical advantage for biological applications. Along with chemical inertness and biocompatibility, diamond substrates permit their safe integration within living cells, thus enabling real-time, noninvasive monitoring of physiological processes [22, 25].

NV centers have been proficiently utilized in the domain of high-resolution magnetic field sensing, encompassing nanoscale MRI (nano-MRI) of singular biomolecules, in addition to nanoscale temperature sensing, thereby yielding significant insights into cellular functionalities and the pathophysiology of diseases [26].

These centers permit the discernment of paramagnetic entities through their engagements with biomolecules that possess magnetic tags. Further refined techniques such as T1 relaxometry and optically detected magnetic resonance (ODMR) have been tested for the biosensing application [27]. NV centers play a pivotal role in facilitating quantum nondemolition (QND) measurements, thereby enabling repeated, noninvasive assessments that preserve the integrity necessary for extended observation of biological systems and enhancing the fidelity of measurements [28]. Recent innovations in technology, including quantum optimal control protocols and the creation of high-density NV ensembles via Ar^+ plasma irradiation, have markedly elevated sensor sensitivity and operational efficacy [29, 30]. Nonetheless, challenges remain in optimizing the stability of NV centers, particularly in sustaining a consistent charge state and prolonging spin coherence times, which are essential for maximizing sensor performance in intricate biological environments [31]. Ongoing investigations aimed at mitigating these challenges are likely to unlock novel opportunities in biological sensing and imaging, thereby establishing NV centers as transformative instruments for the elucidation of complex biological systems [32].

NV centers function as atomic-scale quantum detection devices proficient in recognizing magnetic resonance signals at the level of a proton, a sensitivity benchmark that is indispensable for single-cell biological inquiries where traditional NMR techniques are limited by their requirement for larger sample sizes [33, 34]. Within the scope of single-cell spectroscopy, NV centers provide a sophisticated characterization of cellular heterogeneity and biochemical dynamics, delivering insights that are beyond the reach of traditional techniques and showcasing substantial prospects for biomedical applications, including disease diagnosis and customized therapeutic interventions [2].

14.3.4 Quantum Interferometric Biosensors

Quantum biosensors that employ quantum interferometry epitomize a leading-edge technology in the domain of biological detection, harnessing quantum phenomena such as entanglement and coherence to transcend the limitations inherent in classical sensing methodologies. These sensors function by identifying phase shifts or interference patterns of light waves that interact with biological specimens, thereby facilitating exceptionally sensitive and label-free assessments of molecular interactions [35, 36].

Methods such as Mach–Zehnder interferometers have shown outstanding refractive index resolutions of 1.39×10^{-8} RIU, allowing for the precise measurement of molecules such as glucose and viral loads. Quantum optical coherence tomography provides high-resolution imaging of biological tissues by its use of frequency-entangled photon pairs. Effectively addressing degradation resulting from group-velocity dispersion, which makes it particularly advantageous for fragile specimens [37–39].

Quantum interferometric biosensors are capable of detecting individual biomolecules, evaluating biomolecular binding interactions, and quantifying biological nanoparticles such as viruses, exosomes, and nanoparticle size and morphology without reliance on fluorescent or enzymatic labeling [39, 40].

Complementary quantum sensing techniques-such as quantum plasmonic biosensing, which integrates plasmonic principles with quantum mechanics for the detection of ultralow concentrations, making it suitable for applications requiring high precision utilizing NV centers-provide nanoscale resolution and exceptional precision in the measurement of magnetic and electric fields [41]. Augmented sensitivity and noninvasive measurement capabilities are further attained through the application of quantum metrology, which facilitates high-sensitivity detection at minimal light intensities, thereby reducing photochemical damage to biological specimens [42, 43].

Quantum interferometric radars that employ entangled quantum signals broaden their applications to vital signs monitoring and organ imaging at low power levels, thereby ensuring patient safety. Furthermore, quantum interferometry is under exploration for sub-shot-noise sensing and high-resolution optical spectroscopy, which are crucial for the investigation of photon-sensitive biological molecules and for characterizing biological responses at the single-photon level [44].

Sensor platforms such as superconducting qubits, quantum Hall devices, and optomechanical sensors harness quantum mechanical principles to attain unparalleled sensitivity and precision in biological detection. Superconducting qubits, known for their capacity to sustain quantum coherence over relatively long periods, demonstrate exceptional efficacy in identifying exceedingly weak magnetic fields, thus presenting promising prospects for applications such as magnetoencephalography (MEG) and magnetocardiography (MCG) [45, 46]. Investigations in this domain are concentrated on prolonging coherence times and formulating robust quantum error correction (QEC) protocols to augment the reliability and practicality of these devices for biosensing applications [46].

Quantum Hall devices employ the quantum Hall effect to assess electrical resistance with remarkable accuracy; they are emerging as invaluable instruments for detecting neuronal action potentials and other bioelectrical signals that necessitate high magnetic field sensitivity. Being in the early phases of development for sensing efforts to tackle the challenges linked to miniaturization and the integration of these devices with biological systems have been carried out [45]. Optomechanical sensors, which identify variations in mass or physical properties by evaluating the interaction between light and mechanical vibrations, provide attogram-level sensitivity and are ideally suited for the detection of biomolecules at very minimal concentrations [41].

14.3.5 Quantum Dots-Based Sensors

Quantum dots are semiconductor nanocrystals that display unique optical characteristics, rendering them suitable for biosensing applications due to their distinctive optical properties, which include size-tunable emission, high brightness, and resistance to photobleaching. The wavelength of emission can be systematically adjusted based on size, promoting multiplexed imaging and diagnostic capabilities, while their wide absorption and narrow emission spectra allow for the simultaneous identification of various targets with reduced spectral overlap [47, 48].

The quantum dots have been extensively used as fluorescent markers, in single-molecule imaging, and in multiplexed diagnostics with stability and luminescence surpassing limitations of conventional dyes. Development of water-soluble and bioconjugated quantum dots has utility in various applications like cellular labeling, *in vivo* imaging, and dynamic tracking within intracellular environments, as well as their engagement in Forster resonance energy transfer (FRET)-based biosensing, as well as applications in tissue staining, gene delivery, and immunolabeling [49].

Furthermore, QDs are currently being developed for applications in drug delivery and therapeutics, functioning both as imaging agents and carriers for targeted therapeutic interventions, and can be engineered for multimodal imaging modalities, encompassing magnetic resonance imaging (MRI) and deep tissue applications [50]. Progress in surface modification techniques has enhanced the biocompatibility, stability, and clearance of QDs; however, challenges pertaining to their toxicity and the translation to clinical applications continue to impede progress [51]. Notwithstanding these obstacles, the distinctive characteristics of quantum dots perpetually propel advancements in biomedical imaging, labeling, drug delivery, and the sensitive detection of biomolecules, thereby positioning them at the vanguard of next-generation diagnostic and therapeutic technologies [52].

14.3.6 Optically Pumped Atomic Magnetometers

Optically Pumped Atomic Magnetometers (OPMs) are progressively asserting their role as a distinguished quantum sensing technology for the noninvasive and remarkably sensitive evaluation of magnetic fields, which capitalize on quantum phenomena that are distinctive to atomic vapors, primarily alkali metals like rubidium or cesium, by implementing optical excitation approaches.

This approach allows OPMs to attain sensitivities at the scale of femtotesla (fT), thus making them capable of identifying the remarkably weak biomagnetic signals produced by physiological and metabolic functions within the human body [45, 53]. OPMs operate effectively at or near ambient temperature, which is a significant advantage over traditional superconducting quantum interference devices (SQUIDs), which require bulky and expensive cryogenic cooling systems, reducing operational complexity and associated costs but also encouraging the development of compact, versatile, and portable sensors that are optimally designed for clinical and research purposes [53].

The nonintrusive attributes of OPMs make them particularly advantageous for the continuous and real-time assessment of metabolic functions, including cardiac and neural activities, without requiring direct physical interaction or invasive techniques [53, 54].

14.4 Applications

The application of quantum sensing, particularly through the introduction of NV centers into diamond matrices, has fundamentally modified NMR spectroscopy by making it feasible to detect and characterize metabolites at both nanoscale and microscale with extraordinary sensitivity and spatial resolution. This advancement not only enhances our understanding of metabolic processes but also opens new avenues for real-time monitoring of biochemical interactions in living systems [2].

14.4.1 NV-Center-Based NMR in Metabolite Detection and Structure Determination

NV centers, which are atomic-scale imperfections situated within the diamond crystalline structure, operate as exceptionally sensitive quantum sensors capable of discerning magnetic resonance signals down to the resolution of a single proton. This remarkable ability enables the execution of single-molecule magnetic resonance, yielding profound insights into the structural and dynamic characteristics of individual biomolecules-an accomplishment that transcends the limitations inherent in conventional ensemble measurements, where vital molecular nuances are frequently obscured. Consequently, NV-based NMR spectroscopy has established a pathway for the direct examination of intricate biological systems at the molecular scale, which is critical for enhancing our comprehension of fundamental biological mechanisms and for the innovation of novel therapeutic approaches [55].

A significant advancement attributed to NV centers is the achievement of nanoscale NMR spectroscopy, which accomplishes single-nucleus sensitivity and facilitates the determination of chemical structures within exceedingly diminutive sample volumes, such as those found in living cells. The utilization of nanodiamonds, which are widely recognized for their biocompatibility and their capability to enhance surface functionalization, has markedly augmented the significance of NV-NMR technologies for intracellular investigations, enabling non-invasive assessments of metabolites and molecular interactions within the framework of cellular research [56, 57]. This technological advancement carries a substantial advantage for metabolite detection, as it allows for the identification and quantification of low-abundance molecules that would otherwise be undetectable via traditional methodologies.

In the field of structural elucidation, NV centers have substantially augmented the spectral resolution of nanoscale NMR, permitting the resolution of chemical shifts within magnetic fields and consequently allowing for the identification of distinct molecular structures and this specialization is particularly advantageous for pharmaceutical development and biochemical investigations, as it facilitates the intricate reconstruction of the three-dimensional configurations of nuclear spins across distinct molecules, yielding a sophisticated representation of molecular architecture that is essential for systematic drug design and the analysis of intricate biochemical pathways [58]. Furthermore, NV-based quantum sensors are poised to transform single-cell NMR spectroscopy by enabling the quantification of biologically significant metabolite concentrations within individual cells, thus revealing cellular heterogeneity and dynamic metabolic phenomena [59].

Despite these exceptional developments, a multitude of challenges continues to impede the complete realization of the capabilities of NV-center-based NMR. The present constraints highlight the need for enhanced magnetic sensitivity and spectral and spatial resolution, alongside tackling the limitations concerning the detection volume of contemporary NV-NMR devices, which presently restricts their usage in more substantial biological systems such as whole cells or tissues. Surmounting these challenges will necessitate continuous optimization of technical parameters, the development of innovative sensor architectures, and the integration of NV-NMR with complementary technologies such as advanced imaging techniques [55, 57].

14.4.2 Quantum Sensing in Metabolite Detection

The amalgamation of quantum sensing with computational quantum chemistry has engendered a pivotal transformation within the realm of metabolomics, signaling significant advancements that are reformulating the paradigm of metabolite detection.

The fluorescent quantum dots, specifically CdTe/ZnS core/shell quantum dots, have shown significant proficiency as fluorescence sensors for the identification of crucial metabolites, encompassing folic acid, glucose, and vitamin C, amidst multifaceted biological matrices. These quantum dots are characterized by their elevated quantum yield, exceptional stability, and enhanced biocompatibility, achieving detection limits as minimal as 0.84 μg/mL for folic acid, 0.33 mM for glucose, and 1.15 μg/mL for vitamin C. Their proficiency in sensitive and selective detection is underpinned by a robust FRET-based mechanism, positioning them at the vanguard of advanced biosensing technologies, thereby establishing their necessity for clinical diagnostic endeavors and environmental evaluations [15].

In the field of MCG, OPMs have been proficiently employed to measure cardiac magnetic fields during physical exertion, providing critical understanding of cardiac functionality and the metabolic reactions of cardiac tissue under stress.

Likewise, in MEG, OPMs have been employed to gauge auditory-evoked fields and various cerebral signals, which are intrinsically associated with the underlying metabolic phenomena in neural tissue. The capacity of OPMs to concurrently measure both direct current and radio frequency magnetic fields further amplifies their applicability in intricate metabolic investigations, where monitoring multiple physiological parameters simultaneously is frequently requisite [53, 60].

14.4.3 Computational Quantum Chemistry in Metabolite Identification

By reconfiguring the identification challenge as a maximum colorful subtree (MCS) problem, researchers have suggested utilizing quantum algorithms originally intended for addressing minimum spanning tree problems, potentially allowing for swift and effective metabolite identification independent of pre-existing spectral or metabolite repositories [61].

A noteworthy advancement is the innovation of quantum chemistry-based protocols such as MOSS-DFT, which facilitates the precise prediction of ^{13}C and ^{1}H NMR chemical shifts for metabolites in aqueous environments. This protocol not only enhances traditional methodologies by offering motif-specific error estimations but also boosts the verification and validation processes of metabolite candidates within samples characterized by unknown or complex compositions [62]. Additionally, quantum mechanical simulations have been incorporated to augment metabolite basis sets for high-resolution magic angle spinning (HRMAS) NMR spectroscopy, skillfully addressing variations between empirical and computational signals, thus improving the fidelity of metabolite concentration determinations in biological tissue specimens [63].

The integration of quantum computing into the metabolomics domain remains in its nascent phase, thus necessitating ongoing research to develop robust quantum algorithms adept at addressing intricate identification challenges such as the MCS problem. The fidelity of quantum chemical computations is influenced by several factors, including the choice of computational approaches and the quality of the input data, thus highlighting the need for continual refinement of methodologies and stringent validation. In any case, the synthesis of quantum sensing techniques with computational quantum chemistry provides a remarkable opportunity for redefining metabolomics, thereby augmenting sensitivity, selectivity, and precision in the identification and detection of metabolites. These innovations are paving the way for novel research and applications not only in biomedical sciences but also in environmental surveillance, agricultural biotechnology, and beyond, ultimately enriching our comprehension of intricate biological systems and fostering the advancement of pioneering diagnostic and therapeutic methodologies [64].

Quantum sensing has significantly revolutionized the detection and identification of metabolites, propelling remarkable enhancements in precision, sensitivity, and efficiency that are redefining the domain of analytical biochemistry and clinical diagnostics. By capitalizing on quantum phenomena, investigators have devised pioneering platforms that surmount the constraints of conventional techniques, permitting the identification and examination of metabolites even within complex biological matrices.

14.4.4 Aptamer-Functionalized Platforms

A prominent illustration is the application of aptamer-functionalized fractal nano-plasmonic surfaces to augment laser desorption/ionization mass spectrometry, which facilitates the selective and label-free evaluation of metabolites in biofluids without necessitating pre-enrichment or purification procedures. This method has been successfully validated in clinical scenarios, such as the identification of diabetic individuals through direct glucose quantification in serum, highlighting its potential for swift and precise disease diagnostics [65].

The evolution of graphene quantum dots (GQDs), specifically those that have been altered with tyramine, has contributed to the establishment of fluorometric sensing platforms skilled in identifying a diverse assortment of metabolites with outstanding sensitivity and specificity, and GQD-based sensors enable multiparametric blood analysis, providing vital diagnostic information for various health conditions such as diabetes and hypertension [66]. Together with these developments, hybrid dual-mode sensors utilizing complementary metal-oxide-semiconductor technologies have been engineered for the concurrent detection of vital serum metabolites such as glucose and cholesterol. These platforms offer a cost-effective and user-friendly solution for personal metabolome sensing, establishing a foundation for accessible, point-of-care diagnostics and personalized health monitoring [67].

Innovative methodologies continue to broaden the scope of quantum sensing in metabolomics. Structure-switching aptamer sequencing signifies a significant advancement in multiplexed metabolite quantification, employing structure-switching aptamers that release distinct DNA barcodes upon target binding, which are subsequently quantified through DNA sequencing. This approach allows for the parallel analysis of various metabolites, markedly improving both throughput and analytical efficiency [68, 69].

Optical nanosensors incorporating aptamers converted into signaling molecules facilitate real-time, quantitative assessment of intracellular metabolite concentrations such as ATP, fructose 1,6-bisphosphate, etc., enabling dynamic investigations of cellular metabolism with elevated spatial and temporal resolution [70]. Moreover, progress in mass spectrometry, including ion mobility mass spectrometry and imaging mass spectrometry, has enhanced the robustness, speed, and ease of metabolite identification, while pharmaco-metabolomics now integrates pre-dose metabolite profiling with chemometrics to forecast individual drug responses, thus offering a personalized paradigm to therapeutics [71].

Despite the existence of these transformative innovations, traditional analytical methodologies, such as mass spectrometry and NMR spectroscopy, continue to remain quintessential to the discipline of metabolomics, facilitating high-throughput, systems-level insights into cellular pathways and biological mechanisms. The synthesis of quantum sensing technologies with recent advancements in classical methodologies is projected to significantly extend the range, precision, and utility of metabolite detection and identification. Altogether, these developments are not only reshaping research in the domain of metabolomics but are also expected to significantly affect clinical diagnostics, personalized medicine, and our holistic insight into human health and disease.

Quantum sensing technologies represent a paradigm shift within the discipline of metabolomics by providing extraordinary sensitivity and precision in the detection and assessment of metabolites. These advancements, quantum sensors utilizing NV centers embedded within diamonds have exhibited exceptional sensitivity; however, their practical implementation is hindered by a multitude of significant technical and conceptual challenges that necessitate resolution to enable the transition from experimental frameworks to routine applications. A primary obstacle pertains to the stability of the sensors. NV-center quantum sensors are particularly susceptible to environmental perturbations, including fluctuations in temperature and electromagnetic interference, which can significantly undermine their performance and reliability over prolonged durations [4, 9]. Furthermore, the assurance of biocompatibility and the minimization of potential harm to biological specimens are essential for maintaining sensor stability, especially within biological contexts [9]. The aspect of scalability presents a further notable challenge. The current configurations of quantum sensors are primarily designed for environments characterized by controlled laboratory settings and exhibit limited adaptability for large-scale manufacturing or widespread use. The progression from prototype devices to commercially viable products necessitates the resolution of manufacturing challenges, including the attainment of consistent sensor performance across extensive production batches [72, 73]. Additionally, the reliance on complex and costly materials, such as diamonds doped with NV centers, constrains scalability and general accessibility [74].

Quantum chemistry occupies a crucial and diverse role in this emerging field, and the progress in the establishment of *in silico* libraries is facilitated by quantum simulation frameworks, which allow for the illustration of complex molecular interactions and provide essential insights into metabolic pathways and their regulatory systems [64, 75]. These computational enhancements may significantly elevate the reliability of analytical approaches such as NMR and mass spectrometry, which play a vital role in the identification and quantification of metabolites [64].

Nevertheless, the integration of quantum chemistry into metabolomics is not devoid of challenges. This field requires reproducibility and accuracy. The financial ramifications and intricacies associated with quantum chemistry computations may initially impede widespread adoption; however, as computational technologies advance and interdisciplinary collaborations broaden, these challenges are likely to be effectively addressed.

14.5 Challenges and Considerations

The principal contributor to errors in quantum sensing is decoherence, wherein interactions with the environment provoke bit-flips and phase-flips that compromise the quantum coherence imperative for accurate measurements [76, 77]. Despite the application of QEC methodologies to alleviate decoherence, these techniques may inadvertently introduce systematic biases, which could result in erroneous estimations of the minimal detectable signals if such biases are not adequately addressed [78]. Measurement inaccuracies are inherently restricted by the standard quantum limit (SQL), a phenomenon that emerges from the back-action effects that the quantum system encounters due to the measurement procedure. To achieve an improvement beyond the SQL, it is vital to adopt cutting-edge techniques such as quantum non-demolition measurements and the application of correlated meter variables, which are significant in addressing measurement-induced variances [79].

In resolving these challenges, ideal practices in quantum sensing stress the significance of employing strong error mitigation strategies, including probabilistic error cancellation, which has been validated as effective in reducing local dephasing noise [80]. The principle of quantum reliability has emerged as an essential measure, allowing for the assessment of sensor efficacy without necessitating prior knowledge about the authentic values of the signal, thereby enhancing the nuanced understanding of the interactions among reliability, sensitivity, and systematic errors [81]. The persistent advancement and optimization of quantum sensing practices, including atom interferometry, are critical in supporting the transition of these technologies from strictly regulated experimental contexts to real-world applications, necessitating ongoing initiatives to overcome the scientific and technological challenges that impede the improvement of sensor precision and operational reliability [82].

To confront these issues, best practices in quantum sensing accentuate robust error mitigation strategies, including probabilistic error cancellation, which has demonstrated efficacy in attenuating local dephasing noise [80]. The notion of quantum reliability has emerged as a significant metric, facilitating the assessment of sensor efficacy without necessitating prior knowledge of true signal values, thus enabling a sophisticated comprehension of the interrelations among reliability, sensitivity, and systematic errors [81]. The issue of reproducibility remains a prominent concern attributed to the lack of standardized quantum computing frameworks, thereby complicating the processes of testing and debugging associated with quantum programming.

The establishment of rigorous testing frameworks and debugging strategies tailored to the specific attributes of quantum systems is essential for preserving the uniformity and reproducibility of experimental results [83]. Moreover, the constant presence of noise and the fundamental limitations of current error correction methods result in discrepancies in experimental outcomes, illustrating the essential need for the systematic execution and development of error correction protocols to secure reproducibility [77, 80].

Forecasting forthcoming advancements, the integration of quantum sensing apparatus with established computational frameworks presents promising prospects for the creation of hybrid systems that leverage the inherent advantages of both paradigms, thereby potentially augmenting sensor efficacy and broadening their applications [83]. Advancements in quantum metrology, particularly concerning the dynamics of noisy quantum systems, will play a crucial role in pushing the boundaries of sensor accuracy and dependability [84]. As quantum sensing technologies continue to evolve, it is expected that their commercial utilization will expand markedly across diverse sectors such as metrology, geophysics, and navigation, consequently fostering further innovation aimed at resolving existing challenges and optimizing sensor efficacy [82].

While quantum sensing possesses substantial potential for attaining unparalleled measurement precision, it is essential to address the complex challenges associated with error sources, best practices, and reproducibility through enhanced mitigation strategies, rigorous testing frameworks, and ongoing technological advancement for the future success and widespread adoption of the field. The continuous convergence of quantum and classical technologies, in conjunction with expanding commercial utilization, will further stimulate innovation and practical applicability in the realm of quantum sensing.

A myriad of challenges remain to be addressed in order to fully realize the potential of quantum sensing within biomedical and clinical arenas. Ongoing initiatives are concentrated on the further enhancement of sensitivity and spatial resolution in quantum sensors, the augmentation of their biocompatibility for secure and efficacious application in living systems, and the development of scalable, economically feasible manufacturing methodologies. As the discipline continues to progress, the intersection of quantum physics, materials science, and bioengineering is poised to unveil new avenues for the investigation of biological processes at the quantum level, ultimately revolutionizing metabolomics, disease detection, personalized medicine, and our comprehensive understanding of health and disease.

14.5.1 Integration with Existing Analytical Workflows

The integration of quantum sensing technologies with pre-existing analytical workflows adds an additional layer of complexity to their assimilation. Quantum sensors often require specialized equipment and a high level of technical proficiency, which may pose significant obstacles to their effortless incorporation into traditional metabolomics laboratories [4]. To illustrate, the synthesis of microfluidic platforms with diamond quantum sensors calls for the establishment of creative experimental protocols and comprehensive training for laboratory technicians. Ensuring alignment with conventional analytical methodologies, such as NMR spectroscopy and mass spectrometry, is essential to ascertain that quantum sensors enhance or supplement these techniques without requiring significant alterations to the existing infrastructure [64].

In consideration of these intricate challenges, the forecasted benefits of quantum sensing within metabolomics are notable, offering superior sensitivity and resolution that can yield more precise, detailed, and comprehensive metabolomic analyses. To effectively harness these benefits, subsequent research projects should aim to create robust, scalable, and integrable quantum sensing frameworks that can be smoothly incorporated into current metabolomics workflows, thus propelling the progression of innovative quantum technologies with applicable biomedical significance.

Quantum sensing, which utilizes principles of quantum mechanics to exceed the precision limitations of classical methods, encounters substantial challenges pertaining to errors, optimal methodologies, and reproducibility that necessitate resolution for the progression of the discipline and the assurance of practical implementation.

14.5.2 Interdisciplinary Collaboration

This paradigm shift necessitates robust interdisciplinary collaboration, integrating chemists, biologists, computational scientists, clinicians, and data scientists to devise innovative strategies for metabolite identification and to translate metabolomic discoveries into actionable clinical applications [64]. In biomedical settings, such collaboration is essential for the progress of precision medicine, wherein metabolomics is increasingly employed to identify disease biomarkers and tailor therapeutic interventions for ailments such as cancer and diabetes [85, 86]. In a parallel manner within environmental studies, metabolomics is instrumental in revealing ecological dynamics and genetic transformations, demanding the collective proficiency of ecologists, geneticists, and metabolomics professionals [87–89]. The mutual enhancement of principles and strategies derived from quantum chemistry and metabolomics is consequently poised to facilitate innovation throughout both basic and applied research sectors [90].

Acknowledging the paramount significance of metabolomics, funding entities such as the NIH have prioritized the development of infrastructure, thereby endorsing the establishment of comprehensive metabolomics resource cores, educational programs, and technological advancement initiatives [91]. These investments are vital for surmounting existing limitations and nurturing the forthcoming generation of metabolomics scholars. Important funding factors also involve the formulation of standardized reference materials and the enhancement of measurement consistency, as illustrated by the NIH's alliance with NIST to produce a human plasma pool intended for application as a standardized reference material. Moreover, strategically allocated funding for translational research seeks to close the gap between foundational scientific inquiry and clinical application, thus ensuring that progress in metabolomics translates into concrete advantages for patient care and public health [92].

14.6 Conclusion

Quantum sensors, with their unrivaled sensitivity and precision, offer a transformative leap in metabolite detection and analysis. By leveraging quantum phenomena like spin manipulation and entanglement, these tools overcome key analytical limitations faced by conventional methods. Their integration into metabolomics sets the stage for a new era of ultra-accurate diagnostics and real-time molecular insights. The commercialization and clinical application of quantum sensors in the field of metabolomics represent a revolutionary development within biomedical science, driven by the remarkable sensitivity, specificity, and real-time analytical capabilities that are intrinsic to quantum technologies.

Quantum sensors such as OPMs and NV centers integrated within diamond matrices exhibit distinctive capabilities to detect low-abundance metabolites in complex biological systems with atomic-scale accuracy, thereby providing essential insights into metabolic processes that are critical for diagnostics, therapeutic monitoring, and the clarification of disease mechanisms [2]. Their capacity to produce real-time data facilitates the dynamic observation of metabolic fluctuations, which can substantially augment the evaluation of disease progression and the efficacy of therapeutic interventions [93]. The amalgamation of quantum sensors with existing metabolomics platforms harbors the potential to generate more exhaustive metabolic profiles, thereby propelling the objectives of precision medicine by fostering more accurate discovery of disease biomarkers and personalized patient care [94].

The trajectory toward commercialization is actively being influenced by an emergent ecosystem of start-ups and technology enterprises dedicated to translating laboratory advancements in quantum sensing into commercially viable diagnostic products [2, 95]. These initiatives are reinforced by interdisciplinary collaborations among physicists, biologists, engineers, and clinicians, which are indispensable for surmounting technical challenges associated with sensor stability, reproducibility, and seamless integration into established diagnostic workflows [93].

Technical challenges such as the maintenance of sensor stability within complex biological environments and the management of the inherent intricacies of quantum interactions necessitate systematic resolution to guarantee consistent and reliable performance in authentic clinical contexts. Furthermore, the integration of data generated by quantum sensors with other omics datasets such as genomics and proteomics will be essential for constructing a comprehensive and actionable understanding of metabolic processes, ultimately bolstering the development of personalized medicine strategies [94]. The initial high expenditure associated with quantum sensor technology may pose a barrier to widespread clinical integration, underscoring the necessity for ongoing innovation in cost reduction and scalable manufacturing methodologies [96].

The integration of quantum sensing and quantum chemistry into metabolomics represents a foundational leap in analytical science. These innovations not only overcome critical limitations in detection sensitivity and metabolite identification but also enable standards-free analysis through *in silico* spectral libraries. Their potential to transform clinical diagnostics, therapeutics, and environmental monitoring is substantial; however, successful translation will require coordinated efforts across research, industry, and regulation. As quantum-enabled methodologies evolve, they promise to redefine metabolomics as a high-precision, real-time platform for understanding and improving biological systems.

References

1 Belser, S., Hart, J., Gu, Q. et al. (2023). Opportunities for diamond quantum metrology in biological systems. *Applied Physics Letters* https://doi.org/10.1063/5.0147469.

2 Aslam, N., Zhou, H.-X., Urbach, E. et al. (2023). Quantum sensors for biomedical applications. *Nature Reviews Physics* https://doi.org/10.1038/s42254-023-00558-3.

3 Sarkar, A., Jones, Z.R., Parashar, M. et al. (2024a). High-precision chemical quantum sensing in flowing monodisperse microdroplets. *Science Advances* https://doi.org/10.1126/sciadv.adp4033.

4 Smits, J., Smits, J., Damron, J. et al. (2019). Two-dimensional nuclear magnetic resonance spectroscopy with a microfluidic diamond quantum sensor. *Science Advances* https://doi.org/10.1126/SCIADV.AAW7895.

5 Trofimov, S. and Naydenov, B. (2024). Combined confocal-atomic-force microscope setup for quantum sensing applications with sub-diffractional spatial resolution. *Physica Status Solidi A: Applications and Materials Science* https://doi.org/10.1002/pssa.202400541.

6 Monnappa, A.B., Monnappa, A.B., Manjunatha, J.G., and Bhatt, A.S. (2020). Design of a sensitive and selective voltammetric sensor based on a cationic surfactant-modified carbon paste electrode for the determination of alloxan. *ACS Omega* 5 (36): 23481–23490. https://doi.org/10.1021/ACSOMEGA.0C03517.

7 Raril, C., Manjunatha, J., Ravishankar, D.K. et al. (2020). Validated electrochemical method for simultaneous resolution of tyrosine, uric acid, and ascorbic acid at polymer modified nano-composite paste electrode. *Surface Engineering and Applied Electrochemistry* 56 (4): 415–426. https://doi.org/10.3103/S1068375520040134.

8 Sandeep, C., Padmashali, B., Kulkarni, R. et al. (2014). Synthesis of substituted 5-acetyl-3-benzoylindolizine-1-carboxylate from substituted 2-acetyl pyridinium bromides. *Heterocyclic Letters* 4 (3): 371–376.

9 Shanahan, L., Belser, S., Hart, J.W. et al. (2024). Q-BiC: a biocompatible integrated chip for in vitro and in vivo spin-based quantum sensing. *PRX Life* https://doi.org/10.48550/arxiv.2406.01181.

10 Wang, X., Xu, J., Ge, S. et al. (2023). Recent applications of nanodiamond quantum biosensors: a review. *APL Materials* https://doi.org/10.1063/5.0170145.

11 Hareesha, N. and Manjunatha, J.G. (2021). Electro-oxidation of formoterol fumarate on the surface of novel poly(thiazole yellow-G) layered multi-walled carbon nanotube paste electrode. *Scientific Reports* 11 (1): 12797. https://doi.org/10.1038/S41598-021-92099-X.

12 Manjunatha, J.G., Swamy, K., Mamatha, G. et al. (2011). Poly(maleic acid) modified carbon paste electrode for simultaneous detection of dopamine in the presence of uric acid: a cyclic voltammetric study. *Analytical and Bioanalytical Electrochemistry* 3: 146–159.

13 Pushpanjali, P.A., Manjunatha, J.G., and Hareesha, N. (2021). An overview of recent developments of carbon-based sensors for the analysis of drug molecules. *Journal of Electrochemical Science and Engineering* 11 (3): 161–177. https://doi.org/10.5599/JESE.999.

14 Tigari, G. and Manjunatha, J.G. (2020). Poly(glutamine) film-coated carbon nanotube paste electrode for the determination of curcumin with vanillin: an electroanalytical approach. *Monatshefte Fur Chemie* 151 (11): 1681–1688. https://doi.org/10.1007/S00706-020-02700-8.

15 Ağbulut, M.S.B., Elibol, E., Çadırcı, M., and Demirci, T. (2025). Fluorescent CdTe/ZnS core/shell quantum dots for sensitive metabolite detection in real samples. *Journal of Fluorescence* https://doi.org/10.1007/s10895-025-04138-9.

16 Ganesh, S., Ganesh, S., Venkatakrishnan, K. et al. (2020). Tailoring carbon for single molecule detection – broad spectrum 3D quantum sensor. *Sensors and Actuators B: Chemical* https://doi.org/10.1016/J.SNB.2020.128216.

17 Pushpanjali, P.A., Manjunatha, J.G., Hareesha, N. et al. (2021). Voltammetric analysis of antihistamine drug cetirizine and paracetamol at poly(L-leucine) layered carbon nanotube paste electrode. *Surfaces and Interfaces* 24: 101154. https://doi.org/10.1016/J.SURFIN.2021.101154.

18 Neuling, N.R., Allert, R.D., and Bucher, D.B. (2023b). Prospects of single-cell nuclear magnetic resonance spectroscopy with quantum sensors. *Current Opinion in Biotechnology* https://doi.org/10.1016/j.copbio.2023.102975.

19 Ghamsari, M.S. and Baniasadi, F. (2024). Quantum for biology: spectroscopy and sensing. *Innovation and Emerging Technologies* https://doi.org/10.1142/s2737599424300058.

20 Fujiwara, M. (2023). Diamond quantum sensors in microfluidics technology. *Biomicrofluidics* https://doi.org/10.1063/5.0172795.

21 Chipaux, M., Xavier, S., Tallaire, A. et al. (2015). Nitrogen vacancies (NV) centers in diamond for magnetic sensors and quantum sensing. In: *Quantum Sensing and Nanophotonic Devices XII*, vol. 9370, 400–405. SPIE. doi: 10.1117/12.2084082.

22 Zhang, T., Pramanik, G., Zhang, K. et al. (2021). Toward quantitative bio-sensing with nitrogen-vacancy center in diamond. *ACS Sensors* https://doi.org/10.1021/ACSSENSORS.1C00415.

23 Rondin, L., Tetienne, J.-P., Hingant, T. et al. (2014). Magnetometry with nitrogen-vacancy defects in diamond. *Reports on Progress in Physics* https://doi.org/10.1088/0034-4885/77/5/056503.

24 Wu, Y., Jelezko, F., Plenio, M.B., and Weil, T. (2016). Diamond quantum devices in biology. *Angewandte Chemie* https://doi.org/10.1002/ANIE.201506556.

25 Schirhagl, R., Chang, K., Loretz, M., and Degen, C.L. (2014). Nitrogen-vacancy centers in diamond: nanoscale sensors for physics and biology. *Annual Review of Physical Chemistry* https://doi.org/10.1146/ANNUREV-PHYSCHEM-040513-103659.

26 Hernández-Gómez, S., Balducci, F., Cappellaro, P. et al. (2022). Optimal quantum control of a spin qubit in diamond for biosensing. *2022 IEEE International Workshop on Metrology for Industry 4.0 & IoT (MetroInd4. 0&IoT)*, pp. 115–120. IEEE. https://doi.org/10.1109/metroind4.0iot54413.2022.9831543.

27 Modak, A., Majumder, A., Parashar, M.S. et al. (2023). Few NV center nanodiamonds enable high-speed and high-resolution sensing of paramagnetic species. *2023 Joint Conference of the European Frequency and Time Forum and IEEE International Frequency Control Symposium (EFTF/IFCS)*, pp. 1–3. IEEE. https://doi.org/10.1109/eftf/ifcs57587.2023.10272183.

28 Dhingra, S. and D'Urso, B. (2017). Nitrogen vacancy centers in diamond as angle-squared sensors. *Journal of Physics: Condensed Matter* https://doi.org/10.1088/1361-648X/AA6697.

29 Karki, P.B., Timalsina, R., Dowran, M. et al. (2023a). An efficient and low-cost method to create high-density nitrogen-vacancy centers in CVD diamond for sensing applications. In: *Diamond and Related Materials*, 110472. Elsevier http://dx.doi.org/10.1016/j.diamond.2023.110472.

30 Karki, P.B., Timalsina, R., Dowran, M. et al. (2023b). An efficient method to create high-density nitrogen-vacancy centers in CVD diamond for sensing applications. *Diamond and Related Materials* https://doi.org/10.1016/j.diamond.2023.110472.

31 Sushkov, A.O., Chisholm, N., Lovchinsky, I. et al. (2014). All-optical sensing of a single-molecule electron spin. *Nano Letters* 14 (11): 6443–6448. https://doi.org/10.1021/nl502988n.

32 Luo, T., Lindner, L., Langer, J. et al. (2022). Creation of nitrogen-vacancy centers in chemical vapor deposition diamond for sensing applications. *New Journal of Physics* https://doi.org/10.1088/1367-2630/ac58b6.

33 Allert, R.D., Briegel, K.D., and Bucher, D.B. (2022). Advances in nano- and microscale NMR spectroscopy using diamond quantum sensors. *Chemical Communications* https://doi.org/10.26434/chemrxiv-2022-26s1t-v3.

34 Rizzato, R., Neuling, N.R., and Bucher, D.B. (2023). Quantum sensors in diamonds for magnetic resonance spectroscopy: current applications and future prospects. *Applied Physics Letters* https://doi.org/10.1063/5.0169027.

35 Mauranyapin, N.P., Booth, L., Marinkovic, I., and Bowen, W.P. (2022). Interferometric biosensing. In: *Nanostructure Science and Technology Single Molecule Sensing Beyond Fluorescence*, 5–36. https://doi.org/10.1007/978-3-030-90339-8_1.

36 Campbell, D.P. and McCloskey, C.J. (2002). Chapter 9 - Interferometric biosensors. In: *Optical Biosensors*, 277–304. https://doi.org/10.1016/B978-044450974-1/50009-4.

37 Chalyan, T., Ottevaere, H., and Pasquardini, L. (2024). Optical biosensors: from working principles to detection methods of label-free devices. In: *Biophotonics and Biosensing*, 15–48. Elsevier https://doi.org/10.1016/b978-0-44-318840-4.00009-7.

38 Davé, D.P. (2009). Phase sensitive interferometry for biosensing applications. *Biosensors and Biodetection* 503: https://doi.org/10.1007/978-1-60327-567-5_9.

39 Sevenler, D., Avci, O., and Ünlü, M.S. (2017). Quantitative interferometric reflectance imaging for the detection and measurement of biological nanoparticles. *Biomedical Optics Express* https://doi.org/10.1364/BOE.8.002976.

40 Love, C.A., Kilper, G.K., Antoshak, E. et al. (2024). Biolayer interferometry as a biological detection platform. In: *Chemical, Biological, Radiological, Nuclear, and Explosives (CBRNE) Sensing XXV*, vol. 13056, 43–50. SPIE https://doi.org/10.1117/12.3013538.

41 Mpofu, K. and Mthunzi-Kufa, P.P. (2024). Recent advances in quantum biosensing technologies. https://doi.org/10.5772/intechopen.1007858.

42 Taylor, M.A., Janousek, J., Daria, V.R. et al. (2013). Biological measurement beyond the quantum limit. *Nature Photonics* 7 (3): 229–233.

43 Luong, D., Lam, I.P.Y., Balaji, B., and Rajan, S. (2024). Quantum interferometric radar for biomedical applications. In: *Radar Sensor Technology XXVIII*, vol. 13048, 169–173. SPIE https://doi.org/10.1117/12.3013139.

44 Taylor, M.A., Taylor, M.A., and Bowen, W.P. (2014). Quantum metrology and its application in biology. *arXiv: Quantum Physics.*

45 Bao, B., Hua, Y., Wang, R., and Li, D. (2023). Quantum-based magnetic field sensors for biosensing. *Advanced Quantum Technologies* https://doi.org/10.1002/qute.202200146.

46 Danilin, S. and Weides, M. (2021). Quantum sensing with superconducting circuits. *arXiv: Quantum Physics.*

47 Sarkar, S., Srivastava, T., Sahoo, O.S. et al. (2024). Applications of quantum dots in preventive oncology. *Asian Pacific Journal of Cancer Prevention* https://doi.org/10.31557/apjcp.2024.25.3.747.

48 Sathe, K.P., Garud, N.S., Bangar, V.B., and Gadakh, N.R. (2022). A review on quantum dots (QDS). *Journal of Advanced Scientific Research* https://doi.org/10.55218/jasr.202213603.

49 Li, W. (2013). Quantum dots for biological imaging. In: *Molecular Imaging: Advanced Topics in Science and Technology in China*. Springer https://doi.org/10.1007/978-3-642-34303-2_14.

50 Chow, J.C.L. (2024). Quantum computing in medicine. *Medical Sciences* 12: 67. https://doi.org/10.3390/medsci12040067.

51 Barik, P. and Pradhan, M. (2022). All-optical detection of biocompatible quantum dots. In: *Application of Quantum Dots in Biology and Medicine*, 35–65. https://doi.org/10.1007/978-981-19-3144-4_3.

52 Matea, C., Mocan, T., Tabaran, F. et al. (2017). Quantum dots in imaging, drug delivery and sensor applications. *International Journal of Nanomedicine* https://doi.org/10.2147/IJN.S138624.

53 Sander, T., Jodko-Władzińska, A., Jodko-Władzińska, A. et al. (2020). Optically pumped magnetometers enable a new level of biomagnetic measurements. *Advanced Optical Technologies* https://doi.org/10.1515/AOT-2020-0027.

54 Konstantinovich, V.A., Pavlovich, D.S., Serafimovich, P.A., and Valerevich, P.M. (2020). Quantum sensor and methods for measuring transverse component of weak magnetic field (versions) https://doi.org/10.13140/RG.2.2.17829.86249.

55 Du, J., Shi, F., Jelezko, F., and Wrachtrup, J. (2024). Single-molecule scale magnetic resonance spectroscopy using quantum diamond sensors. *Reviews of Modern Physics* https://doi.org/10.1103/revmodphys.96.025001.

56 Holzgrafe, J., Holzgrafe, J., Gu, Q. et al. (2020). Nanoscale NMR spectroscopy using nanodiamond quantum sensors. *Physical Review Applied* https://doi.org/10.1103/PHYSREVAPPLIED.13.044004.

57 Holzgrafe, J., Gu, Q., Beitner, J. et al. (2019). Nanoscale NMR spectroscopy using self-calibrating nanodiamond quantum sensors. *arXiv: Mesoscale and Nanoscale Physics*.

58 Kong, X., Kong, X., Stark, A. et al. (2015). Towards chemical structure resolution with nanoscale nuclear magnetic resonance spectroscopy. *Physical Review Applied* https://doi.org/10.1103/PHYSREVAPPLIED.4.024004.

59 Neuling, N.R., Allert, R.D., and Bucher, D.B. (2023). Prospects of single-cell NMR spectroscopy with quantum sensors. *Current Opinion in Biotechnology* 83: 102975. https://doi.org/10.1016/j.copbio.2023.102975.

60 Lipka, M., Sierant, A., Troullinou, C., and Mitchell, M. (2023). Multiparameter quantum sensing and magnetic communication with a hybrid dc and rf optically pumped magnetometer. *Physical Review Applied* https://doi.org/10.1103/physrevapplied.21.034054.

61 Tsai, L.-A., Nuckels, E., and Wang, Y. (2023). Integrating quantum computing into de novo metabolite identification. *Journal on Systemics, Cybernetics and Informatics* https://doi.org/10.54808/jsci.21.02.83.

62 Hoffmann, F., Li, D., Sebastiani, D., and Brüschweiler, R. (2017). Improved quantum chemical NMR chemical shift prediction of metabolites in aqueous solution toward the validation of unknowns. *Journal of Physical Chemistry A* https://doi.org/10.1021/ACS.JPCA.7B01954.

63 Lazariev, A., Aubert-Frécon, M., Fauvelle, F. et al. (2011). A quantum mechanics-based approach for optimization of metabolite basis-sets. Application to quantitation of HRMAS-NMR signals Optimisation des bases de métabolites fondée sur une approche par mécanique quantique. Application à la quantification de signaux RMN-HRMAS. *IRBM* 32 (5): 279–283.

64 Borges, R.M., Colby, S.M., Das, S. et al. (2021). Quantum chemistry calculations for metabolomics. *Chemical Reviews* https://doi.org/10.1021/ACS.CHEMREV.0C00901.

65 Wang, X., Yan, L., Yu, Z. et al. (2022). Aptamer-functionalized fractal nanoplasmonics-assisted laser desorption/ionization mass spectrometry for metabolite detection. *Collection of Czechoslovak Chemical Communications* https://doi.org/10.1002/cplu.202100479.

66 Li, N., Than, A., Wang, X. et al. (2016). Ultrasensitive profiling of metabolites using tyramine-functionalized graphene quantum dots. *ACS Nano* https://doi.org/10.1021/ACSNANO.5B08103.

67 Hu, C., Al-Rawhani, M.A., Cheah, B.C. et al. (2018). Hybrid dual mode sensor for simultaneous detection of two serum metabolites. *IEEE Sensors Journal* https://doi.org/10.1109/JSEN.2017.2774359.

68 Tan, J.H. and Fraser, A.G. (n.d.). Quantifying metabolites using structure-switching aptamers coupled to DNA sequencing. *Nature Biotechnology* https://doi.org/10.1038/s41587-025-02554-7.

69 Tan, J. H., Mercado, M. P., & Fraser, A. G. (2023). A novel platform for metabolomics using barcoded structure-switching aptamers. *bioRxiv*. https://doi.org/10.1101/2023.06.09.544402.

70 Long, Y., Long, Y., Pfeiffer, F. et al. (2016). Selection of aptamers for metabolite sensing and construction of optical nanosensors. In: *Nucleic Acid Aptamers: Selection, Characterization, and Application*, 3–19. New York: Springer New York https://doi.org/10.1007/978-1-4939-3197-2_1.

71 Lu, W., Xu, Y., Zhao, Y., and Cen, X. (2015). Emerging technologies, recent developments, and novel applications for drug metabolite identification. *Current Drug Metabolism* https://doi.org/10.2174/1389200216666141230105649.

72 Oh, E., Gregoire, M. D., Black, A., Hughes, K. J., Kunz, P. D., Larsen, M. D., Lautier-Gaud, J., Lee, J., Schwindt, P., Mouradian, S., Narducci, F. A., & Sackett, C. A. (2024). A perspective on quantum sensors from basic research to commercial applications. *AIAA Journal*. https://doi.org/10.48550/arxiv.2407.00689.

73 Kantsepolsky, B., Aviv, I., Weitzfeld, R., and Bordo, E. (2023). Exploring quantum sensing potential for systems applications. *IEEE Access* https://doi.org/10.1109/access.2023.3262506.

74 Moreva, E.V., Cimini, V., Gianani, I. et al. (2025). Quantum photonics sensing in biosystems. *APL Photonics* https://doi.org/10.1063/5.0232183.

75 Njoki, W.A. (2024). Transformative potential of quantum simulation in chemistry and materials science: advances, applications, and future directions. https://doi.org/10.59298/jcas/2024/91.152732001.

76 Hanzo, L., Babar, Z., Cai, Z. et al. (2025). Quantum information processing, sensing, and communications: their myths, realities, and futures. *Proceedings of the IEEE* https://doi.org/10.1109/jproc.2024.3510394.

77 Rojkov, I., Layden, D., Cappellaro, P. et al. (2022). Bias in error-corrected quantum sensing. *Physical Review Letters* https://doi.org/10.1103/physrevlett.128.140503.

78 Rojkov, I., Layden, D., Cappellaro, P. et al. (2021). Bias in error-corrected quantum sensing. *arXiv: Quantum Physics*.

79 Vorontsov, Y.I. (1994). Standard quantum limits of measurement error and methods of overcoming them. *Physics-Uspekhi* https://doi.org/10.1070/PU1994V037N01ABEH000004.

80 Takagi, R., Minagawa, S., Oláh, V.P., and Reyes, J.M. (2022). Fundamental limits of quantum error mitigation. *npj Quantum Information* https://doi.org/10.1038/s41534-022-00618-z.

81 Cui, L., Du, Y.-M., and Sun, C.-P. (2025). On quantum reliability characterizing systematic errors in quantum sensing. *Journal of Reliability Science and Engineering* https://doi.org/10.1088/3050-2454/ada8ac.

82 Bongs, K., Holynski, M., Vovrosh, J. et al. (2019). Taking atom interferometric quantum sensors from the laboratory to real-world applications. *Nature Reviews Physics* https://doi.org/10.1038/S42254-019-0117-4.

83 Ramalho, N.C.L., de Souza, H.A., and Chaim, M.L. (2025). Testing and debugging quantum programs: the road to 2030. *ACM Transactions on Software Engineering and Methodology* https://doi.org/10.1145/3715106.

84 Davidovich, L. (2024). Quantum sensing: beyond the classical limits of precision. https://doi.org/10.48550/arxiv.2401.13658.

85 Gonzalez-Covarrubias, V., Martínez-Martínez, E., and del Bosque-Plata, L. (2022). The potential of metabolomics in biomedical applications. *Metabolites* https://doi.org/10.3390/metabo12020194.

86 Siddesh, M.B., Padmashali, B., and Thriveni, K. (2014). Synthesis of thiophene-linked pyrimidopyrimidines as pharmaceutical leads. *Journal of Chemical Sciences* 126: 821–826.

87 Thriveni, K., Padmashali, B., Siddesh, M., and Sandeep, C. (2014). Synthesis of pyrimidine incorporated piperazine derivatives and their antimicrobial activity. *Indian Journal of Pharmaceutical Sciences* 76 (4): 332–338.

88 Mallikarjuna, S. and Chandrashekharappa, S. (2017). Acid amine coupling of 1*H*-indole-6-yl(ppiperazin-1-yl) methanone with substituted acids using HATU coupling reagent and their antimicrobial and antioxidant activity. *International Journal of Pharmaceutical Sciences and Research* 8 (7): 2879–2885.

89 Sandeep, C., Basavaraj, P., and Rashmi, S. (2013). Synthesis of isomeric subtituted 6-acetyl-3-benzoylindolizine-1-carboxylate and 8-acetyl-3-benzoylindolizine-1-carboxylate from subtituteded 3-acetyl pyridinium bromides and their antimicrobial activity. *Journal of Applied Chemistry* 2: 1049.

90 Munir, A. (2024). The transformative future of metabolomics from diagnostics to therapeutics. *International Journal of Pharmacy & Integrated Health Sciences* https://doi.org/10.56536/ijpihs.v5i2.213.

91 Zanetti, K.A., Mette, E., Maruvada, P. et al. (2014). The future of metabolomic profiling in population-based research: opportunities and challenges. *Journal of Analytical and Bioanalytical Techniques* https://doi.org/10.4172/2155-9872.1000203.

92 Kaddurah-Daouk, R. (2012). Interest in metabolomics continues to grow within US National Institutes of Health (NIH). *Metabolomics* https://doi.org/10.1007/S11306-012-0444-2.

93 Das, S., Mazumdar, H., Khondakar, K.R. et al. (2024). Quantum biosensors: principles and applications in medical diagnostics. *ECS Sensors Plus* https://doi.org/10.1149/2754-2726/ad47e2.

94 Beger, R.D., Dunn, W.B., Schmidt, M.A. et al. (2016). Metabolomics enables precision medicine: "a white paper, community perspective." *Metabolomics* https://doi.org/10.1007/S11306-016-1094-6.

95 Choi, C.Q., Fairley, P., Perry, T.S., and Patel, P. (2022). Sensors: a guide to the quantum-sensor boom: atomic scale bolsters sensing revolutions in medicine, tech, and engineering. *IEEE Spectrum* https://doi.org/10.1109/MSPEC.2022.9792186.

96 Gschwendtner, M., Bormuth, Y., Soller, H. et al. (2024). Quantum sensing can already make a difference. But where? *Journal of Innovation Management* https://doi.org/10.24840/2183-0606_012.001_l001.

15

Quantum Plasmonic in Biosensing

Raril Chenthattil, Sree Lekshmi, Aswathy S. Murali, Leona R. Varghese, Anuja Sudarsanan, and Beena Saraswathyamma

Department of Chemistry, Amrita Vishwa Vidyapeetham, Kollam, Kerala, India

15.1 Introduction

Biosensing is an emerging analytical field for the detection of biological indicators. These processes convert biological variation into optical, thermal, or electrochemical signals. Several biological targets, including small biomolecules, biomacromolecules, cells, bacteria, and viruses, can be detected via biosensing. High selectivity, sensitivity, and quick response biosensors offer a wide range of applications in diagnostics and disease treatment. Developments in the biomedical sciences are often driven by the development of instruments with enhanced sensitivity and resolution, which allow the imaging and detection of increasingly weaker and more localized signals. Diagnostics and therapy have greatly advanced as a result of improvement in many analytical techniques; conventional methods seem to limit further progress in sensitivity and resolution. However, developments in quantum science and technology offer a promising path toward a new generation of biological sensors with significantly improved performance. With the ongoing rise in pandemics and epidemics, more sensitive and accurate biosensing technologies have to be developed [1–3].

Quantum biosensing (QB) refers to the application of quantum phenomena or quantum mechanics principles in biological sensing and detection concepts to biological sensing and detecting [4–6]. It involves using the special properties of quantum systems to develop accurate and sensitive methods for identifying infections, biological molecules, or other biological entities. QB enables the detection of specific molecules or even single atoms within biological samples with remarkable sensitivity and precision. This emerging field combines nanoscale materials and quantum mechanics to improve biosensor performance, with improved sensitivity and functionality compared to conventional methods. Quantum biosensors (QBs) use quantum phenomena to improve sensitivity, accuracy, and selectivity in detecting biological entities. A wide range of sensing materials developed using quantum processes measures various properties, including temperature, strain, thermal conductivity, electric and magnetic field, and pH. QBs are expected to find increasing applications in the

Quantum Biosensing in Medical Diagnostics, First Edition. Edited by J. G. Manjunatha.

real world as technology develops and understanding of quantum phenomena deepens. Biotechnology and healthcare are expected to be greatly impacted by this advancement [7].

Various fields of research and technology make use of optical sensors. Notable examples include electrochemical devices for evaluating renewable energy storage, density sensors for analyzing saltwater in climate change research, gyroscopes for navigation [8], accelerometers for tracking structural deformations [9], and seismic wave activity [10]. Sensors that detect and monitor the presence and behavior of bacteria, viruses, and proteins are of particular interest to the biological and chemical sciences [11–19]. Plasmonic effects have been successfully exploited in practical photonic sensors, offering significantly higher sensitivities compared to conventional photonic sensing platform [20].

The sensitivity of surface plasmon resonances (SPRs), which are employed in photonic sensors, varies depending on the geometrical and material configuration, leading to a diverse range of sensing platforms [21–23]. A new path for integrated nanophotonics and miniaturized optoelectronic devices has been made possible by the ability to enhance and confine light at dimensions far smaller than the incident wavelength. At the interface between two with dielectric constants of opposite signs—usually a metal and a dielectric—with the light intensity rapidly decaying as it moves away from the interface. These properties make plasmonic platforms well-suited for point-of-care (POC) diagnostic devices [24, 25], owing to their integration into microfluidic systems and high sensitivity to changes at the interface. Despite the substantial research efforts on SPR biosensing since its first demonstration around four decades ago, detecting small analytes or those dispersed in solutions at very low concentration remains challenging. Minimal changes in dielectric properties near the metal/dielectric interface may be overcome by improving sensing performance. In this chapter, we aim to introduce the fundamentals of quantum plasmonics, the mechanism of biosensing enhancement through plasmonics, and its application in biosensing.

15.2 Fundamentals of Quantum Plasmonic

15.2.1 Surface Plasmon Resonances (SPRs) and Localized Surface Plasmons (LSPs)

Surface plasmons are collective oscillations of free electrons at the interface between a metal and a dielectric substance, such as air or water [26]. The ability of a metal to excite surface plasmons is determined by its electromagnetic properties, specifically its dielectric function or complex refractive index. Nobel metals such as gold, silver, and copper are particularly well suited for surface plasmon excitation due to their negative real portion of the dielectric function at optical frequencies. This enables strong coupling between incident light and the collective oscillations of electrons at the metal surface, known as surface plasmons. On the other hand, because of their weak coupling with light, metals such as aluminum and titanium that have a low density of free electrons cannot support surface plasmons at optical frequencies. The ability to excite surface plasmons can also be influenced by the dimensions and form of the metal nanostructures. Gold or silver are typically used to excite surface plasmons because their electromagnetic properties, such as charge arrangement and conductivity, are the most appropriate for obtaining an excitation response. Antibacterial and

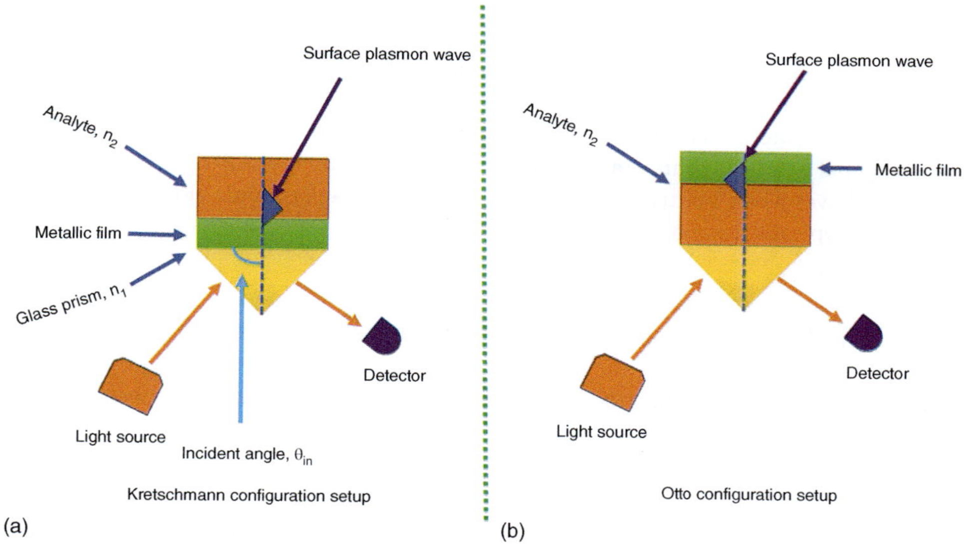

Figure 15.1 Exciting surface plasmons in prism configuration setups. (a) Krestchmann configuration setup and (b) Otto configuration setup. Figures (a) and (b) show how surface plasmons are generated in a prism-based SPR setup. There are two common configurations that can be used to excite surface plasmons, that is, the Krestchmann configuration shown in (a) and the Otto configuration shown in (b). Adapted from Hindawi, *International Journal of Optics*, 2023.

antiviral properties make silver not to be used for biological experiments. Figure 15.1 shows the surface plasmon waves whose propagation is confined to a metal-dielectric interface.

LSPs, or localized surface plasmons, are the oscillations of the electromagnetic field and free electrons in metallic nanoparticles that are merged. For nanostructures made of noble metals such as gold (Au), silver (Ag), and copper (Cu), the resonance typically occurs in the visible to near-infrared region and is dependent on the particle size, shape, composition, and optical environment [27]. LSPs are an attractive alternative for propagating surface plasmons (PSPs) in a variety of applications. Since LSPs do not have translational symmetry, they can be excited without momentum matching. This makes them more adaptable and simpler to integrate into practical devices. Furthermore, LSPs benefit from relatively easy production methods for metal nanoparticles [28] and for nanoparticle arrays. LSPs exhibit a number of special features and capabilities, enabling advanced light and transforming light at the nanoscale level [29–31]. Sub-wavelength imaging [32], nano-lensing, nano-lasing, field concentration [33], and nanotweezers are important uses. Additionally, LSPs are essential for the development of supersensitive plasmonic nanosensors [34, 35], enhanced photovoltaic devices, and active optical elements. However, compared to PSPs, LSPs typically have greater spectral width. For example, the resonance width of LSPs for Au nanostructures typically exceeds ~80–100 nm (Full Width at Half Maximum whereas PSPs have a spectral width of ~50 nm. For many applications, the potential of LSPs is severely limited by this increased width. Aluminum can also be employed as a feasible plasmonic material that supports resonances in the UV-visible range, despite not being on the noble metal list.

15.2.2 Evaluation of Plasmonic Biosensors

To evaluate the performance of biosensors, a number of figures of merit are widely used. The most widely used of these are sensitivity, selectivity, specificity, and limit of detection (LOD). Two important criteria that can be used to compare biosensors from various platforms are sensitivity and detection limits. In the context of virus detection, sensitivity describes how a sensor behaves when a virus is present. The sensitivity of a plasmonic biosensor for viral detection is typically described as:

$$S = \frac{\Delta\lambda}{\Delta C}$$

Here, $\Delta\lambda$ and ΔC are the change in peak reflection/transmission wavelength and change in analyte concentration.

The lowest concentration that the sensor can detect is known as the LOD. A formula that is frequently used to determine LOD is

$$\text{LOD} = \frac{3\sigma}{S}$$

Here σ and S are the standard deviation of the blank solution (without analyte), and S is the slope of the linear fit.

15.2.3 Mechanism of Plasmonic Biosensors

Plasmonic optical biosensors fall into two categories: those that employ thin metal-based films and those that employ inorganic plasmon resonance based on nanostructures [36]. The SPR sensor, a metal-based film sensor primarily composed of gold, is the most widely used plasmonic biosensor for describing biomolecular interactions [37]. The beam of light is focused on the metal film and reflected onto a detector to produce the SPR. The energized plasmonic electrons that result from the collective refractive index (RI) of the conduction band electron oscillations and the light-induced electric field oscillations are responsible for the output that was gathered. The RI of the substance affixed to the metal film determines the angle at which the SPR is measured. Therefore, any change in the material's RI or angle of incidence has an impact on the resonance observed during analyte detection [38]. A red shift that holds a low frequency is produced by raising the surrounding dielectric constant, whereas a blueshift that holds a high frequency is produced by decreasing the size [39]. As illustrated in Figure 15.2, targeted analytes in a sample attach to the biological receptor that is immobilized on the metal film, altering the local RI. This change in RI produces a measurable shift in the resonance angle, which indicates the presence of the target analyte [40]. Moreover, SPR biosensors can be reused after proper chemical treatment to remove the bound analytes, maintaining sensor performance for subsequent measurements [41].

15.2.4 Quantum Plasmonic in Biosensing

In today's biomedical and global healthcare fields, they play a critical role in diagnosing diseases, identifying multidrug-resistant organisms, detecting epidemic, and detecting extremely low levels of toxins and microorganisms in food or drink. However, significant challenges remain for scientists and engineers working to improve biosensor performance.

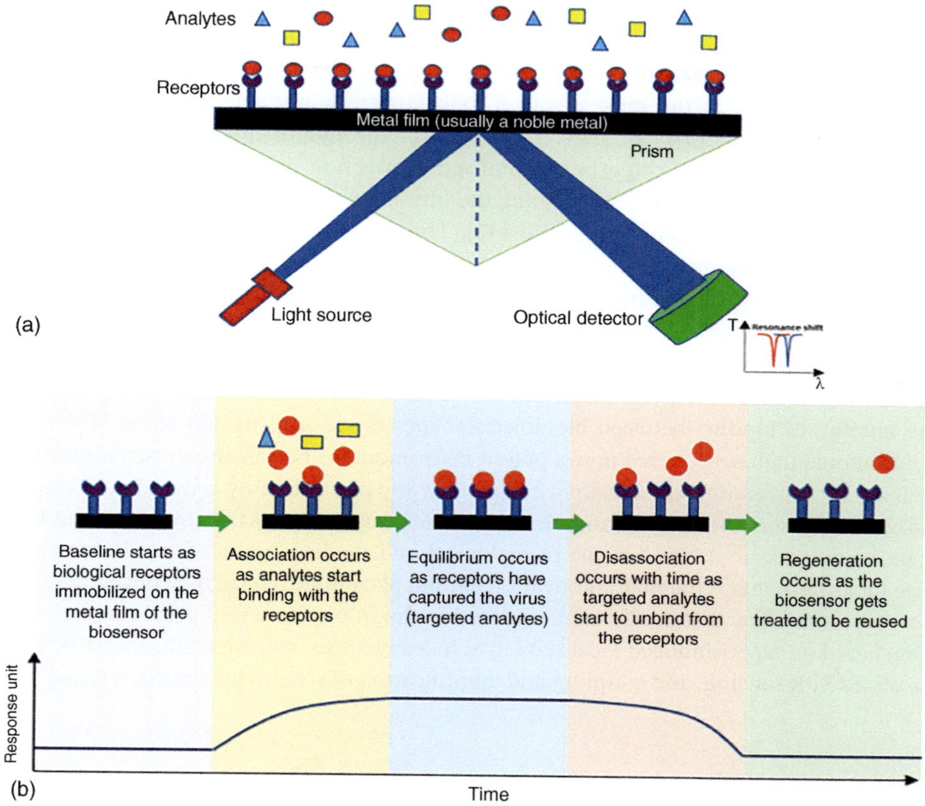

Figure 15.2 (a) A schematic diagram of the mechanism of plasmonic optical biosensors. (b) Stages of SPR sensor from detecting analytes to detaching to be reused. Adapted from MDPI, Biology, 2022.

These include sensitivity and selectivity, reducing fabrication costs, and improving stability and reproducibility in complex biological samples. For example, biosensors need to be sensitive enough to identify even minute levels of bacteria in blood or other biological samples if they are to improve the accuracy and efficacy of medical diagnosis. The QB can be used to precisely outline tumor borders during treatment or follow a drug across the cytoplasm and membrane of a cell, among other possible applications. It is difficult to see with the naked eye, but the QB may be able to detect the electrical tensions at a neural synapse, follow a single ion moving across a cell membrane, or record the movement of peptides among microscopic organelles inside a cell. Quantum engineering and biology-related technologies have the power to drastically change how we think about medicine. The informatics obtained from QB provides increased precision, which can enhance the use and application of existing sensing technologies by collecting and using finer statistics to produce better data outputs. Although quantum optics usually depends on measurements utilizing different features of photons or light, QB can be produced, for example, by solid-phase devices and atoms and molecules that reside in free space. Clinical diagnostic scans from QB are more accurate, less expensive, and less likely to have unfavorable clinical consequences for healthcare practitioners. The use of biosensors to diagnose diseases, identify organisms that are resistant to multiple drugs, identify new epidemics, and detect incredibly low concentrations of toxins and microorganisms in food and drink is one of the most practical tools currently available in the biomedical and international fields of healthcare. Regardless of the modulation technique (angle, wavelength, amplitude) and the experimental setup (prism or grating-coupler) for phase and/or intensity measurements, noise limits the ultimate performance in SPR biosensing because of the quantum qualities of light [42]. The shot-noise limit (SNL) is the name given to this limit. Quantum metrology takes advantage of this limitation by employing nonclassical light states [43], often referred to as squeezed states, to improve the accuracy of measurements from gravitational wave interferometry to particle tracking in a biological sample [44]. With rising uncertainty in the other variance, the goal is to reduce the uncertainty by squeezing one of the variances.

The kinetics of binding between biomolecular species are investigated using SPR sensors. Plasmonic nanosensors are now a potent instrument for biosensing uses. This is due to their ability to overcome the sensitivity, flexibility, and photostability constraints of traditional optical sensors. New detection techniques have been created as the area has expanded and are being used in a broad range of biological and medical applications. Enhancing precise measurements is the main motivation behind plasmonic nanosensor development. Some of the strategies that are being considered in the improvement of precision include sensing based on target-induced local refractive index changes, colorimetric sensing based on localized SPR sensing, and coupling and amplification of sensitivity detection based on

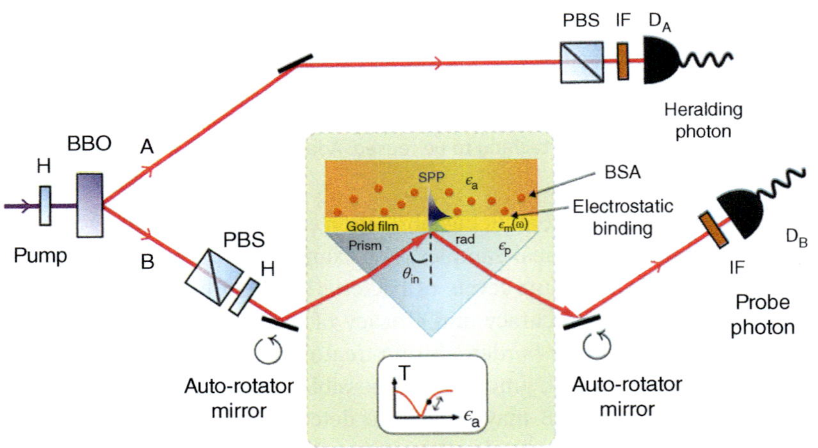

Figure 15.3 A schematic representation of a quantum plasmonic sensor setup. The system utilizes a pump source to generate photon pairs through a beta barium borate (BBO) crystal. The heralding photon (path A) is directed toward detector D_A, while the probe photon (path B) interacts with a surface plasmon polariton (SPP) mode on a gold film prism, where biotin-streptavidin binding is used for sensing. Polarizing beam splitters (PBS) and interference filters (IF) are used to manage the photonic paths, with auto-rotator mirrors adjusting the polarization states of the probe photon. The transmitted signal is detected by D_B, providing information about the target analyte's presence and concentration. Adapted from *Journal of Applied Physics*, 2022.

nanoparticle growth. Commercially available SPR-based plasmonic nanosensors exist, but the accuracy of current SPR equipment in determining the reflectivity of light from the sensor is constrained. Such restrictions cause issues in applications. For instance, it is crucial that the precision be greater than what is currently possible with commercially available SPR-based nanosensors while researching the pharmacological kinetics for medications used to treat different HIV-1 virus variants. To get around this, scientists must examine the sensor's physics in greater detail and take into account more basic optical resources, such as quantum mechanical resources. Understanding the quantum nature of light can help overcome the drawbacks of the SPR-based nanosensors that are now on the market. Specifically, the precision can be improved by preparing light in entangled multi-photon states and compressed states that utilize the uncertainty principle. Recent efforts to demonstrate that both quantum and plasmonic resonance features can be combined to produce several advantageous aspects have sparked interest in the possible improvement of plasmonic devices' sensing performance using quantum optical resources. A schematic representation of a quantum plasmonic sensor setup is shown in Figure 15.3.

15.3 Quantum Plasmonic Sensing and Microsystems

15.3.1 General Concept of Quantum Plasmonic Sensing

One interesting method is quantum plasmonic biosensing, which has the benefit of using less optical power and lowering the possibility of causing harm to the biological samples being studied [45]. The degree of interaction with the environment, which was meant to serve as a gauge of how the system regulates and safeguards quantum information, has been used to categorize quantum technology for encoding quantum information in recent years, with reports of evolution across many frameworks [46]. Photons play a crucial role as they serve as the scaffolding that better protects the quantum information during its propagation, whether in vacuum or in matter [47]. In addition, quantum photonic processors enable the creation and manipulation of qubits to encode the photons for universal two-qubit quantum computation [48]. The drawbacks include a quantum communication rate that decreases exponentially with distance and inevitable photon losses due to the tiny interactions between photons and their surroundings.

15.3.2 Light/Matter Coupling for Quantum Plasmonic Biosensing

By improving the interaction between light and matter at the nanoscale, which is connected with the sensitivity of plasmonic biosensors, it is still conceivable to build quantum networks at high speeds despite the intrinsic issue with photons. Then, a strong coupling may result in increased sensitivity with less optical power since light is confined and amplified inside a cavity or nanogap structure. Due to the mechanism of generation, the interaction of light and matter in plasmon polaritons relies on light-matter coupling and exhibits a quantum nature, which depends on the coupling between light and matter and has a quantum character that emerges in energy losses associated with a mechanism of hot carrier generation. The fundamental idea of quantum electrodynamic cavities is the interplay between light and matter inherent in plasmon polaritons. The efficiency of light-matter coupling finds itself in several factors characterized by its strength. The photon generation rate, which accounts for the decay rate of the radiative excited state, is correlated with the mode volume V/V_0, which accounts for the spatial distribution of electromagnetic energy surrounding the matter, and the quality factor Q, which measures the duration of interaction between photons and emitter. When the generation rate is greater than the decay rate, the spectrum breadth narrows, which is indicative of a strong light/matter connection. The coupling between the resonator mode and the emission modes is made possible by placing the atoms and emitters in a cavity that resembles a resonator. This causes the emitters' spectral width to increase and surpass the cavity's spectral width to shrink and its emission linewidth to narrow, both of which affect the radiation's directionality. The light/matter interaction regimes are schematized [49]. An optical resonator with a single two-level atom, a qubit (left), is shown on the left side of the top panel, while the same optical resonator is connected to a number N of quantum emitters on the right. The coupling strength is estimated using parameters like the cavity mode resonance frequency (ωc), the qubit transition frequency (ωq), the coupling strength (g), and the cavity and qubit loss rate (κ and γ). In contrast, for a system with N emitters, the coupling strength is dependent on the coupling with the single qubit and all emitters ($g\sqrt{N}$). From a weak condition derived by tuning the electromagnetic environment according to a Purcell effect to a deep strong coupling generated by the effect of THz metamaterial coupled to a cyclotron resonance of a 2D electron gas [50], the light/matter coupling regimes are shown in the figure's bottom panel. Because strong and deep strong coupling results are common in quantum phenomena and their quantification, there is a purpose to quantify the coupling strength. An illustration of this idea is provided by Raman spectroscopy, and some recent research will be shown in the following section.

15.3.3 Quantum Plasmonic Microsystem Biosensors

Following a conceptual premise pertaining to the idea of light/matter coupling, this section goes on to describe quantum plasmonic biosensing and emphasizes integration in nano- and microsystems, keeping in mind that the suggested solutions could become increasingly complex with the help of thriving micro- and nanotechnology. The basic metrology architecture comprises a classical detector for the quantum states of light, while the detection principle integrates the coherent photon-plasmon conversion processes or the signal from a noncoherent source. There are various configurations for quantum plasmonic design for metrology. Numerous solutions, ranging from a multilayer to a cavity/antenna structure, have been presented in light of the cavity and hybrid cavity modes. In order to obtain a propagation mode within the diffraction limit and to encourage strong light/matter interactions down to the single photon, metal-dielectric combinations have been devised [51]. The optical architecture showed an improvement over the integration in fluidic microsystems, allowing for a 90% increase in photon collection efficiency. As a result, molecular analysis has been performed in recent applications. For instance, a cavity mode has been effectively paired with an optical fiber composed of a clever and flexible microgel to create a glucose sensor [52]. Because an interwoven interaction between the plasmonic resonances and cavity modes is turned on and off by the swelling/shrinking gel mechanism, the fiber/cavity integration gave the cavity a tunable nature. It is important to note that even with a high measuring system efficiency, the aqueous system needed for biological samples cannot obtain the phase singularity that depends on the cavity surface topology and allows for tremendous efficiency when reached. For the design of fluidic microsystems, achieving a single mode is a significant problem [53] because doing so would lead to an improvement in sensitivity for both single molecules and single cells. One way to get around the challenges associated with integrating the optical couplers is to use materials like silicon, which has the dual benefits of being biocompatible and allowing for the achievement of a quality factor (e.g., Q > 105) and a high step change in resonance. The binding activity of individual adenoviruses with a typical diameter of 100 nm can actually be monitored thanks to a confined cavity mode proven by similarly built devices. Systems with low radiation rates, such as photonic cavities, exhibit decreased efficiency. An effective electromagnetic field confinement and photon emission faster than the lasting time must be triggered inside the matter. They are made possible by hybridizing the cavity with optical antennae, which overcomes the decrease of the radiation rate of a photonic cavity and improves the measurement efficiency. Systems with an antenna mode and a cavity mode have accomplished this dual behavior; the antenna enhances the directionality of the emission by coupling the system to the free space, while the cavity achieves a high efficiency in photon collection by retaining the radiation and improving the light/matter interaction [54]. By causing the matching of two impedances and pushing the detection limit down to a THz regime, the hybridization of the two modes can reduce the lossy nature of the metal and linewidth, which also has a major effect on the radiative behavior. The device is strain tuned and was demonstrated when combined with a quantum emitter and a planar integrated optical resonator made of a silicon nitride ring resonator fabricated on a piezoelectric substrate (panel ii). Figure 15.4a presents an example of a waveguide (panel i) made using a combination of bottom-up/top-down processes. A voltage was applied to the piezoelectric substrate to evaluate the devices' tunability, and the focusing optical fiber (panel iii) was used to record the quantum dot emission that was captured through the waveguide [55]. The two conditions have been implemented in the photonic cavity like a whispering gallery mode that integrates plasmonic structures in an opportune architecture for achieving the enhancement [56]. Ideally, an adiabatic coupling between the plasmon polaritons into a dielectric waveguide allows for an efficient collection and concurrently a significant enhancement of the sensitivity. Figure 15.4b(i) depicts a simplified version of the whispering gallery mode in microfluidics, where the whispering mode is created between a sphere and an Au nanorod that is permanently adsorbed on the spherical surface; Figure 15.4b(ii) illustrates the expected shift of the transmitted signal resulting from the hybrid mode. Compared to photonic waveguides, which can only convert between photons and plasmons, the SPP modes that cause electromagnetic waves to propagate via metal-dielectric or metal-air interfaces carry a lot more high-quality information. Using a unique optical arrangement that includes nanohole and nanodisk structures, it has been observed that the detection limit can be reduced down to zeptomoles with a corresponding increase in sensitivity. With its small linewidth spectral properties, high peak-to-dip signal ratio, and considerable near-field electromagnetic amplification, a nanoring resonator array is a great platform for high-performance chemical and biological sensing. The light actually creates powerful electric fields within a few nanometers, which intensify the scattering of adsorbed molecules and result in a rich fingerprint spectrum that contains all of the chemical information that is known. In order to merge quantum light sources and nonlinear elements, ridge waveguides and grating far-field couplers have been used to integrate quantum emitters built of organic molecules. This is a step toward photon-based quantum information and communication. A molecule contained in anthracene is deposited on the waveguide with a coupling effectiveness of up to $42 \pm 2\%$ over both propagation directions, according to the device's animation, which is displayed in Figure 15.4c(i) [57]. Consequently, the signal analysis shows that the waveguide/molecule interaction (panel ii) causes the intensity to increase more than it would if the molecule were off-guide.

Importantly, in these complex systems, which lacked a quantum intrinsic nature, the interaction between light and matter was boosted and intensified by the instantaneous electric field enhancement imposed on the photons by short pulses, as well as by nonlinear effects and interactions of photons with the surrounding matter. It has been shown that other methods of single-photon emission can be achieved by linking optical fields with quantum emitters such as quantum dots, atoms, solids, or ions. Although a single photon can be captured and released by these sources, the isotropic emission diminishes sensor effectiveness and makes photon collecting more challenging. Nonclassical sources have drawn more attention than promising sources recently obtained for quantum noise reduction, even if coherent control over light/matter interactions at the single-photon level was a driving force for quantum plasmonic sensing, consistent with the process being pursued for information technology. Squeezed states [5] and entangled states [58] are the most preferred quantum states of nonclassical light for quantum photon detection. Photon entanglement, the latter, permits a quantum state of photon pairs that is not a single one. According to this correlation, $\Delta\Phi\Delta I = \hbar/2$, the statistical noise of the phase $\Delta\Phi$ or of the amplitude ΔI of a quantum field is reduced at the expense of an increased noise in the conjugate variable. This redistribution of the observable uncertainty leads to the squeezed states of light, and in fact, the advantage arises by reducing the uncertainty in the variance of one quadrature. When physical occurrences that transduce the squeezed variable are detected, the signal-to-noise ratio increases. The goal of integrating quantum plasmonic sensing into microsystems is to offer a strong foundation for guaranteeing sample stability and, in turn, reducing all noise and signal contamination sources. It was shown that devices with a similar optomechanical design offer the benefit of combining the stability of integrated optics with the high-precision handling of liquid samples. Examples of parametric single-photon sources that use microfluidics and quantum sensing to measure protein concentration were reported. According to recent studies, micro- and nanosystems also have the benefit of lowering measurement noise by minimizing the uncertainties and oscillations brought on by sample handling.

An example of an integrated measurement system with an idler and a coincidence point for the signal comparison is provided by the optomechanical setup shown in Figure 15.4d(i). The estimation of the statistical error was used to calculate the variation of the analyte refractive index, and Figure 15.4d(ii) (panels 1–3 from left to right) illustrates the trend of the signal-to-noise ratio vs the refractive index. Although the idler and source have a Poissonian distribution, the signal and the idler have the same photon number. In order to estimate the standard deviation and show that estimated errors surpass the signal noise limit that would be achievable by a coherent state of light with the same

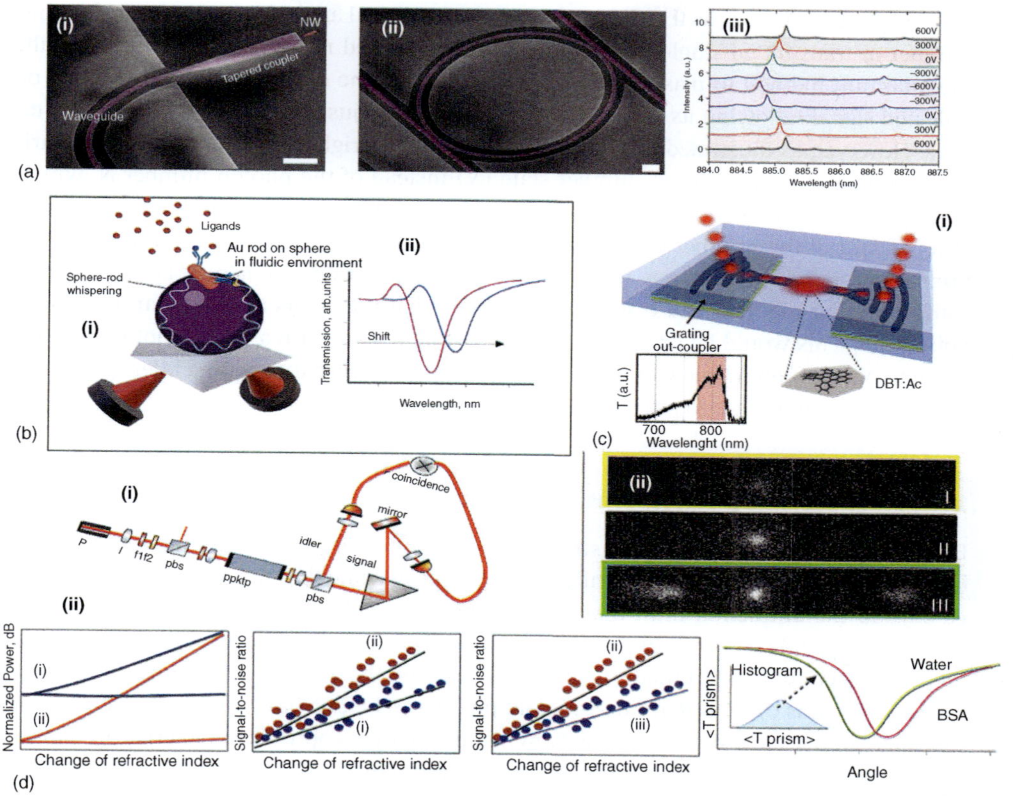

Figure 15.4 Microsystems and quantum sensing. (a) (i) Scanning electron microscope image of a silicon nitride waveguide in purple (scale bar 2 μm); (ii) scanning electron microscope image of a silicon nitride ring resonator (scale bar 2 μm); (iii) emission spectra of the nanowire quantum dot collected from the waveguide as a function of the applied voltage to the piezoelectric substrate. (b) Whispering principle: (i) schematic diagram of an optoplasmonic sensor made of spheres and Au rods; (ii) shift enhancement. (c) Waveguide/molecule coupling: (i) cartoon of the waveguide with two grating couplers. A molecule embedded in anthracene is deposited on the waveguide. Inset: grating optical response of the molecule (dibenzoterrylene [DBT]); (ii) image of the fluorescent signal by excitation with a diffraction-limited laser spot. Scheme I: no molecule on waveguide; II: molecule off guide; III: coupling waveguide/molecule. (d) Signal/idler quantum sensing. (i) A single photon is the input in a plasmonic ATR sensor with the Kretschmann configuration, where the actual reflection of a single photon is treated as transmission through the attenuated total reflectance (ATR) setup. The generation of the single-photon state (the signal) is heralded by the detection of its twin photon (the idler) by a quantum correlation of photon pairs initially produced via spontaneous parametric down-conversion. (ii) Sensitivity enhancement of plasmonic sensor with quantum resources when probing with coherent states. Left: measured signal while linearly ramping the driving voltage of the ultrasonic transducer that changes the refractive index of air; Middle panels: signal-to-noise ratio when probing the plasmonic sensor vs the change of the refractive index (plasmonic sensor with coherent states trace (i), with twin beams trace (ii). Trace (iii) gives the estimated SNR for the single coherent-state configuration). Right panel: Measured transmittances ⟨*T* prism⟩ over the incident angles for BSA and high-purity water. The error of the plot is measured as a standard deviation of *T* prism in the histogram embedded in the figure. Reproduced by permission of Taylor & Francis, Copyright 2024.

average photon number as the single photon, the statistical analysis was performed on the number of transmitted single photons for a predetermined number of times. As a result, the quantum method to plasmonic detection examines the change in the transmitted or reflected signal following its passage through the sensor, just like the coherent technique does. However, as illustrated in Figure 15.4d(ii) (panel 4 right), the quantum benchmark takes into account the photon number state $|N\rangle$ instead of the photon number N, which makes the optimal state for single-mode transmission and the magnitude of quantum gain independent of the number of photons. The statistical analysis is then conducted on the number of individual photons transmitted. The standard deviation may be determined using a similar method, and it was demonstrated that the errors are less than the signal noise limit. This would be feasible for a coherent light state with a mean photon number equal to one photon. In view of these findings, it makes sense to combine idler and signal analysis on a single chip to provide a compact solution.

15.3.4 Intensity and Phase-Sensitive Sensing

By specifically creating plasmonic structures, plasmonic sensors seek to improve sensitivity, LOD, and other parameters while simultaneously achieving sensing capabilities that go beyond the diffraction limit that plagues traditional photonic sensors [59]. Generally speaking, photonic sensors—both conventional and plasmonic—like those discussed in the preceding sections can be broadly categorized based on the quantity whose change is being tracked [60]. It is evident that intensity and phase are the two basic conjugate parameters that characterize electromagnetic fields. When the analyte's optical characteristics change, these values also change. Depending on the sensing scheme and the material, the change in the analyte leads to a change in both quantities or only one of them. If a sensor is based on tracking changes in intensity, it is referred to as "intensity-sensitive" sensing; if it tracks changes in phase, it is referred to as "phase-sensitive" sensing. While such a categorization may or may not be required in conventional sensing, it is essential in quantum sensing since a suitable light input state needs to be designed. To put it another way, unlike classical sensing, the two sensing kinds require entirely distinct light input states as an ideal signal. The following section goes into greater detail about this topic.

15.3.5 Other Plasmonic Sensors

Plasmonic sensors have proven to be helpful for a number of additional applications, such as assessing medical diagnostics [12, 61] or detecting food safety and quality [62], in addition to the previously mentioned refractive index sensing. In addition, throughout the past few decades, several plasmonic sensor types have been proposed. Plasmonic metamaterials, for instance, have been thoroughly researched for biosensing [63]. Circular dichroism spectroscopy and the identification of amyloid fibrils in Parkinson's disease at nanomolar concentrations are two examples of how chiral plasmonic devices have lately become a valuable sensing platform for determining the handedness of electromagnetic fields [64]. Another platform where developments have been made for biosensing and bioimaging applications is magneto-plasmonic structures [65].

15.4 Conclusions and Future Perspectives

The nanooptics foundation of plasmonic technology makes it theoretically suitable for point-of-care diagnostics and downsizing. Among other things, plasmonic biosensing—which inherently depends on subwavelength-scale optical phenomena—has discovered that microsystem integration provides the ideal setting for stimulating the sense mechanism. However, in recent years, micro- and nanotechnologies have reached a critical stage in their development, enabling the rapid development of bioanalytical techniques and the fabrication of devices as small as a few hundred nanometers that are meant for sensor integration. The adaptability of the microsystems environment has made it a good place to use cutting-edge analytical tools to research cellomics and biomolecular behavior. Since the initial instance of plasmonic sensor integration into a microsystem, arrays, gratings, waveguides, nanometric particles, and features have all been integrated into increasingly intricate and condensed systems. Due to their superior compatibility with the microfluidic environment, certain plasmon excitation strategies—such as gratings, waveguides, arrays, and meta-surfaces—have shown the most promise; nevertheless, other optomechanical topologies require greater work to align the transducers and the source.

Several benefits have been brought about by this combination; some are solely dependent on the inherent characteristics of the miniature environment, while others are the result of the overlap between plasmonics and microfluidics. These benefits include real-time and label-free monitoring of molecular binding events, as well as intriguing benefits like cost savings, high sensitivity and specificity, quick response, and the potential for multiplex detection. Plasmonic sensors have the potential to become a point-of-care diagnostic tool and make up a significant amount of the label-free sensor market, despite current limitations on their development. For the technology to be widely adopted, it must be small, precise, dependable, repeatable, inexpensive, and disposable, and it must be able to compete with other outstanding technologies. While these achievements are not negligible, they are unquestionably achievable. Plasmonic sensors for point-of-care have not yet reached a high level of industrialization, which may be explained from a critical perspective by two main issues: the prices and the sensitivity of the miniaturized integrated systems. This later phase, which involves a number of characters and procedures that determine whether the product will succeed on the market, is what brings the product to market. Plasmonic point-of-care sensors have not yet found a profitable market for a number of reasons. Signal interference still affects several of the suggested plasmonic devices, lowering the specificity of the binding signal. This contrasts with the high specificity of label-based methods (such as immunoassays). Furthermore, because plasmonic equipment is still costly and bulky, it is not yet ready for mass-market manufacture. The numerous benefits that plasmonic sensors exhibit drive efforts to promote the expansion of plasmonic measurement in therapeutic applications. Among these are the capacity to identify the equilibrium constants of the kinetic rates of biomolecular interactions, which are essential for basic molecular analysis; nevertheless, the benefits of direct detection and the speed of plasmonic sensors are much more significant.

This class of sensors is consequently highly suited for applications involving extensive and large-scale deployment due to benefits like flexibility and quick testing. These characteristics do not appear to be the key that enables plasmonic sensors to take the place of other bioanalytical techniques, such as dipstick tests, but they can provide assistance in a market segment that does not currently have a useful answer. In order to differentiate itself from the previously mentioned needs, the development perspective of plasmonic sensors must aspire to become a fully integrated and wearable sensor platform in addition to lowering the production cost of the batch devices and improving reproducibility. A promising outcome for basic research is the use of quantum plasmonics, which has demonstrated reduced optical power consumption and a decreased risk of causing harm to the biological material being studied. Cavity sensing and surface-enhanced Raman scattering are made possible by quantum plasmonic biosensing, which depends on a strong light/matter connection. This interaction shows a frequency range rich in Raman transitions that are seldom perceptible in a conventional frame. Quantum plasmonic sensing may develop into a realistic and useful application that fully utilizes quantum phenomena with additional research into various kinds of plasmonic sensors and quantum methodologies. Additionally, it would encourage multidisciplinary research in a range of engineering and scientific domains.

Acknowledgment

The authors gratefully acknowledge Amrita Vishwa Vidyapeetham, Amritapuri Campus, for all the support and infrastructure.

References

1. De Araujo, W.R., Lukas, H., Torres, M.D. et al. (2024). Low-cost biosensor technologies for rapid detection of covid-19 and future pandemics. *ACS Nano* 18: 1757–1777.
2. Kabay, G., De Castro, J., Altay, A. et al. (2022). Emerging biosensing technologies for the diagnostics of viral infectious diseases. *Advanced Materials* 34: 2201085.
3. Franklin, R.K. (2010) in vivo electrochemical sensors.
4. Peng, Y., Zhao, Y.X., and Hu. and Yang, Y. (2020). Optical fiber quantum biosensor based on surface plasmon polaritons for the label-free measurement of protein. *Sensors and Actuators B* 316: 128097.
5. Degen, C.L., Reinhard, F., and Cappellaro, P. (2017). Quantum sensing. *Reviews of Modern Physics* 89: 035002.
6. Kaushik, A., Srinivasan, S.S., and Mishra, Y.K. (2023). *Analytical Techniques for Biomedical Nanotechnology*. IOP Science https://doi.org/10.1088/978-0-7503-3379-5.
7. Nabeel, A., Hengyun, Z., Elana, K.U. et al. (2023). Quantum sensors for biomedical applications. *Nature Reviews Physics* 5: 157–169.
8. Lai, Y.H., Suh, M.G., Lu, Y.K. et al. (2020). Earth rotation measured by a chip-scale ring laser gyroscope. *Nature Photonics* 14: 345–349.
9. Na, Y., Jeon, C.G., Ahn, C. et al. (2020). Ultrafast, sub-nanometre-precision and multifunctional time-of-flight detection. *Nature Photonics* 14: 355–360.
10. Rong, Q., Guo, T., Bao, W. et al. (2017). Highly sensitive fiber-optic accelerometer by grating inscription in specific core dip fiber. *Scientific Reports* 7: 11856.
11. Leonard, P., Hearty, S., Brennan, J. et al. (2003). Advances in biosensors for detection of pathogens in food and water. *Enzyme and Microbial Technology* 32: 03–13.

12 Andreescu, S. and Sadik, O.A. (2004). Trends and challenges in biochemical sensors for clinical and environmental monitoring. *Pure and Applied Chemistry* 76: 861–878.

13 Hareesha, N. and Manjunatha, J.G. (2021). Electro-oxidation of formoterol fumarate on the surface of novel poly(thiazole yellow-G) layered multi-walled carbon nanotube paste electrode. *Scientific Reports* 11: 12797.

14 Pushpanjali, P.A., Manjunatha, J.G., Hareesha, N. et al. (2021). Voltammetric analysis of antihistamine drug cetirizine and paracetamol at poly(L-Leucine) layered carbon nanotube paste electrode. *Surfaces and Interfaces* 24: 101154.

15 Manjunatha, J.G., Kumara Swamy, B.E., Mamatha, G.P. et al. (2011). Poly (maleic acid) modified carbon paste electrode for simultaneous detection of dopamine in the presence of uric acid: a cyclic voltammetric study. *Analytical and Bioanalytical Electrochemistry* 3: 146–159.

16 Amrutha, B.M., Manjunatha, J.G., and Aarti, S.B. (2020). Design of a sensitive and selective voltammetric sensor based on a cationic surfactant-modified carbon paste electrode for the determination of alloxan. *ACS Omega* 5: 23481–23490.

17 Raril, C., Manjunatha, J.G., Ravishankar, D.K. et al. (2020). Validated electrochemical method for simultaneous resolution of tyrosine, uric acid, and ascorbic acid at polymer modified nano-composite paste electrode. *Surface Engineering and Applied Electrochemistry* 56: 415–426.

18 Pushpanjali, P.A., Manjunatha, J.G., and Hareesha, N. (2021). An overview of recent developments of carbon-based sensors for the analysis of drug molecules. *Journal of Electrochemical Science and Engineering* 11: 161–177.

19 Girish, T., Manjunatha, J.G., Hareesha, N., and Prinith, N.S. (2020). Research developments in carbon materials-based sensors for determination of hormones. *Monatshefte für Chemie* 151: 1681–1688.

20 Anker, J.N., Hall, W.P., Lyandres, O. et al. (2008). Biosensing with plasmonic nanosensors. *Nature Materials* 7: 442–453.

21 Rothenhausler, B. and Knoll, W. (1988). Surface plasmon microscopy. *Nature* 332: 615–617.

22 Jorgenson, R.C. and Yee, S.S. (1993). A fiber-optic chemical sensor based on surface plasmon resonance. *Sensors and Actuators B* 12: 213–220.

23 Homola, J., Koudela, I., and Yee, S. (1999). Surface plasmon resonance sensors based on diffraction gratings and prism couplers: sensitivity comparison. *Sensors and Actuators B: Chemical* 54: 16–24.

24 Brolo, A.G. (2012). Plasmonics for future biosensors. *Nature Photonics* 6: 709–713.

25 Liedberg, B., Nylander, C., and Lundstrom, I. (1983). Surface plasmons resonance for gas detection and biosensing. *Sensors and Actuators* 4: 299–304.

26 Maier, S.A. (2007). *Localized Surface Plasmons, Plasmonics: Fundamentals and Applications*, 65–88. Springer Nature.

27 Kreibig, U. and Vollmer, M. (2013). *Optical Properties of Metal Clusters*. Springer Science & Business Media.

28 Cosgrove, T. (2010). *Colloid Science: Principles, Methods and Applications*. John Wiley & Sons.

29 Pendry, J.B. (2000). Negative refraction makes a perfect lens. *Physical Review Letters* 85: 3966–3969.

30 Grigorenko, A., Geim, A., Gleeson, H. et al. (2005). Nanofabricated media with negative permeability at visible frequencies. *Nature* 438: 335–338.

31 Shalaev, V.M. (2007). Optical negative-index metamaterials. *Nature Photonics* 1: 41–48.

32 Lal, S., Link, S., and Halas, N.J. (2007). Nano-optics from sensing to waveguiding. *Nature Photonics* 1: 641–648.

33 Li, K., Stockman, M.I., and Bergman, D.J. (2003). Self-similar chain of metal nano spheres as an efficient nanolens. *Physical Review Letters* 91: 227402.

34 Kravets, V.G., Schedin, F., Jalil, R. et al. (2013). Singular phase nano-optics in plasmonic metamaterials for label-free single-molecule detection. *Nature Materials* 12: 304–309.

35 Sreekanth, K.V., Alapan, Y., ElKabbash, M. et al. (2016). Extreme sensitivity biosensing platform based on hyperbolic metamaterials. *Nature Materials* 5: 621–627.

36 Mondal, B. and Zeng, S. (2020). Recent advances in surface plasmon resonance for biosensing applications and future prospects. In: *Nanophotonics in Biomedical Engineering* (ed. X. Zhao and M. Lu), 21–48. Springer Nature.

37 Miyazaki, C., Shimizu, F., and Ferreira, M. (2017). Surface plasmon resonance for sensors and biosensors. In: *Nano Characterization Techniques* (ed. L.D.R. Alessandra, M. Ferreira, and N.O. Osvaldo Jr.), 183–200. Norwich: William Andrew Publishing.

38 Kravets, V.G., Wu, F., Yu, T., and Grigorenko, A.N. (2022). Metal-dielectric-graphene hybrid heterostructures with enhanced surface plasmon resonance sensitivity based on amplitude and phase measurements. *Plasmonics* 17: 973–987.

39 Derkachova, A., Kolwas, K., and Demchenko, I. (2016). Dielectric function for gold in plasmonics applications: size dependence of plasmon resonance frequencies and damping rates for nanospheres. *Plasmonics* 11: 941–951.

40 Zhu, X. and Gao, T. (2019). Spectrometry. In: *Nano-Inspired Biosensors for Protein Assay with Clinical Applications* (ed. G. Li), 237–264. Amsterdam, The Netherlands: Elsevier.

41 Zhou, X., Chen, K., Li, L. et al. (2017). Angle modulated surface plasmon resonance spectrometer for refractive index sensing with enhanced detection resolution. *Optics Communications* 382: 610–614.

42 Piliarik, M. and Homola, J. (2009). Surface plasmon resonance sensors: approaching their limits. *Optics Express* 17: 16505–16517.

43 Caves, C.M. (1981). Quantum-mechanical noise in an interferometer. *Physical Review D* 23: 1693.

44 Taylor, M.A., Janousek, J., Daria, V. et al. (2013). Biological measurement beyond the quantum limit. *Nature Photonics* 7: 229–233.

45 Fan, W., Lawrie, B.J., and Pooser, R.C. (2015). Quantum plasmonic sensing. *Physical Review A* 92: 53812.

46 Georgieva, H., Lopez, M., Hofer, H. et al. (2021). Absolute calibration of a single-photon avalanche detector using a bright triggered single-photon source based on an InGaAs quantum dot. *Optical Express* 29: 23500–23507.

47 Riemensberger, J., Kuznetsov, N., Liu, J. et al. (2022). Photonic integrated continuous-travelling-wave parametric amplifier. *Nature* 612: 56–61.

48 Qiang, X., Zhou, X., Wang, J. et al. (2018). Large-scale silicon quantum photonics implementing arbitrary two-qubit processing. *Nature Photonics* 12: 534–539.

49 Frisk, K.A., Miranowicz, A., De Liberto, S. et al. (2019). Ultrastrong coupling between light and matter. *Nature Reviews Physics* 1: 19–40.

50 Le Roux, F., Taylor, R.A., and Bradley, D.D.C. (2020). Enhanced and polarization-dependent coupling for photoaligned liquid crystalline conjugated polymer microcavities. *ACS Photonics* 7: 746–758.

51 Fehler, K.G., Antoniuk, L., Lettner, N. et al. (2021). Hybrid quantum photonics based on artificial atoms placed inside one hole of a photonic crystal cavity. *ACS Photonics* 8: 2635–2641.

52 Peng, Y., Zhao, Y., Hu, X., and Yang, Y. (2020). Optical fiber quantum biosensor based on surface plasmon polaritons for the label-free measurement of protein. *Sensors and Actuators B: Chemical* 316: 128097.

53 Sreekanth, K.V., Sreejith, S., Han, S. et al. (2018). Biosensing with the singular phase of an ultrathin metal-dielectric nanophotonic cavity. *Nature Communications* 9: 369.

54 Watanabe, K., Wu, H.Y., Xavier, J. et al. (2022). Single virus detection on silicon photonic crystal random cavities. *Small* 18: 2107597.

55 Fehler, K.G., Ovvyan, A.P., Gruhler, N. et al. (2019). Efficient coupling of an ensemble of nitrogen vacancy center to the mode of a High-Q, Si_3N_4 photonic crystal cavity. *ACS Nano* 13: 6891–6898.

56 Wang, Y., Zeng, S., Humbert, G., and Ho, H.-P. (2020). Microfluidic whispering gallery mode optical sensors for biological applications. *Laser & Photonics Reviews* 14: 2000135.

57 Lombardi, P., Ovvyan, A.P., Pazzagli, S. et al. (2018). Photostable molecules on chip: integrated sources of nonclassical light. *ACS Photonics* 5: 126–132.

58 Crespi, A., Lobino, M., Matthews, J.C.F. et al. (2012). Measuring protein concentration with entangled photons. *Applied Physics Letters* 100: 233704.

59 Takahara, J. (2009). Negative dielectric optical waveguides for nano-optical guiding. In: *Plasmonic Nanoguides and Circuits* (ed. S. Bozhevolnyi). Pan Stanford Publishing, Singapore https://doi.org/10.1201/9780429066429.

60 Homola, J. (2008). Surface plasmon resonance sensors for detection of chemical and biological species. *Chemical Reviews* 108: 462–493.

61 Pejcic, B., Marco, R.D., and Parkinson, G. (2006). The role of biosensors in the detection of emerging infectious diseases. *Analyst* 131: 1079–1090.

62 Mello, L.D. and Kubota, L.T. (2022). Review of the use of biosensors as analytical tools in the food and drink industries. *Food Chemistry* 77: 237–256.

63 Kabashin, A.V., Evans, P., Pastkovsky, S. et al. (2009). Plasmonic nanorod metamaterials for biosensing. *Nature Materials* 8: 867–871.

64 Kumar, J., Erana, H., Lopez-Martınez, E. et al. (2018). Detection of amyloid fibrils in Parkinson's disease using plasmonic chirality. *Proceedings of the National Academy of Sciences* 115: 3225–3230.

65 Armelles, G., Cebollada, A., Garcıa-Martın, A., and Gonzales, M.U. (2013). Magnetoplasmonics: combining magnetic and plasmonic functionalities. *Advanced Optical Materials* 1: 10–35.

Index

Quantum Biosensing in Medical Diagnostics, First Edition. Edited by J. G. Manjunatha.

Printed and bound by CPI Group (UK) Ltd, Croydon, CR0 4YY

17/09/2026

14964269-0004